Lecture Notes in Computer Science 14837

Founding Editors

Gerhard Goos
Juris Hartmanis

The series Lecture Notes in Computer Science (LNCS), including its subseries Lecture Notes in Artificial Intelligence (LNAI) and Lecture Notes in Bioinformatics (LNBI), has established itself as a medium for the publication of new developments in computer science and information technology research, teaching, and education.

LNCS enjoys close cooperation with the computer science R & D community, the series counts many renowned academics among its volume editors and paper authors, and collaborates with prestigious societies. Its mission is to serve this international community by providing an invaluable service, mainly focused on the publication of conference and workshop proceedings and postproceedings. LNCS commenced publication in 1973.

Leonardo Franco · Clélia de Mulatier ·
Maciej Paszynski · Valeria V. Krzhizhanovskaya ·
Jack J. Dongarra · Peter M. A. Sloot
Editors

Computational Science – ICCS 2024

24th International Conference
Malaga, Spain, July 2–4, 2024
Proceedings, Part VI

 Springer

Editors
Leonardo Franco 🆔
University of Malaga
Malaga, Spain

Clélia de Mulatier 🆔
University of Amsterdam
Amsterdam, The Netherlands

Maciej Paszynski 🆔
AGH University of Science and Technology
Krakow, Poland

Valeria V. Krzhizhanovskaya 🆔
University of Amsterdam
Amsterdam, The Netherlands

Jack J. Dongarra 🆔
University of Tennessee
Knoxville, TN, USA

Peter M. A. Sloot 🆔
University of Amsterdam
Amsterdam, The Netherlands

ISSN 0302-9743   ISSN 1611-3349 (electronic)
Lecture Notes in Computer Science
ISBN 978-3-031-63777-3   ISBN 978-3-031-63778-0 (eBook)
https://doi.org/10.1007/978-3-031-63778-0

This Springer imprint is published by the registered company Springer Nature Switzerland AG
The registered company address is: Gewerbestrasse 11, 6330 Cham, Switzerland

If disposing of this product, please recycle the paper.

Preface

Welcome to the proceedings of the 24th International Conference on Computational Science (https://www.iccs-meeting.org/iccs2024/), held on July 2–4, 2024 at the University of Málaga, Spain.

In keeping with the new normal of our times, ICCS featured both in-person and online sessions. Although the challenges of such a hybrid format are manifold, we have always tried our best to keep the ICCS community as dynamic, creative, and productive as possible. We are proud to present the proceedings you are reading as a result.

ICCS 2024 was jointly organized by the University of Málaga, the University of Amsterdam, and the University of Tennessee.

Facing the Mediterranean in Spain's Costa del Sol, Málaga is the country's sixth-largest city, and a major hub for finance, tourism, and technology in the region.

The University of Málaga (Universidad de Málaga, UMA) is a modern, public university, offering 63 degrees and 120 postgraduate degrees. Close to 40,000 students study at UMA, taught by 2500 lecturers, distributed over 81 departments and 19 centers. The UMA has 278 research groups, which are involved in 80 national projects and 30 European and international projects. ICCS took place at the Teatinos Campus, home to the School of Computer Science and Engineering (ETSI Informática), which is a pioneer in its field and offers the widest range of IT-related subjects in the region of Andalusia.

The International Conference on Computational Science is an annual conference that brings together researchers and scientists from mathematics and computer science as basic computing disciplines, as well as researchers from various application areas who are pioneering computational methods in sciences such as physics, chemistry, life sciences, engineering, arts, and the humanities, to discuss problems and solutions in the area, identify new issues, and shape future directions for research.

The ICCS proceedings series have become a primary intellectual resource for computational science researchers, defining and advancing the state of the art in this field.

We are proud to note that this 24th edition, with 17 tracks (16 thematic tracks and one main track) and close to 300 participants, has kept to the tradition and high standards of previous editions.

The theme for 2024, "Computational Science: Guiding the Way Towards a Sustainable Society", highlights the role of Computational Science in assisting multidisciplinary research on sustainable solutions. This conference was a unique event focusing on recent developments in scalable scientific algorithms; advanced software tools; computational grids; advanced numerical methods; and novel application areas. These innovative novel models, algorithms, and tools drive new science through efficient application in physical systems, computational and systems biology, environmental systems, finance, and others.

ICCS is well known for its excellent lineup of keynote speakers. The keynotes for 2024 were:

- David Abramson, University of Queensland, Australia
- Manuel Castro Díaz, University of Málaga, Spain
- Jiří Mikyška, Czech Technical University in Prague, Czechia
- Takemasa Miyoshi, RIKEN, Japan
- Coral Calero Muñoz, University of Castilla-La Mancha, Spain
- Petra Ritter, Berlin Institute of Health & Charité University Hospital Berlin, Germany

This year we had 430 submissions (152 to the main track and 278 to the thematic tracks). In the main track, 51 full papers were accepted (33.5%); in the thematic tracks, 104 full papers (37.4%). The higher acceptance rate in the thematic tracks is explained by their particular nature, whereby track organizers personally invite many experts in the field to participate. Each submission received at least 2 single-blind reviews (2.6 reviews per paper on average).

ICCS relies strongly on our thematic track organizers' vital contributions to attract high-quality papers in many subject areas. We would like to thank all committee members from the main and thematic tracks for their contribution to ensuring a high standard for the accepted papers. We would also like to thank Springer, Elsevier, and Intellegibilis for their support. Finally, we appreciate all the local organizing committee members for their hard work in preparing this conference.

We hope the attendees enjoyed the conference, whether virtually or in person.

July 2024

Leonardo Franco
Clélia de Mulatier
Maciej Paszynski
Valeria V. Krzhizhanovskaya
Jack J. Dongarra
Peter M. A. Sloot

Organization

Conference Chairs

General Chair

Valeria Krzhizhanovskaya   University of Amsterdam, The Netherlands

Main Track Chair

Clélia de Mulatier   University of Amsterdam, The Netherlands

Thematic Tracks Chair

Maciej Paszynski   AGH University of Krakow, Poland

Thematic Tracks Vice Chair

Michael Harold Lees   University of Amsterdam, The Netherlands

Scientific Chairs

Peter M. A. Sloot   University of Amsterdam, The Netherlands
Jack Dongarra   University of Tennessee, USA

Local Organizing Committee

Leonardo Franco (Chair)   University of Malaga, Spain
Francisco Ortega-Zamorano   University of Malaga, Spain
Francisco J. Moreno-Barea   University of Malaga, Spain
José L. Subirats-Contreras   University of Malaga, Spain

Thematic Tracks and Organizers

Advances in High-Performance Computational Earth Sciences: Numerical Methods, Frameworks & Applications (IHPCES)

Takashi Shimokawabe	University of Tokyo, Japan
Kohei Fujita	University of Tokyo, Japan
Dominik Bartuschat	FAU Erlangen-Nürnberg, Germany

Artificial Intelligence and High-Performance Computing for Advanced Simulations (AIHPC4AS)

Maciej Paszynski	AGH University of Krakow, Poland

Biomedical and Bioinformatics Challenges for Computer Science (BBC)

Mario Cannataro	University Magna Graecia of Catanzaro, Italy
Giuseppe Agapito	University Magna Graecia of Catanzaro, Italy
Mauro Castelli	Universidade Nova de Lisboa, Portugal
Riccardo Dondi	University of Bergamo, Italy
Rodrigo Weber dos Santos	Federal University of Juiz de Fora, Brazil
Italo Zoppis	University of Milano-Bicocca, Italy

Computational Diplomacy and Policy (CoDiP)

Roland Bouffanais	University of Geneva, Switzerland
Michael Lees	University of Amsterdam, The Netherlands
Brian Castellani	Durham University, UK

Computational Health (CompHealth)

Sergey Kovalchuk	Huawei, Russia
Georgiy Bobashev	RTI International, USA
Anastasia Angelopoulou	University of Westminster, UK
Jude Hemanth	Karunya University, India

Computational Optimization, Modelling, and Simulation (COMS)

Xin-She Yang   Middlesex University London, UK
Slawomir Koziel   Reykjavik University, Iceland
Leifur Leifsson   Purdue University, USA

Generative AI and Large Language Models (LLMs) in Advancing Computational Medicine (CMGAI)

Ahmed Abdeen Hamed   State University of New York at Binghamton,
                              USA
Qiao Jin   National Institutes of Health, USA
Xindong Wu   Hefei University of Technology, China
Byung Lee   University of Vermont, USA
Zhiyong Lu   National Institutes of Health, USA
Karin Verspoor   RMIT University, Australia
Christopher Savoie   Zapata AI, USA

Machine Learning and Data Assimilation for Dynamical Systems (MLDADS)

Rossella Arcucci   Imperial College London, UK
Cesar Quilodran-Casas   Imperial College London, UK

Multiscale Modelling and Simulation (MMS)

Derek Groen   Brunel University London, UK
Diana Suleimenova   Brunel University London, UK

Network Models and Analysis: From Foundations to Artificial Intelligence (NMAI)

Marianna Milano   Università Magna Graecia of Catanzaro, Italy
Giuseppe Agapito   University Magna Graecia of Catanzaro, Italy
Pietro Cinaglia   University Magna Graecia of Catanzaro, Italy
Chiara Zucco   University Magna Graecia of Catanzaro, Italy

Numerical Algorithms and Computer Arithmetic for Computational Science (NACA)

Pawel Gepner   Warsaw Technical University, Poland
Ewa Deelman   University of Southern California, Marina del
                            Rey, USA
Hatem Ltaief   KAUST, Saudi Arabia

Quantum Computing (QCW)

Katarzyna Rycerz   AGH University of Krakow, Poland
Marian Bubak   Sano and AGH University of Krakow, Poland

Simulations of Flow and Transport: Modeling, Algorithms, and Computation (SOFTMAC)

Shuyu Sun   King Abdullah University of Science and
                            Technology, Saudi Arabia
Jingfa Li   Beijing Institute of Petrochemical Technology,
                            China
James Liu   Colorado State University, USA

Smart Systems: Bringing Together Computer Vision, Sensor Networks and Artificial Intelligence (SmartSys)

Pedro Cardoso   University of Algarve, Portugal
João Rodrigues   University of Algarve, Portugal
Jânio Monteiro   University of Algarve, Portugal
Roberto Lam   University of Algarve, Portugal

Solving Problems with Uncertainties (SPU)

Vassil Alexandrov   Hartree Centre – STFC, UK
Aneta Karaivanova   IICT – Bulgarian Academy of Science, Bulgaria

Teaching Computational Science (WTCS)

Evguenia Alexandrova   Hartree Centre – STFC, UK
Tseden Taddese   UK Research and Innovation, UK

Reviewers

Ahmed Abdelgawad   Central Michigan University, USA
Samaneh Abolpour Mofrad   Imperial College London, UK
Tesfamariam Mulugeta Abuhay Queen's University, Canada
Giuseppe Agapito   University of Catanzaro, Italy
Elisabete Alberdi   University of the Basque Country, Spain
Luis Alexandre   UBI and NOVA LINCS, Portugal
Vassil Alexandrov   Hartree Centre – STFC, UK
Evguenia Alexandrova   Hartree Centre – STFC, UK
Julen Alvarez-Aramberri   Basque Center for Applied Mathematics, Spain
Domingos Alves   Ribeirão Preto Medical School, University of São
                             Paulo, Brazil
Sergey Alyaev   NORCE, Norway
Anastasia Anagnostou   Brunel University London, UK
Anastasia Angelopoulou   University of Westminster, UK
Rossella Arcucci   Imperial College London, UK
Emanouil Atanasov   IICT – Bulgarian Academy of Sciences, Bulgaria
Krzysztof Banaś   AGH University of Krakow, Poland
Luca Barillaro   Magna Graecia University of Catanzaro, Italy
Dominik Bartuschat   FAU Erlangen-Nürnberg, Germany
Pouria Behnodfaur   Curtin University, Australia
Jörn Behrens   University of Hamburg, Germany
Adrian Bekasiewicz   Gdansk University of Technology, Poland
Gebrail Bekdas   Istanbul University, Turkey
Mehmet Ali Belen   Iskenderun Technical University, Turkey
Stefano Beretta   San Raffaele Telethon Institute for Gene Therapy,
                             Italy
Anabela Moreira Bernardino   Polytechnic Institute of Leiria, Portugal
Eugénia Bernardino   Polytechnic Institute of Leiria, Portugal
Daniel Berrar   Tokyo Institute of Technology, Japan
Piotr Biskupski   IBM, Poland
Georgiy Bobashev   RTI International, USA
Carlos Bordons   University of Seville, Spain
Bartosz Bosak   PSNC, Poland
Lorella Bottino   University Magna Graecia of Catanzaro, Italy

Bhaskar Dasgupta	University of Illinois at Chicago, USA
Clélia de Mulatier	University of Amsterdam, The Netherlands
Ewa Deelman	University of Southern California, Marina del Rey, USA
Quanling Deng	Australian National University, Australia
Eric Dignum	University of Amsterdam, The Netherlands
Riccardo Dondi	University of Bergamo, Italy
Rafal Drezewski	AGH University of Krakow, Poland
Simon Driscoll	University of Reading, UK
Hans du Buf	University of the Algarve, Portugal
Vitor Duarte	Universidade NOVA de Lisboa, Portugal
Jacek Długopolski	AGH University of Krakow, Poland
Wouter Edeling	Vrije Universiteit Amsterdam, The Netherlands
Nahid Emad	University of Paris Saclay, France
Christian Engelmann	ORNL, USA
August Ernstsson	Linköping University, Sweden
Aniello Esposito	Hewlett Packard Enterprise, Switzerland
Roberto R. Expósito	Universidade da Coruna, Spain
Hongwei Fan	Imperial College London, UK
Tamer Fandy	University of Charleston, USA
Giuseppe Fedele	University of Calabria, Italy
Christos Filelis-Papadopoulos	Democritus University of Thrace, Greece
Alberto Freitas	University of Porto, Portugal
Ruy Freitas Reis	Universidade Federal de Juiz de Fora, Brazil
Kohei Fujita	University of Tokyo, Japan
Takeshi Fukaya	Hokkaido University, Japan
Wlodzimierz Funika	AGH University of Krakow, Poland
Takashi Furumura	University of Tokyo, Japan
Teresa Galvão	University of Porto, Portugal
Luis Garcia-Castillo	Carlos III University of Madrid, Spain
Bartłomiej Gardas	Institute of Theoretical and Applied Informatics, Polish Academy of Sciences, Poland
Victoria Garibay	University of Amsterdam, The Netherlands
Frédéric Gava	Paris-East Créteil University, France
Piotr Gawron	Nicolaus Copernicus Astronomical Centre, Polish Academy of Sciences, Poland
Bernhard Geiger	Know-Center GmbH, Austria
Pawel Gepner	Warsaw Technical University, Poland
Alex Gerbessiotis	NJIT, USA
Maziar Ghorbani	Brunel University London, UK
Konstantinos Giannoutakis	University of Macedonia, Greece
Alfonso Gijón	University of Granada, Spain

Jorge González-Domínguez	Universidade da Coruña, Spain
Alexandrino Gonçalves	CIIC – ESTG – Polytechnic University of Leiria, Portugal
Yuriy Gorbachev	Soft-Impact LLC, Russia
Pawel Gorecki	University of Warsaw, Poland
Michael Gowanlock	Northern Arizona University, USA
George Gravvanis	Democritus University of Thrace, Greece
Derek Groen	Brunel University London, UK
Loïc Guégan	UiT the Arctic University of Norway, Norway
Tobias Guggemos	University of Vienna, Austria
Serge Guillas	University College London, UK
Manish Gupta	Harish-Chandra Research Institute, India
Piotr Gurgul	SnapChat, Switzerland
Oscar Gustafsson	Linköping University, Sweden
Ahmed Abdeen Hamed	State University of New York at Binghamton, USA
Laura Harbach	Brunel University London, UK
Agus Hartoyo	TU Kaiserslautern, Germany
Ali Hashemian	Basque Center for Applied Mathematics, Spain
Mohamed Hassan	Virginia Tech, USA
Alexander Heinecke	Intel Parallel Computing Lab, USA
Jude Hemanth	Karunya University, India
Aochi Hideo	BRGM, France
Alfons Hoekstra	University of Amsterdam, The Netherlands
George Holt	UK Research and Innovation, UK
Maximilian Höb	Leibniz-Rechenzentrum der Bayerischen Akademie der Wissenschaften, Germany
Huda Ibeid	Intel Corporation, USA
Alireza Jahani	Brunel University London, UK
Jiří Jaroš	Brno University of Technology, Czechia
Qiao Jin	National Institutes of Health, USA
Zhong Jin	Computer Network Information Center, Chinese Academy of Sciences, China
David Johnson	Uppsala University, Sweden
Eleda Johnson	Imperial College London, UK
Piotr Kalita	Jagiellonian University, Poland
Drona Kandhai	University of Amsterdam, The Netherlands
Aneta Karaivanova	IICT-Bulgarian Academy of Science, Bulgaria
Sven Karbach	University of Amsterdam, The Netherlands
Takahiro Katagiri	Nagoya University, Japan
Haruo Kobayashi	Gunma University, Japan
Marcel Koch	KIT, Germany

Harald Koestler	University of Erlangen-Nuremberg, Germany
Georgy Kopanitsa	Tomsk Polytechnic University, Russia
Sotiris Kotsiantis	University of Patras, Greece
Remous-Aris Koutsiamanis	IMT Atlantique/DAPI, STACK (LS2N/Inria), France
Sergey Kovalchuk	Huawei, Russia
Slawomir Koziel	Reykjavik University, Iceland
Ronald Kriemann	MPI MIS Leipzig, Germany
Valeria Krzhizhanovskaya	University of Amsterdam, The Netherlands
Sebastian Kuckuk	Friedrich-Alexander-Universität Erlangen-Nürnberg, Germany
Michael Kuhn	Otto von Guericke University Magdeburg, Germany
Ryszard Kukulski	Institute of Theoretical and Applied Informatics, Polish Academy of Sciences, Poland
Krzysztof Kurowski	PSNC, Poland
Marcin Kuta	AGH University of Krakow, Poland
Marcin Łoś	AGH University of Krakow, Poland
Roberto Lam	Universidade do Algarve, Portugal
Tomasz Lamża	ACK Cyfronet, Poland
Ilaria Lazzaro	Università degli studi Magna Graecia di Catanzaro, Italy
Paola Lecca	Free University of Bozen-Bolzano, Italy
Byung Lee	University of Vermont, USA
Mike Lees	University of Amsterdam, The Netherlands
Leifur Leifsson	Purdue University, USA
Kenneth Leiter	U.S. Army Research Laboratory, USA
Paulina Lewandowska	IT4Innovations National Supercomputing Center, Czechia
Jingfa Li	Beijing Institute of Petrochemical Technology, China
Siyi Li	Imperial College London, UK
Che Liu	Imperial College London, UK
James Liu	Colorado State University, USA
Zhao Liu	National Supercomputing Center in Wuxi, China
Marcelo Lobosco	UFJF, Brazil
Jay F. Lofstead	Sandia National Laboratories, USA
Chu Kiong Loo	University of Malaya, Malaysia
Stephane Louise	CEA, LIST, France
Frédéric Loulergue	University of Orléans, INSA CVL, LIFO EA 4022, France
Hatem Ltaief	KAUST, Saudi Arabia
Zhiyong Lu	National Institutes of Health, USA

Stefan Luding	University of Twente, The Netherlands
Lukasz Madej	AGH University of Krakow, Poland
Luca Magri	Imperial College London, UK
Anirban Mandal	Renaissance Computing Institute, USA
Soheil Mansouri	Technical University of Denmark, Denmark
Tomas Margalef	Universitat Autònoma de Barcelona, Spain
Arbitrio Mariamena	Consiglio Nazionale delle Ricerche, Italy
Osni Marques	Lawrence Berkeley National Laboratory, USA
Maria Chiara Martinis	Università Magna Graecia di Catanzaro, Italy
Jaime A. Martins	University of Algarve, Portugal
Paula Martins	CinTurs – Research Centre for Tourism Sustainability and Well-being; FCT-University of Algarve, Portugal
Michele Martone	Max-Planck-Institut für Plasmaphysik, Germany
Pawel Matuszyk	Baker-Hughes, USA
Francesca Mazzia	University di Bari, Italy
Jon McCullough	University College London, UK
Pedro Medeiros	Universidade Nova de Lisboa, Portugal
Wen Mei	National University of Defense Technology, China
Wagner Meira	Universidade Federal de Minas Gerais, Brazil
Roderick Melnik	Wilfrid Laurier University, Canada
Pedro Mendes Guerreiro	Universidade do Algarve, Portugal
Isaak Mengesha	University of Amsterdam, The Netherlands
Wout Merbis	University of Amsterdam, The Netherlands
Ivan Merelli	ITB-CNR, Italy
Marianna Milano	Università Magna Graecia di Catanzaro, Italy
Magdalena Misiak	Howard University College of Medicine, USA
Jaroslaw Miszczak	Institute of Theoretical and Applied Informatics, Polish Academy of Sciences, Poland
Dhruv Mittal	University of Amsterdam, The Netherlands
Fernando Monteiro	Polytechnic Institute of Bragança, Portugal
Jânio Monteiro	University of Algarve, Portugal
Andrew Moore	University of California Santa Cruz, USA
Francisco J. Moreno-Barea	Universidad de Málaga, Spain
Leonid Moroz	Warsaw University of Technology, Poland
Peter Mueller	IBM Zurich Research Laboratory, Switzerland
Judit Munoz-Matute	Basque Center for Applied Mathematics, Spain
Hiromichi Nagao	University of Tokyo, Japan
Kengo Nakajima	University of Tokyo, Japan
Philipp Neumann	Helmut-Schmidt-Universität, Germany
Sinan Melih Nigdeli	Istanbul University – Cerrahpasa, Turkey

Fernando Nobrega Santos	University of Amsterdam, The Netherlands
Joseph O'Connor	University of Edinburgh, UK
Frederike Oetker	University of Amsterdam, The Netherlands
Arianna Olivelli	Imperial College London, UK
Ángel Omella	Basque Center for Applied Mathematics, Spain
Kenji Ono	Kyushu University, Japan
Hiroyuki Ootomo	Tokyo Institute of Technology, Japan
Eneko Osaba	TECNALIA Research & Innovation, Spain
George Papadimitriou	University of Southern California, USA
Nikela Papadopoulou	University of Glasgow, UK
Marcin Paprzycki	IBS PAN and WSM, Poland
David Pardo	Basque Center for Applied Mathematics, Spain
Anna Paszynska	Jagiellonian University, Poland
Maciej Paszynski	AGH University of Krakow, Poland
Łukasz Pawela	Institute of Theoretical and Applied Informatics, Polish Academy of Sciences, Poland
Giulia Pederzani	Universiteit van Amsterdam, The Netherlands
Alberto Percz de Alba Ortiz	University of Amsterdam, The Netherlands
Dana Petcu	West University of Timisoara, Romania
Beáta Petrovski	University of Oslo, Norway
Frank Phillipson	TNO, The Netherlands
Eugenio Piasini	International School for Advanced Studies (SISSA), Italy
Juan C. Pichel	Universidade de Santiago de Compostela, Spain
Anna Pietrenko-Dabrowska	Gdansk University of Technology, Poland
Armando Pinho	University of Aveiro, Portugal
Pietro Pinoli	Politecnico di Milano, Italy
Yuri Pirola	Università degli Studi di Milano-Bicocca, Italy
Ollie Pitts	Imperial College London, UK
Robert Platt	Imperial College London, UK
Dirk Pleiter	KTH/Forschungszentrum Jülich, Germany
Paweł Poczekajło	Koszalin University of Technology, Poland
Cristina Portalés Ricart	Universidad de Valencia, Spain
Simon Portegies Zwart	Leiden University, The Netherlands
Anna Procopio	Università Magna Graecia di Catanzaro, Italy
Ela Pustulka-Hunt	FHNW Olten, Switzerland
Marcin Płodzień	ICFO, Spain
Ubaid Qadri	Hartree Centre – STFC, UK
Rick Quax	University of Amsterdam, The Netherlands
Cesar Quilodran Casas	Imperial College London, UK
Andrianirina Rakotoharisoa	Imperial College London, UK
Celia Ramos	University of the Algarve, Portugal

Robin Richardson	Netherlands eScience Center, The Netherlands
Sophie Robert	University of Orléans, France
João Rodrigues	Universidade do Algarve, Portugal
Daniel Rodriguez	University of Alcalá, Spain
Marcin Rogowski	Saudi Aramco, Saudi Arabia
Sergio Rojas	Pontifical Catholic University of Valparaiso, Chile
Diego Romano	ICAR-CNR, Italy
Albert Romkes	South Dakota School of Mines and Technology, USA
Juan Ruiz	University of Buenos Aires, Argentina
Tomasz Rybotycki	IBS PAN, CAMK PAN, AGH, Poland
Katarzyna Rycerz	AGH University of Krakow, Poland
Grażyna Ślusarczyk	Jagiellonian University, Poland
Emre Sahin	Science and Technology Facilities Council, UK
Ozlem Salehi	Özyeğin University, Turkey
Ayşin Sancı	Altinay, Turkey
Christopher Savoie	Zapata Computing, USA
Ileana Scarpino	University "Magna Graecia" of Catanzaro, Italy
Robert Schaefer	AGH University of Krakow, Poland
Ulf D. Schiller	University of Delaware, USA
Bertil Schmidt	University of Mainz, Germany
Karen Scholz	Fraunhofer MEVIS, Germany
Martin Schreiber	Université Grenoble Alpes, France
Paulina Sepúlveda-Salas	Pontifical Catholic University of Valparaiso, Chile
Marzia Settino	Università Magna Graecia di Catanzaro, Italy
Mostafa Shahriari	Basque Center for Applied Mathematics, Spain
Takashi Shimokawabe	University of Tokyo, Japan
Alexander Shukhman	Orenburg State University, Russia
Marcin Sieniek	Google, USA
Joaquim Silva	Nova School of Science and Technology – NOVA LINCS, Portugal
Mateusz Sitko	AGH University of Krakow, Poland
Haozhen Situ	South China Agricultural University, China
Leszek Siwik	AGH University of Krakow, Poland
Peter Sloot	University of Amsterdam, The Netherlands
Oskar Slowik	Center for Theoretical Physics PAS, Poland
Sucha Smanchat	King Mongkut's University of Technology North Bangkok, Thailand
Alexander Smirnovsky	SPbPU, Russia
Maciej Smołka	AGH University of Krakow, Poland
Isabel Sofia	Instituto Politécnico de Beja, Portugal
Robert Staszewski	University College Dublin, Ireland

Magdalena Stobińska	University of Warsaw, Poland
Tomasz Stopa	IBM, Poland
Achim Streit	KIT, Germany
Barbara Strug	Jagiellonian University, Poland
Diana Suleimenova	Brunel University London, UK
Shuyu Sun	King Abdullah University of Science and Technology, Saudi Arabia
Martin Swain	Aberystwyth University, UK
Renata G. Słota	AGH University of Krakow, Poland
Tseden Taddese	UK Research and Innovation, UK
Ryszard Tadeusiewicz	AGH University of Krakow, Poland
Claude Tadonki	Mines ParisTech/CRI – Centre de Recherche en Informatique, France
Daisuke Takahashi	University of Tsukuba, Japan
Osamu Tatebe	University of Tsukuba, Japan
Michela Taufer	University of Tennessee, USA
Andrei Tchernykh	CICESE, Mexico
Kasim Terzic	University of St Andrews, UK
Jannis Teunissen	KU Leuven, Belgium
Sue Thorne	Hartree Centre – STFC, UK
Ed Threlfall	United Kingdom Atomic Energy Authority, UK
Vinod Tipparaju	AMD, USA
Pawel Topa	AGH University of Krakow, Poland
Paolo Trunfio	University of Calabria, Italy
Ola Tørudbakken	Meta, Norway
Carlos Uriarte	University of the Basque Country, BCAM – Basque Center for Applied Mathematics, Spain
Eirik Valseth	University of Life Sciences & Simula, Norway
Rein van den Boomgaard	University of Amsterdam, The Netherlands
Vítor V. Vasconcelos	University of Amsterdam, The Netherlands
Aleksandra Vatian	ITMO University, Russia
Francesc Verdugo	Vrije Universiteit Amsterdam, The Netherlands
Karin Verspoor	RMIT University, Australia
Salvatore Vitabile	University of Palermo, Italy
Milana Vuckovic	European Centre for Medium-Range Weather Forecasts, UK
Kun Wang	Imperial College London, UK
Peng Wang	NVIDIA, China
Rodrigo Weber dos Santos	Federal University of Juiz de Fora, Brazil
Markus Wenzel	Fraunhofer Institute for Digital Medicine MEVIS, Germany

Lars Wienbrandt	Kiel University, Germany
Wendy Winnard	UKRI STFC, UK
Maciej Woźniak	AGH University of Krakow, Poland
Xindong Wu	Hefei University of Technology, China
Dunhui Xiao	Tongji University, China
Huilin Xing	University of Queensland, Australia
Yani Xue	Brunel University, UK
Abuzer Yakaryilmaz	University of Latvia, Latvia
Xin-She Yang	Middlesex University London, UK
Dongwei Ye	University of Amsterdam, The Netherlands
Karol Życzkowski	Jagiellonian University, Poland
Gabor Závodszky	University of Amsterdam, Hungary
Sebastian Zając	SGH Warsaw School of Economics, Poland
Małgorzata Zajęcka	AGH University of Krakow, Poland
Justyna Zawalska	ACC Cyfronet AGH, Poland
Wei Zhang	Huazhong University of Science and Technology, China
Yao Zhang	Google, USA
Jinghui Zhong	South China University of Technology, China
Sotirios Ziavras	New Jersey Institute of Technology, USA
Zoltan Zimboras	Wigner Research Center, Hungary
Italo Zoppis	University of Milano-Bicocca, Italy
Chiara Zucco	University Magna Graecia of Catanzaro, Italy
Pavel Zun	ITMO University, Russia

Contents – Part VI

Quantum Computing

Network Models and Analysis: From Foundations to Artificial Intelligence

Representation Learning in Multiplex Graphs: Where and How to Fuse Information?

Piotr Bielak[(✉)][ID] and Tomasz Kajdanowicz[ID]

Department of Artificial Intelligence, Wroclaw University of Science and Technology,
Wrocław, Poland
`piotr.bielak@pwr.edu.pl`

Abstract. In recent years, unsupervised and self-supervised graph representation learning has gained popularity in the research community. However, most proposed methods are focused on homogeneous networks, whereas real-world graphs often contain multiple node and edge types. Multiplex graphs, a special type of heterogeneous graphs, possess richer information, provide better modeling capabilities and integrate more detailed data from potentially different sources. The diverse edge types in multiplex graphs provide more context and insights into the underlying processes of representation learning. In this paper, we tackle the problem of learning representations for nodes in multiplex networks in an unsupervised or self-supervised manner. To that end, we explore diverse information fusion schemes performed at different levels of the graph processing pipeline. The detailed analysis and experimental evaluation of various scenarios inspired us to propose improvements in how to construct GNN architectures that deal with multiplex graphs.

Keywords: representation learning · graph neural networks ·
multiplex graphs · unsupervised learning · self-supervised learning

1 Introduction

Real-world data often exhibits a complex and heterogeneous nature, leading to challenges and opportunities in graph representation learning. From social networks, where multiple relationship types exist among users, to biological networks, where various interaction types occur among molecules, a multiplex view is often more appropriate to capture the entirety of the network structure. Multiplex networks are characterized by the presence of multiple edge types (graph layers), span across a common set of nodes. They provide richer information, better modeling capabilities, and facilitate the integration of more detailed data from diverse sources. A multiplex network is not simply a combination of homogeneous graphs but a fundamentally more complex system encapsulating a range of various interaction types. Despite the rich information inherent to multiplex

© The Author(s), under exclusive license to Springer Nature Switzerland AG 2024
L. Franco et al. (Eds.): ICCS 2024, LNCS 14837, pp. 3–18, 2024.
https://doi.org/10.1007/978-3-031-63778-0_1

networks, the research focus in graph representation learning has been over-whelmingly concentrated on methods for homogeneous graphs. This has left a substantial gap in understanding how to leverage the unique features of multiplex networks effectively.

To this end, we examine the problem of learning representations for nodes in multiplex networks from the perspective of unsupervised and self-supervised learning. Unlike supervised learning, these approaches do not require labeled data and, hence, offer a flexible and scalable method for learning node representations. In particular, we investigate various information fusion schemes that can be performed at different stages of the representation learning pipeline to harness the richness of multiplex networks. These schemes are designed to integrate the diverse edge types in multiplex graphs, providing more context and insights into the underlying processes.

This paper details an extensive analysis and experimental evaluation of various information fusion scenarios. We hope our work advances our understanding of multiplex network representation learning and paves the way for developing more robust, efficient, and versatile multiplex network methods.

Our contributions can be summarized as follows:

1. We propose an **information fusion taxonomy** for multiplex networks.
2. We provide an **extensive experimental evaluation** of existing representation learning approaches in multiplex networks. We categorize them according to our proposed taxonomy. We evaluate the node embeddings in three tasks: node classification, node clustering and similarity search.
3. We **identify research gaps** in each fusion strategy category and provide preliminary implementations of those methods. We evaluate their performance and compare them to existing approaches. We **provide a limitations analysis of existing methods** and identify future directions in representation learning for multiplex networks.

2   Related Work

Graph Representation Learning. It has emerged as a vital area of research with applications spanning various domains, such as social networks, bioinformatics, or recommender systems. The essence lies in learning continuous, low-dimensional representations of nodes that capture their structural and attribute information within the graph. Plenty of methods have been developed, each focused on a different aspect of network embeddings, such as structure, attributes, learning paradigm or scalability. Shallow methods, such as Deep-Walk [9], LINE [11] use a simple notion of graph coding through random walks or objectives that optimize first and second-order node similarity. More complex graph neural networks, such as GCN [5] or GAT [13], implement the message-passing algorithm over graph edges combined with various message aggregation schemes.

Un- and Self-supervised Graph Representation Learning. Inspired by the success of contrastive methods in other domains, the procedures were adapted to graphs. Early approaches were based on the autoencoder architecture (GAE [4]). Another method, DGI [14], employed a GNN to learn node embeddings and maximized the mutual information between node embeddings and the graph embedding (readout function) by discriminating nodes in the original graph from nodes in a corrupted graph. GRACE [20] and GraphCL [18] utilized contrastive learning on graphs. All the previous methods rely on negative sampling which yields a high complexity. Negative-sample-free methods, such as BGRL [12] or GBT [1] use graph augmentations combined with either an asymmetric pipeline architecture or a decorrelation-based loss to prevent representation collapse and learn node embeddings in a self-supervised manner.

Multiplex Graph Representation Learning. Despite the advances in representation learning for homogeneous graphs, research on multiplex graphs is relatively limited. MHGCN [19] utilizes a weighted sum of adjacency matrices of the multiplex graph layers, where the combination weights are learnable. The resulting graph is passed into a standard GCN backbone while the whole model is trained using a link prediction objective with negative sampling. DMGI [8], extends the idea of DGI to multiplex graphs. It applies a GCN backbone with the DGI objective for each graph layer. Node embeddings are fused using a trainable lookup embedding that optimized via a loss function that minimizes the distance layer-wise node embeddings of the original graph and maximizes the distance to embeddings from the corrupted graph. HDGI [10] also builds upon the DGI method. It applies a GCN backbone at each graph layer and fuses the resulting vectors into a single node embedding using a semantic attention mechanism. Such scheme is applied to the original graph and its corrupted version. The two node embedding sets are processed by DGI's original loss function. Contrary to previous approaches, S^2MGRL [7] extends the idea of GBT [1] to multiplex networks. For each graph layer, the model first applies an MLP followed by a GCN block. The loss function applies the Barlow Twins objective between (a) the outputs of the MLPs and GCNs, and (b) all pairs of GCN outputs to ensure correlation between embeddings from all graph layers. This training step (and all previously introduced multiplex graph representation learning approaches) is carried out in an un- or self-supervised manner. However, S^2MGRL trains an attention mechanism on the frozen embedding from all layers to obtain the final node embeddings, utilizing node labels on the output.

3   Information Fusion in Multiplex Graphs

Let us start with basic definitions and the introduction of the multiplex graph representation learning pipeline (see: Fig. 1). Next, we will introduce a taxonomy of information fusion methods in multiplex graphs, followed by a detailed discussion of existing methods from each category and our proposed extensions. *Multiplex Graph.* We define a multiplex graph $\mathcal{G} = (\mathcal{V}, \mathcal{E}_1, \mathcal{E}_2, \ldots, \mathcal{E}_K, \mathbf{X})$, where $\mathcal{V}$ denotes a set of nodes, $\mathbf{X} \in \mathbb{R}^{|\mathcal{V}| \times d_{\text{in}}}$ is a matrix whose rows contain d_{in}-dimensional node's features. The K layers of the multiplex graph are defined by the K sets of edges: $\mathcal{E}_1, \mathcal{E}_2, \ldots, \mathcal{E}_K$, where each set $\mathcal{E}_i \in \mathcal{V} \times \mathcal{V}$ contains edges between pairs of nodes. Note that the nodes and their features are shared across all graph layers.

Multiplex Graph Representation Learning. When learning embeddings for nodes in multiplex graphs, the input graph is processed by a representation learning model f_θ (parametrized by θ), which computes embedding vectors $\mathbf{Z} \in \mathbb{R}^{|\mathcal{V}| \times d}$ for all nodes in the graph. This representation learning model is often a graph neural network (GNN) based on the message-passing paradigm. In particular, each graph neural network layer is comprised of two phases – message generation and message aggregation (see Eq. 1):

$$
\begin{aligned}
\mathbf{m}_i &= \text{MESSAGE}(\mathbf{h}_i^{(in)}) \\
\mathbf{h}_u^{(out)} &= \text{UPDATE}\left(\mathbf{h}_u^{(in)}, \text{AGG}\left(\{\mathbf{m}_v, v \in \mathcal{N}(u)\}\right)\right)
\end{aligned}
\tag{1}
$$

where $\text{MESSAGE}(\cdot)$ is a function that converts node i's input features $h_i^{(in)}$ into a message vector $\mathbf{m}_i$ (for the first GNN layer, we use $H^{(in)} = \mathbf{X}$). Incoming messages from node u's neighbors $\mathcal{N}(u)$ are aggregated using a permutation-invariant function $\text{AGG}(\cdot)$. Finally, the u's output representation (embedding) is computed via the $\text{UPDATE}(\cdot)$ function, which takes into account both the node's input features $h_u^{(in)}$ and the aggregated neighbor messages.

GNN layers are modified to accommodate the multiple edge types found in multiplex graphs, or separate GNNs are applied on each multiplex graph layer. In the latter case, which is often used in existing multiplex representation learning methods, the output of those GNNs must be fused into a single representation vector for each node.

The resulting node embedding vectors can be applied in various downstream tasks, such as node classification or link prediction. For instance, a node classifier can be trained in a supervised manner on the frozen node representation vectors and corresponding labels.

Multiplex Graph Information Fusion Taxonomy. The complex and rich structure of multiplex graphs and their representation learning and processing pipeline enables us to identify several locations for applying information fusion. We propose the following taxonomy of information fusion in multiplex graphs:

- **Graph-level fusion** – methods that directly modify the graph structure and fuse the information from several multiplex graph layers together,
- **GNN-level fusion** – methods that either (a) utilize a dedicated GNN architecture that is specifically designed for multiplex graphs, or (b) models that utilize standard GNNs in combination with trainable fusion mechanisms (such as attention) to produce fused embedding vectors for nodes,
- **Embedding-level fusion** – methods that firstly precompute node embeddings in each multiplex graph layer and then fuse the frozen embeddings into a single embedding per node (post-hoc methods),
- **Prediction-level fusion** – methods that utilize precomputed node embeddings at each multiplex graph layer and then train downstream task models

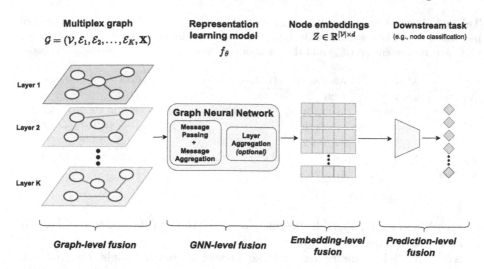

Fig. 1. Multiplex graph processing pipeline and information fusion taxonomy.

to obtain and fuse their predictions (note that these methods are not part of the unsupervised or self-supervised representation learning pipeline).

Such taxonomy can capture various multiplex graph representation learning methods and categorize them according to where the critical information fusion step is applied. Moreover, one could build another categorization based on the actual fusion mechanism. However, to keep things clearer, we introduce them in detail in Sects. 3.3–3.6.

3.1 Experimental Setup and Evaluation

In the following sections, we will introduce the existing methods for each category in our proposed taxonomy, along with an experimental evaluation and comparison of the methods' performance in selected downstream tasks. We implemented all models using the PyTorch-Geometric [2] library, and we utilize the DVC package [6] to ensure reproducibility. We make our code and data publicly available at https://github.com/graphml-lab-pwr/multiplex-fusion. Due to page limits, we leave the details about the hyperparameter search ranges and the final chosen ones in the code repository. Let us now introduce the utilized datasets and downstream tasks with the appropriate evaluation protocols.

Datasets. We selected six real-world datasets frequently utilized in representation learning papers for multiplex graphs [7,8,10,19]. We summarize the statistics of those datasets in Table 1.

- ACM [15] is an academic paper dataset with nodes assigned to one of 3 classes: database, wireless communication, and data mining. The multiplex layers are built from 2 meta-paths obtained from this graph: Paper-Author-Paper (PAP) and Paper-Subject-Paper (PSP). Node features are TF-IDF

Table 1. Dataset statistics. We summarize the number of nodes ($|\mathcal{V}|$), the names of particular graph layers along with the number of edges in each layer ((Layer i) $|\mathcal{E}_i|$), the node feature dimension (d_{in}) and the number of classes in the node classification task.

	ACM	Amazon	Freebase	IMDB	Cora	CiteSeer
$|\mathcal{V}|$	3,025	7,621	3,492	3,550	2,708	3,327
(Layer i) $|\mathcal{E}_i|$	(PAP) 29,281	(IBI) 1,104,257	(MAM) 254,702	(MAM) 66,428	(CIT) 10,556	(CIT) 9,104
	(PSP) 2,210,761	(IOI) 14,305	(MDM) 8,404	(MDM) 13,788	(KNN) 54,160	(KNN) 66,540
		(IVI) 266,237	(MWM) 10,706			
d_{in}	1,870	2,000	3,492	2,000	1,433	3,703
#classes	3	4	3	3	7	6

encodings of the papers' keywords. Note that the PSP layer has significantly more edges than the PAP layer.

- Amazon [3] is a multiplex network dataset of items bought on Amazon. The graph layers refer to common activities of users, i.e., also-bought (IBI), bought-together (IOI) and also-viewed (IVI). Each item can be categorized into one of 4 classes - Beauty, Automotive, Patio Lawn and Garden, and Baby. Note that there is a disproportion in the size of particular graph layers. Node features are built as bag-of-words encodings of the item descriptions.
- Freebase [16] is a movies graph dataset (from the Freebase KG) with nodes divided into three classes – Action, Comedy and Drama. Graph layers are built upon the following meta-paths: Movie-Actor-Movie (MAM), Movie-Director-Movie (MDM) and Movie-Writer-Movie (MWM). No inherent node features are available, so a one-hot encoding is utilized.
- IMDB [17] is another movie graph dataset extracted from the IMDB knowledge graph. Nodes are movies, and each is assigned one of 3 classes - Action, Comedy and Drama. Graph layers are built upon the following meta-paths: Movie-Actor-Movie (MAM) and Movie-Director-Movie (MDM). The node features are TF-IDF encodings of movie metadata, such as title or language.
- Cora, CiteSeer [5] are two citation graphs where nodes are research papers divided into seven and six different research areas, respectively. Node features are bag-of-words encodings of paper abstracts. These datasets are inherently homogeneous, and the original edges are put into the CIT layer. We build the multiplex graph by extending the original networks by another graph layer – a paper similarity layer. We find each node's k-nearest neighbors (KNN) based on the cosine similarity of node features and select the top 10 similar papers (i.e., $k = 10$).

Downstream Tasks. The utilized datasets are equipped with node labels, so a node classification is a natural choice for the downstream evaluation of node representations. However, to fully explore the generalization abilities of the learned embedding vectors, we employ two additional tasks – node clustering and node similarity search [8]. For all tasks, we collect various metrics (provided in the code repository). To ensure clarity, we focus only on one metric per task in the paper's result tables. Before executing the downstream evaluation, we freeze the representation backbone model and extract the node embeddings.

- node classification (**Clf**) – we follow the standard evaluation protocol for unsupervised and self-supervised node representation learning methods [14]. We train a logistic regression classifier on the node embeddings and corresponding label pairs from the training set. We provide the Macro-F1 (**MaF1**) metric on the test set. Additionally, we use the this task for the models' hyper-parameter searches. We select the models with the highest Macro-F1 score on the validation set.
- node clustering (**Clu**) – we train a K-Means model on the node embeddings and measure the clustering performance using the Normalized Mutual Information (**NMI**) score on the test set. We train the clustering model ten times with different seeds and average the scores to obtain the final NMI score.
- node similarity search – we compute the cosine similarity scores of the node embeddings between all pairs of nodes and build a ranking for each node according to the similarity score. Next, we calculate the ratio of the nodes that belong to the same class within top-5 ranked nodes (**Sim@5**).

Table Notation. In all result tables (Table 2, 3, 4, 5, 6 and 7), we utilize the same notations and present the mean and std (given in parentheses) of the Macro-F1, NMI and Sim@5 scores as percentages in the node classification, node clustering and similarity search tasks, respectively. Please refer to the "***Methods.***" paragraphs in corresponding sections for explanations of model names. Shaded rows denote our proposed methods and extensions.

3.2 Baseline Approaches (no Fusion)

Methods. We examine two baseline approaches – **Layers** and **Features**. In the first approach, we train the DGI [14] model on each graph layer separately and evaluate the resulting embeddings. We selected DGI due to its popularity in multiplex representation learning method design and its ability to learn node embeddings in an unsupervised way. The **Features** approach does not involve any learning but evaluates the initial node features as the node representations. ***Discussion.*** Results are given in Table 2. In most cases, we observe that different graph layers perform differently in the downstream tasks. For the ACM dataset, the smaller layer PAP performs much better in the node classification and similarity search tasks than the much larger PSP layer. However, the clustering performance is better on the PSP layer. Utilizing the node features also yields

Table 2. Downstream tasks performance of methods **without information fusion**.

	ACM			Amazon			Freebase		
	Clf (MaF1)	Clu (NMI)	Sim@5	Clf (MaF1)	Clu (NMI)	Sim@5	Clf (MaF1)	Clu (NMI)	Sim@5
Layers	PAP			IBI			MAM		
	82.33 (6.51)	37.01 (6.06)	85.26 (0.56)	26.50 (0.79)	0.15 (0.04)	31.06 (0.25)	51.79 (1.24)	15.82 (4.44)	55.72 (0.28)
	PSP			IOI			MDM		
	56.17 (5.16)	50.42 (1.57)	66.06 (2.88)	35.63 (1.09)	1.01 (0.53)	48.83 (1.08)	35.97 (1.28)	0.87 (0.08)	41.28 (1.06)
				IVI			MWM		
				27.08 (0.63)	0.14 (0.03)	29.94 (0.33)	25.26 (1.00)	0.27 (0.09)	41.54 (1.12)
Features	73.30 (0.00)	14.83 (4.14)	71.75 (0.00)	71.27 (0.00)	4.11 (1.80)	84.09 (0.00)	20.18 (0.00)	0.41 (0.03)	32.04 (0.00)

	IMDB			Cora			CiteSeer		
	Clf (MaF1)	Clu (NMI)	Sim@5	Clf (MaF1)	Clu (NMI)	Sim@5	Clf (MaF1)	Clu (NMI)	Sim@5
Layers	MAM			CIT			CIT		
	41.99 (2.92)	0.48 (0.28)	45.98 (0.30)	65.32 (3.37)	25.70 (5.26)	75.49 (1.37)	58.50 (3.74)	27.36 (4.74)	63.47 (0.82)
	MDM			KNN			KNN		
	51.50 (2.01)	5.09 (1.36)	51.68 (0.36)	51.87 (4.57)	16.99 (1.01)	57.67 (1.17)	55.93 (1.89)	24.08 (1.49)	61.23 (0.32)
Features	56.59 (0.00)	6.75 (2.01)	51.64 (0.00)	55.42 (0.00)	13.47 (0.63)	50.10 (0.00)	58.73 (0.00)	18.64 (2.22)	53.48 (0.00)

a decent performance; however, it is not as good as for the PAP layer. A similar situation happens for the IMDB dataset, where the larger MAM layer performs worse than the MDM layer – this time in all three downstream tasks. The node features yield the best overall performance. Another case worth noting occurs in the Amazon dataset – the node features outperform all layer information with about 40 pp difference in the node classification task, 50 pp for the similarity search task, and about 4–40 times better performance on the clustering task. Note that the IBI and IVI layers achieve only 0.15% and 0.14% NMI, which are inferior scores. In the case of the artificially constructed Cora and CiteSeer datasets, we observe that the added KNN layers achieve similar performance to the node features, which is expected as these datasets are highly homophilic.

3.3   Graph-Level Fusion

Methods. Information fusion in multiplex graph representation learning can be applied in the graphs themselves. Not much research has been carried out, and the most prominent existing method for graph-level fusion is MHGCN [19]. The method consists of a GCN backbone trained using a link prediction objective with negative edge sampling. For a K-layer multiplex graph, the method jointly learns K edge weights: $\beta_1, \beta_2, \ldots, \beta_K$, which are assigned to each edge in the corresponding graph layer. Finally, the multiplex graph is converted into a homogeneous weighted graph, where the edge weights are summed over the graph layers. For instance, if edge (u, v) exists in three graph layers: i, j and k, the final edge weight is given as: $\beta^{(u,v)} = \beta_i + \beta_j + \beta_k$. We compare the MHGCN method with an approach where we flatten (**Flattened**) the input graph into a non-weighted homogeneous graph with multi-edges, i.e., $\mathcal{G} = (\mathcal{V}, \mathcal{E}_1, \mathcal{E}_2, \ldots, \mathcal{E}_K, \mathbf{X}) \rightarrow G_{\text{flattened}} = (\mathcal{V}, \mathcal{E}_\Sigma, \mathbf{X})$, where $\mathcal{E}_\Sigma = \mathcal{E}_1 \cup \mathcal{E}_2 \cup \ldots \cup \mathcal{E}_K$ is the multi-set of edges (an edge between two nodes might occur in several graph layers). We evaluate the performance of several popular node representation learning methods for homogenous graphs – the supervised GCN and GAT models, the unsupervised DeepWalk (DW) and DGI models.

Discussion. Results are given in Table 3. The MHGCN method did not perform well on the ACM, Amazon and IMDB datasets compared to baseline approaches (no fusion). This might be related to the overall high number of edges (approx. 2M and 1.3M, respectively) combined with the link prediction objective – mining the negative edges might be troublesome. We leave the evaluation of a modified model training objective as future work. On the contrary, for the Freebase dataset, we observe an approx. 11 pp improvement in the node classification task along with substantial improvement in the similarity search task and comparable results for the clustering task. Similarly, Cora and CiteSeer also benefit from the learnable graph-level fusion approach, which increases performance by up to 15 pp. When it comes to the Flattened graph methods, we firstly observe a consistently better performance for the supervised models compared to unsupervised ones, which is related to the direct access to node labels. However, the DW and DGI methods achieve similar results to MHGCN on ACM and Amazon, while for other datasets, the results are 5–10 pp worse.

Table 3. Downstream tasks performance of **graph-level fusion methods**.

		ACM		Amazon			Freebase		
	Clf (MaF1)	Clu (NMI)	Sim@5	Clf (MaF1)	Clu (NMI)	Sim@5	Clf (MaF1)	Clu (NMI)	Sim@5
GCN	70.19 (0.22)	50.12 (0.35)	73.38 (0.23)	27.84 (0.17)	0.81 (0.11)	27.88 (0.20)	60.70 (0.55)	18.71 (1.20)	56.91 (0.43)
GAT	70.34 (2.89)	44.35 (2.10)	72.17 (1.16)	25.88 (2.06)	0.20 (0.10)	30.08 (1.02)	57.22 (1.24)	16.54 (1.26)	56.80 (0.42)
DW	64.25 (0.64)	37.53 (1.82)	62.98 (0.79)	24.45 (0.39)	0.04 (0.01)	26.79 (0.39)	49.20 (1.55)	17.01 (0.47)	53.67 (0.79)
DGI	59.54 (0.78)	36.07 (7.00)	72.85 (0.55)	25.60 (0.92)	0.12 (0.02)	28.38 (0.15)	51.22 (0.57)	17.86 (0.79)	55.72 (0.51)
MHGCN	63.35 (0.88)	38.89 (0.00)	71.05 (3.13)	23.49 (1.69)	0.04 (0.01)	29.44 (0.17)	62.75 (0.20)	14.48 (1.59)	60.84 (0.61)

		IMDB		Cora			CiteSeer		
	Clf (MaF1)	Clu (NMI)	Sim@5	Clf (MaF1)	Clu (NMI)	Sim@5	Clf (MaF1)	Clu (NMI)	Sim@5
GCN	57.04 (0.50)	10.69 (0.21)	49.01 (1.03)	68.76 (0.45)	41.11 (0.82)	66.65 (0.33)	64.27 (0.22)	39.93 (0.75)	62.28 (0.73)
GAT	54.81 (0.67)	10.11 (0.49)	46.90 (0.61)	65.29 (1.58)	36.99 (1.94)	59.94 (1.18)	60.53 (1.84)	32.84 (3.10)	60.17 (1.37)
DW	38.09 (0.78)	0.94 (0.19)	38.11 (0.60)	61.50 (1.26)	39.71 (1.00)	53.86 (1.24)	54.16 (1.35)	40.38 (1.07)	53.74 (0.32)
DGI	40.51 (4.44)	0.27 (0.08)	48.84 (0.49)	58.09 (3.60)	21.53 (3.48)	62.57 (0.64)	60.43 (2.25)	29.08 (2.38)	63.38 (0.71)
MHGCN	49.77 (0.83)	2.69 (0.78)	50.07 (0.56)	71.22 (1.17)	35.61 (1.70)	67.09 (0.25)	62.55 (0.84)	39.64 (2.13)	63.02 (0.71)

3.4 Prediction-Level Fusion

Methods. In this scenario, we assume that precomputed node embedding vectors exist for each multiplex graph layer. We utilize the layer-wise node representations of the DGI method from the baseline experiments in Sect. 3.2. Based on those embeddings and the node labels, we train logistic regression classifiers and combine their predictions to obtain a prediction-level information fusion scheme. In particular, we validate two approaches – soft and hard ensemble voting. In the hard voting case, we first obtain predicted node labels from each classifier and find the most occurring label (also known as majority voting). In the `soft` case, we first obtain the class probabilities, average them over all classifiers, and select the label with the highest probability. Note that such a scheme is applicable only in the node classification task.

Discussion. Results are given in Table 4. We observe that the `soft` approach performs consistently better on all datasets compared to `hard`. One might lose too much information about the actual class distribution and model confidence by discarding the actual probabilities. Compared to baseline methods and graph-level fusion, we do not observe a significant improvement, yet sometimes, the performance is worse than just utilizing a single graph layer. The information fusion occurs too late in the graph processing pipeline, making discovering complex relationships in the multiplex data harder.

Table 4. Downstream tasks performance of **prediction-level fusion methods**.

		ACM Clf (MaF1)	Amazon Clf (MaF1)	Freebase Clf (MaF1)	IMDB Clf (MaF1)	Cora Clf (MaF1)	CiteSeer Clf (MaF1)
Vote	soft	81.95 (6.23)	36.20 (1.01)	51.08 (1.17)	49.73 (3.02)	65.38 (4.12)	60.40 (2.71)
	hard	58.32 (6.81)	31.34 (0.89)	45.18 (1.04)	42.13 (3.51)	61.68 (2.77)	57.32 (1.98)

3.5   GNN-Level Fusion

Methods. This case is where most existing methods apply the fusion mechanism. It seems a natural choice to perform information fusion as a part of the representation learning model. `DMGI` [8] applies a GCN backbone with the DGI objective at each graph layer and learns the fused node representations as a trainable embedding lookup matrix. The additional loss term minimizes the distance to positive layer-wise embeddings (original graph) and maximizes the distance to negative ones (corrupted graph). `HDGI` [10] employs a similar approach but uses a semantic attention mechanism to fuse the layer-wise node embeddings. Instead of applying the DGI loss to each graph layer, the model applies it to the fused embeddings. S^2`MGRL` [7] builds upon GBT [1] and extends it to multiplex networks. Each graph layer is processed by an MLP, followed by a GCN block. The loss function is twofold: (1) a Barlow Twins (BT) loss term between the outputs of the MLPs and GCNs, and (2) the BT loss computed between all pairs of GCN outputs. Contrary to previous approaches, S^2MGRL is not fully un-/self-supervised. It trains an attention mechanism on pairs of frozen layer-wise embeddings and node labels to fuse node embeddings. We propose extensions to the above methods: `F(GBT, *)` and `F(DGI, *)`, where * denotes the fusion method (attention `Att` or concatenation with linear projection `CL`). For `F(GBT, *)` we employ a similar scheme as S^2`MGRL`, but instead of computing the BT loss for all graph layer pairs, we compute the BT loss between each graph layer output and the fused node embedding (which scales linearly). For the `F(DGI,`

·), we apply the DGI loss at each graph layer, and additionally on the fused node embeddings (of the original and corrupted graph). Such settings are both inductive and self-supervised.

Table 5. Downstream tasks performance of **GNN-level fusion methods**.

	ACM			Amazon			Freebase		
	Clf (MaF1)	Clu (NMI)	Sim@5	Clf (MaF1)	Clu (NMI)	Sim@5	Clf (MaF1)	Clu (NMI)	Sim@5
DMGI	87.76 (0.82)	57.29 (4.47)	87.94 (0.52)	73.15 (1.85)	36.11 (2.60)	79.27 (0.59)	49.79 (1.97)	15.36 (0.49)	54.83 (2.16)
HDGI	85.01 (2.47)	56.33 (9.53)	87.28 (0.64)	37.16 (3.33)	1.21 (0.96)	44.40 (2.25)	47.00 (2.55)	16.58 (0.51)	55.29 (0.53)
S^{2}MGRL	86.14 (2.14)	43.02 (25.10)	85.79 (1.04)	50.99 (3.78)	3.91 (1.86)	58.27 (2.80)	47.63 (7.63)	7.22 (3.79)	57.41 (1.46)
F(GBT,Att)	74.52 (3.22)	22.90 (12.59)	78.45 (2.04)	48.06 (2.64)	0.76 (0.17)	52.51 (0.85)	49.20 (2.48)	2.01 (1.21)	55.53 (1.57)
F(GBT,CL)	81.37 (2.18)	30.54 (23.06)	83.41 (1.51)	50.54 (4.17)	1.41 (1.66)	52.22 (3.07)	48.26 (2.07)	3.02 (2.53)	55.51 (1.12)
F(DGI,Att)	86.68 (2.16)	67.57 (4.36)	89.05 (0.45)	31.17 (3.54)	0.39 (0.16)	44.86 (2.97)	49.79 (0.98)	16.09 (1.17)	57.50 (0.59)
F(DGI,CL)	84.11 (4.30)	61.13 (6.28)	89.21 (0.29)	34.24 (2.68)	0.28 (0.10)	42.28 (3.38)	49.21 (0.97)	0.80 (0.57)	52.97 (0.95)

	IMDB			Cora			CiteSeer		
	Clf (MaF1)	Clu (NMI)	Sim@5	Clf (MaF1)	Clu (NMI)	Sim@5	Clf (MaF1)	Clu (NMI)	Sim@5
DMGI	60.81 (0.66)	19.70 (0.49)	60.05 (0.63)	73.55 (0.56)	42.30 (2.70)	72.78 (0.83)	66.13 (1.75)	43.37 (0.59)	66.84 (0.56)
HDGI	52.60 (1.70)	13.66 (1.42)	55.45 (1.36)	61.45 (2.53)	34.18 (3.00)	67.43 (2.80)	60.54 (1.18)	40.01 (2.18)	64.06 (1.54)
S^{2}MGRL	44.23 (8.44)	1.36 (1.76)	48.84 (2.55)	61.03 (7.64)	19.99 (11.76)	67.86 (1.71)	52.66 (3.98)	20.96 (8.81)	62.30 (0.61)
F(GBT,Att)	53.70 (0.95)	2.00 (2.06)	50.39 (0.42)	68.70 (2.22)	22.01 (3.50)	68.60 (1.01)	59.30 (2.40)	19.95 (6.32)	61.36 (1.20)
F(GBT,CL)	56.30 (1.08)	0.53 (0.90)	52.59 (0.48)	70.57 (1.81)	21.61 (1.48)	69.87 (1.21)	60.39 (2.24)	18.61 (5.33)	62.54 (0.98)
F(DGI,Att)	55.56 (1.53)	11.85 (3.88)	56.59 (0.74)	66.44 (1.01)	35.18 (0.48)	68.18 (0.71)	62.25 (2.04)	36.58 (3.17)	62.58 (0.82)
F(DGI,CL)	52.29 (3.40)	1.03 (1.58)	53.83 (1.02)	62.33 (3.68)	13.51 (4.36)	66.40 (1.15)	60.88 (1.49)	17.55 (1.49)	64.08 (0.74)

Discussion. Results are given in Table 5. We observe that the lookup embedding-based fusion approach of the DMGI method consistently delivers the best results on the node classification task compared to the other two approaches. A similar observation can be made for the clustering and similarity search task except for the Freebase dataset, where the performance is alike. Compared to previous fusion scenarios, the results of DMGI are best overall on all datasets except for Freebase, where the model achieves slightly worse results. One could argue that such lookup embedding is the best fusion mechanism. However, note that it does not allow us to easily obtain embeddings for newly added nodes or changes in the graph structure (it is transductive). HDGI's attention mechanism proves to perform similarly to DMGI on ACM and Freebase, worse on IMDB, Cora and CiteSeer, but on the Amazon dataset, we observe an approx. 35 pp difference in the node classification and similarity search tasks. The utilization of the BT objective in S^{2}MGRL achieves comparable performance to the aforementioned methods on ACM, Freebase and Cora, but on Amazon and IMDB, the difference is more substantial. For our proposed extensions, F(GBT, ∗) and F(DGI, ∗), we observe that both variants provide competitive performance to existing multiplex approaches. If we compare the results to HDGI, which is the only model that is both inductive and self-supervised, the performance gains are even greater. For instance, 56% vs. 52% MaF1 on IMDB, 70% vs. 61% MaF1 on Cora, or even 50% vs. 37% MaF1 on Amazon.

3.6 Embedding-Level Fusion

Methods. A basic fusion approach often presented in multiplex representation learning papers is first to precompute node embeddings for each graph layer separately and then fuse them by computing the average vectors (over all graph layers). We denote that approach as `Mean` and use it in combination with a supervised `GCN`, `GAT` and unsupervised DeepWalk (`DW`) and `DGI` models (same set of models as in the flattened graph case – see Sect. 3.3). We extend these embedding-level (post-hoc) fusion methods by examining other non-trainable fusion operators (i.e., `Concat`, `Min`, `Max`, `Sum`). We also build models with trainable fusion mechanisms – i.e., attention (`Att`), concatenation with a linear projection (`CL`) and a lookup embedding (`Lk`). We utilize an MSE or BT loss computed between the layer-wise embedding vectors and the fused output representation.

Table 6. Downstream tasks performance of **embedding-level fusion methods**.

		ACM			Amazon			Freebase	
		Clf (MaF1)	Clu (NMI)	Sim@5	Clf (MaF1)	Clu (NMI)	Sim@5	Clf (MaF1)	Clu (NMI)	Sim@5
GCN		84.66 (0.16)	62.53 (0.53)	85.61 (0.11)	45.31 (11.90)	9.06 (8.14)	46.05 (5.44)	58.32 (2.69)	20.24 (1.05)	58.07 (0.66)
GAT	Mean	85.03 (0.98)	59.55 (4.51)	85.36 (0.16)	35.83 (9.66)	4.18 (6.28)	41.13 (4.05)	50.38 (6.87)	15.81 (3.89)	57.44 (0.65)
DW		69.21 (0.96)	35.63 (3.20)	64.04 (3.23)	24.33 (0.59)	0.04 (0.02)	26.74 (0.53)	41.61 (2.48)	8.75 (3.98)	43.28 (2.37)
DGI		80.14 (8.60)	41.12 (4.15)	85.95 (1.04)	31.93 (0.43)	0.36 (0.09)	42.93 (0.58)	49.22 (2.47)	6.57 (8.05)	54.36 (1.42)
	Concat	82.42 (5.51)	42.02 (4.81)	85.90 (1.09)	36.21 (1.02)	0.45 (0.13)	46.42 (0.81)	50.90 (1.22)	7.52 (7.62)	54.82 (1.56)
DGI	Min	77.80 (9.57)	42.40 (7.76)	85.69 (0.98)	28.22 (2.20)	0.72 (0.04)	39.45 (0.33)	50.60 (1.44)	8.51 (7.60)	55.35 (0.80)
	Max	82.21 (6.30)	40.35 (4.88)	85.94 (1.06)	28.49 (1.00)	0.72 (0.09)	37.45 (0.51)	49.89 (2.58)	9.08 (8.84)	54.66 (0.82)
	Sum	82.44 (6.42)	41.12 (4.15)	85.95 (1.04)	31.96 (0.44)	0.36 (0.09)	42.93 (0.58)	50.53 (1.53)	6.57 (8.05)	54.36 (1.42)
	Att,BT	80.37 (9.83)	50.97 (6.63)	86.17 (0.90)	31.60 (0.48)	0.31 (0.06)	42.04 (1.25)	49.67 (2.03)	8.07 (7.38)	54.48 (1.13)
	Att,MSE	80.12 (8.57)	41.48 (4.60)	85.95 (1.02)	31.95 (0.43)	0.39 (0.08)	42.92 (0.58)	49.24 (2.50)	6.57 (8.05)	54.37 (1.38)
DGI	CL,BT	85.39 (1.79)	36.26 (6.87)	86.10 (0.92)	30.25 (1.05)	0.21 (0.15)	34.17 (2.03)	50.24 (2.01)	7.77 (3.56)	54.23 (0.66)
	CL,MSE	80.17 (9.18)	39.82 (3.85)	85.85 (0.73)	32.43 (0.93)	0.31 (0.14)	42.30 (1.00)	49.04 (2.55)	6.36 (8.04)	54.00 (1.08)
	Lk,BT	82.71 (1.03)	43.93 (7.08)	83.65 (1.83)	27.86 (0.65)	0.13 (0.06)	33.75 (1.35)	45.18 (1.12)	8.44 (6.00)	53.12 (0.87)
	Lk,MSE	33.59 (0.64)	0.09 (0.01)	32.67 (0.28)	24.67 (0.51)	0.05 (0.01)	26.75 (0.18)	33.08 (2.63)	0.22 (0.05)	36.98 (0.84)

		IMDB			Cora			CiteSeer	
		Clf (MaF1)	Clu (NMI)	Sim@5	Clf (MaF1)	Clu (NMI)	Sim@5	Clf (MaF1)	Clu (NMI)	Sim@5
GCN		61.63 (0.83)	16.39 (0.82)	55.41 (0.47)	76.88 (1.44)	48.40 (4.13)	74.73 (1.50)	63.50 (0.86)	39.51 (1.79)	63.12 (0.77)
GAT	Mean	54.57 (4.41)	6.46 (2.00)	51.14 (1.28)	73.94 (3.18)	46.98 (5.84)	72.52 (2.01)	64.10 (0.72)	39.20 (2.42)	64.52 (1.09)
DW		34.02 (0.80)	0.11 (0.05)	35.03 (0.54)	54.89 (2.84)	34.13 (1.98)	49.82 (1.11)	44.32 (1.41)	27.10 (2.49)	43.33 (1.30)
DGI		49.05 (3.22)	0.80 (0.86)	51.64 (0.59)	62.63 (4.33)	22.96 (1.42)	69.18 (2.55)	58.37 (2.34)	27.61 (1.48)	64.01 (0.87)
	Concat	49.72 (2.93)	0.61 (0.50)	52.15 (0.43)	65.51 (4.09)	22.84 (0.72)	70.41 (2.44)	60.62 (2.49)	28.82 (2.29)	64.47 (1.04)
DGI	Min	49.20 (2.74)	0.63 (0.51)	51.48 (0.86)	63.81 (6.87)	23.90 (3.20)	09.86 (1.75)	58.96 (4.59)	30.47 (3.45)	64.36 (0.59)
	Max	47.27 (3.78)	0.47 (0.25)	51.12 (0.75)	60.24 (4.91)	19.37 (1.81)	68.53 (3.38)	57.62 (1.97)	23.70 (1.83)	63.24 (0.94)
	Sum	49.17 (3.09)	0.80 (0.86)	51.64 (0.59)	63.42 (4.16)	22.96 (1.42)	69.18 (2.55)	59.00 (2.24)	27.61 (1.48)	64.01 (0.87)
	Att,BT	50.76 (0.81)	1.46 (1.63)	51.66 (0.73)	66.34 (3.20)	29.10 (3.10)	70.99 (1.11)	60.71 (2.89)	31.49 (4.04)	64.36 (0.93)
	Att,MSE	49.05 (3.31)	0.81 (0.59)	51.62 (0.62)	62.56 (4.29)	22.49 (1.45)	69.15 (2.56)	58.37 (2.34)	27.73 (1.43)	64.01 (0.89)
DGI	CL,BT	52.82 (1.67)	1.51 (0.74)	50.88 (1.33)	68.16 (2.22)	32.67 (3.13)	70.40 (0.78)	61.19 (2.11)	33.92 (2.93)	62.88 (0.47)
	CL,MSE	46.95 (2.50)	0.57 (0.55)	50.62 (1.51)	61.61 (4.11)	22.27 (1.65)	68.11 (2.22)	57.91 (3.24)	27.57 (0.71)	63.52 (0.82)
	Lk,BT	50.50 (1.78)	2.26 (2.28)	49.40 (1.32)	68.43 (2.03)	28.44 (3.30)	66.39 (0.47)	54.76 (1.54)	31.86 (0.37)	60.92 (0.94)
	Lk,MSE	33.11 (0.86)	0.06 (0.01)	34.03 (0.53)	14.12 (1.16)	1.09 (0.08)	18.22 (0.34)	16.85 (1.17)	0.78 (0.12)	18.22 (0.65)

Discussion. Results are given in Table 6. The supervised `GCN` and `GAT` models perform better than the unsupervised ones. However, in some cases, the performance gap is small (e.g., `GAT` and `DGI` on Freebase in the node classification task).

Regarding other non-trainable operators in conjunction with the DGI model, we observe that Concat often delivers the best performance, however, at the cost of higher-dimensional node embedding vectors. This approach would be infeasible when dealing with multiplex networks with many layers (edge types). The other operators (Min, Max, Sum) perform quite similarly to each other on all datasets. However, on the ACM dataset, we observe that using either Max or Sum aggregation leads to a performance increase of about 2 pp compared to Mean along at a reduced std. Trainable operators also allow for an increase in the performance in downstream tasks in most cases – on ACM, the best choice is the CL, BT model with approx. 85% MaF1 and 86% Sim@5, while for clustering the best score is achieved by Att, BT. On Amazon and Freebase, the results are not significantly different. On IMDB and CiteSeer, we obtain about 2-3pp increase in node classification, and on Cora CL, BT achieves about 68% MaF1, while outperforming the results of DGI, Mean on clustering and similarity search tasks.

4   Conclusions, Research Gaps and Future Work

In this paper, we tackled the problem of node representation learning in multiplex graphs. We proposed a novel taxonomy for categorizing embedding methods based on the location where the information fusion from different graph layers (edge types) occurs. In particular, we identified graph-level, GNN-level, posthoc embedding-level and prediction-level fusion approaches. These are based either on non-trainable basic operators (e.g., averaging) or more complex trainable ones (e.g., attention). We summarize a detailed comparison of methods and fusion approaches in Table 7. Combined with our experimental evaluation on three downstream tasks and six multiplex graph datasets, we find the following limitations of existing approaches and propose guidelines for future research:

Model Architecture. Most un-/self-supervised representation learning methods for multiplex graphs are based on the DGI method. Self-supervised learning has proved to achieve SoTA performance on homogeneous graph datasets. We propose to evaluate other base architectures. The attempt to utilize the G-BT architecture on multiplex graphs, i.e., S^2MGRL, revealed promising results. However, the final proposed model does not work in a fully unsupervised fashion and requires node labels to train the fusion mechanism.

Inductive Learning. DMGI achieves the best results on most datasets and downstream tasks. However, the utilized lookup embedding-based fusion mechanism does not provide inductive capabilities. This is especially important in real-world applications where the graph data is subject to change.

Table 7. Comparison of multiplex graph representation learning and fusion methods.

Name		Unsuperivsed?	Inductive?		Fusion location/level				Fusion type
			Node/edge	Layer	Graph	GNN	Emb.	Pred.	
Flattened	GCN	✗	✓	✓	✓	✗	✗	✗	graph flattening
	GAT	✗	✓	✓	✓	✗	✗	✗	
	DW		✗	✗	✓	✗	✗	✗	
	DGI		✓	✓	✓	✗	✗	✗	
MHGCN		✓	✓	✗	✓	✗	✗	✗	weighted sum of adj. matrices
DMGI		✓	✗	✗	✗	✓	✗	✗	lookup
HDGI		✓	✗	✗	✗	✓	✗	✗	semantic attention
S^2MGRL		✓ (fusion: ✗)	✗	✗	✗	✓	✗	✗	supervised attention
F(GBT, *)			✓	✗	✗	✓	✗	✗	attention / concat with linear proj.
F(DGI, *)			✓	✗	✗	✓	✗	✗	
GCN	Mean	✗	✓	✗	✗	✗	✓	✗	mean
GAT		✗	✓	✗	✗	✗	✓	✗	
DW		✓	✗	✗	✗	✗	✓	✗	
DGI		✓	✓	✗	✗	✗	✓	✗	
DGI	Concat		✓	✗	✗	✗	✓	✗	concat
	Min		✓	✗	✗	✗	✓	✗	min
	Max		✓	✗	✗	✗	✓	✗	max
	Sum		✓	✗	✗	✗	✓	✗	sum
DGI	Att,BT		✓	✗	✗	✗	✓	✗	attention
	Att,MSE		✓	✗	✗	✗	✓	✗	
	CL,BT		✓	✗	✗	✗	✓	✗	concat with linear proj.
	CL,MSE		✓	✗	✗	✗	✓	✗	
	Lk,BT		✗	✗	✗	✗	✓	✗	lookup
	Lk,MSE		✗	✗	✗	✗	✓	✗	
Vote	soft	✗	✓	✗	✗	✗	✗	✓	ensemble voting
	hard	✗	✓	✗	✗	✗	✗	✓	

New Type of Inductive Models. Inductivity in terms of multiplex graphs can be viewed twofold: (A) modification of a single graph layer (addition of nodes or edges), where most GNN-based methods are capable of inferring new embeddings in such scenarios, (B) addition of new multiplex graph layers. Virtually all methods requires a whole model retraining when adding a new graph layer. Methods based on graph flattening are capable of layer-inductivity at the cost of rather poor performance in downstream tasks.

Multiplex GNN Layer. We did not find any attempts to build dedicated GNN layer architectures designed for multiplex graph data. Proposed models instead utilize existing ones and apply them to each graph layer separately, relying on information fusion to be executed in a later step.

Model Scalability. Moreover, applying GNNs at each layer separately yields a high number of model parameters. We propose to explore the application of GNN sharing. For instance, this could be achieved by designing new positional encodings for multiplex graph layers.

Acknowledgments. The project was partially supported by the National Science Centre, Poland (grant number 2021/41/N/ST6/03694), the European Union under the Horizon Europe grant OMINO (grant number 101086321) and Polish Ministry of Science, as well as statutory funds from the Department of Artificial Intelligence.

References

1. Bielak, P., Kajdanowicz, T., Chawla, N.V.: Graph Barlow twins: a self-supervised representation learning framework for graphs. Knowl.-Based Syst. **256**, 109631 (2022). https://doi.org/10.1016/j.knosys.2022.109631
2. Fey, M., Lenssen, J.E.: Fast graph representation learning with PyTorch Geometric. In: ICLR Workshop on Representation Learning on Graphs and Manifolds (2019)
3. He, R., McAuley, J.: Ups and downs: modeling the visual evolution of fashion trends with one-class collaborative filtering. In: WWW 2016 (2016). https://doi.org/10.1145/2872427.2883037
4. Kipf, T.N., Welling, M.: Variational graph auto-encoders. In: NIPS Workshop on Bayesian Deep Learning (2016)
5. Kipf, T.N., Welling, M.: Semi-supervised classification with graph convolutional networks. In: International Conference on Learning Representations (ICLR) (2017)
6. Kuprieiev, R., et al.: DVC: data version control - git for data & models, May 2021. https://doi.org/10.5281/zenodo.4733984
7. Mo, Y., Chen, Y., Peng, L., Shi, X., Zhu, X.: Simple self-supervised multiplex graph representation learning. In: Proceedings of the 30th ACM International Conference on Multimedia, MM 2022 (2022). https://doi.org/10.1145/3503161.3547949
8. Park, C., Kim, D., Han, J., Yu, H.: Unsupervised attributed multiplex network embedding. In: AAAI 2020 (2020). https://doi.org/10.1609/aaai.v34i04.5985
9. Perozzi, B., Al-Rfou, R., Skiena, S.: DeepWalk: online learning of social representations. In: KDD 2014 (2014). https://doi.org/10.1145/2623330.2623732
10. Ren, Y., Liu, B., Huang, C., Dai, P., Bo, L., Zhang, J.: Heterogeneous deep graph infomax. ArXiv **abs/1911.08538** (2019)
11. Tang, J., Qu, M., Wang, M., Zhang, M., Yan, J., Mei, Q.: Line: large-scale information network embedding. In: WWW. ACM (2015)
12. Thakoor, S., Tallec, C., Azar, M.G., Munos, R., Veličković, P., Valko, M.: Bootstrapped representation learning on graphs. In: ICLR 2021 Workshop on Geometrical and Topological Representation Learning (2021)
13. Veličković, P., Cucurull, G., Casanova, A., Romero, A., Liò, P., Bengio, Y.: Graph attention networks. In: International Conference on Learning Representations (2018)
14. Veličković, P., Fedus, W., Hamilton, W.L., Liò, P., Bengio, Y., Hjelm, R.D.: Deep graph infomax. In: International Conference on Learning Representations (2019)
15. Wang, X., et al.: Heterogeneous graph attention network. In: WWW 2019 (2019). https://doi.org/10.1145/3308558.3313562
16. Wang, X., Liu, N., Han, H., Shi, C.: Self-supervised heterogeneous graph neural network with co-contrastive learning. In: KDD 2021 (2021). https://doi.org/10.1145/3447548.3467415
17. Wu, C.Y., Beutel, A., Ahmed, A., Smola, A.J.: Explaining reviews and ratings with PACO: poisson additive co-clustering. In: WWW 2016 (2016). https://doi.org/10.1145/2872518.2889400
18. You, Y., Chen, T., Sui, Y., Chen, T., Wang, Z., Shen, Y.: Graph contrastive learning with augmentations. In: NeurIPS 2020 (2020)

19. Yu, P., Fu, C., Yu, Y., Huang, C., Zhao, Z., Dong, J.: Multiplex heterogeneous graph convolutional network. In: KDD 2022 (2022). https://doi.org/10.1145/3534678.3539482
20. Zhu, Y., Xu, Y., Yu, F., Liu, Q., Wu, S., Wang, L.: Deep graph contrastive representation learning. In: ICML Workshop on Graph Representation Learning and Beyond (2020)

Data Augmentation to Improve Molecular Subtype Prognosis Prediction in Breast Cancer

Francisco J. Moreno-Barea[1]([✉])(iD), José M. Jerez[1](iD), Nuria Ribelles[2](iD), Emilio Alba[2](iD), and Leonardo Franco[1](iD)

[1] Departamento de Lenguajes y Ciencias de la Computación, Escuela Técnica Superior de Ingeniería Informática, Universidad de Málaga, Málaga, Spain
fjmoreno@lcc.uma.es
[2] Unidad de Gestión Clínica Intercentros de Oncología, Hospitales Universitarios Regional y Virgen de la Victoria, Málaga, Spain

Abstract. Breast cancer is a major public health problem, with 2.3M new cases diagnosed each year. Immunotherapy is an effective treatment for breast cancer depending on several factors like subtype of tumours or associated prognosis. However, the immune system's efficiency depends on the local microenvironment and requires region-specific trials with a reduced number of samples. To minimise this drawback and improve the accuracy of patient prognosis predictions, we explore several data augmentation methods, i.e. noise injection, oversampling techniques and generative adversarial networks. The experiment was conducted through a set of immune system gene expression samples donated by 165 breast cancer patients from the Málaga region. Results showed a 5% increase in AUC and a 23-36% increase in F_1 score for subtype prediction.

Keywords: Data augmentation · Breast Cancer · Cancer Prognosis · Data Mining · E-Health

1 Introduction

Despite several advances, cancer remains a serious medical problem. Cancer is currently the second leading cause of death in developed countries, after major cardiovascular diseases. Breast cancer in particular is one of the world's major health problems, with a high number of cases diagnosed each year. It is the most common tumour in women, with an estimated 2.3 million cases worldwide in 2022 (11.7% of all diagnosed cancers). Fortunately, breast cancer has a lower mortality rate due to improvements in early detection and less aggressive therapies.

In recent decades, immunotherapy has emerged as a potent and less aggressive strategy for treating cancer, and countless studies have clearly provided a boost to understanding the effects of the immune system on tumour development and establishment [9]. One way to monitor the immune response of patients is to

L. Franco et al. (Eds.): ICCS 2024, LNCS 14837, pp. 19–27, 2024.
https://doi.org/10.1007/978-3-031-63778-0_2

analyse the molecular variables encoded by the major histocompatibility complex (MHC). The MHC is a system of interrelated genes whose main role is to control the expression of cell surface molecules acting as markers of the immune response. This paper examines a cohort of breast cancer patients from the region of Málaga, Spain. The dataset represents the MHC gene expression of patients associated with the molecular subtype of breast cancer [15] considered as the labelled class in a supervised learning analysis.

The disadvantage of applying machine learning (ML) methods to these regional bioinformatics datasets is the scarcity of data. The application of data augmentation (DA) methods, which allow the addition of synthetically generated samples, has become a relevant topic. DA has proven to be very effective for improving the performance of ML models, and recent results from generative artificial intelligence are demonstrating the potential of these models, even in the medicine and healthcare field [7,17]. But applying DA techniques to non-structured datasets is far more complex. DA techniques available to deal with this type of dataset include noise injection techniques [13], oversampling techniques [3], and recently the Generative Adversarial Networks (GAN) [5]. GAN models have shown an impressive level of success in generating synthetic samples, and recently they have shown good results as a DA method for datasets without spatial or temporal structure [11], also in the biomedical domain [4,6,8,12].

Considering all the above aspects, the main objective of this work is to apply state-of-the-art DA methods to a small MHC gene expression data set expecting an increase in the prediction performance of the prognosis associated with the molecular subtype of breast cancer patients.

2 Related Works

Works on the application of DA to bioinformatics problems have mainly focused on the treatment of medical images and tasks involving time series. However, DA with -omic data is recent and challenging, but recently works shows that DA methods can be beneficial to improve prediction performance. Among the DA studies using classical methods, Beinecke and Heider [2] applied Gaussian noise, SMOTE and ADASYN methods to clinical data from the UCI ML repository covering different medical fields. Related to the application of deep learning-based models to unstructured data, the work of Marouf et al. [8] used a GAN for realistic generation of single cell RNA-Seq data and detection of marker genes. García-Ordás et al. [4] built a variational autoencoder to predict diabetes in pima indians. Barile et al. [1] employed a Generative Adversarial Autoencoder for the generation of synthetic structural brain network with sclerosis.

To the best of our knowledge, this paper is the first work that proposes the application of DA to improve the predictive performance of prognosis associated with molecular subtypes of breast cancer using genomic data. Only a few papers applied DA to genomic cancer samples. Moreno-Barea et al. [12] used DA to improve the prediction of fixed-time events in cancer given 18 different cancer types from The TCGA database; Wei et al. [16] developed a GAN-based model to expand 12 cancer datasets, improving the accuracy of cancer diagnosis; and Gutta et al. [6] used a GAN model and image processing to improve survival prediction in breast cancer patients.

3   Breast Cancer Cohort

According to studies of gene expression patterns associated with prognosis or metastatic risk [15], breast cancer can be divided into two main groups based on estrogen receptor positivity. A first group are low-grade neoplasms or estrogen receptor-positive tumours, also called luminal subtypes; and a second group are high-grade neoplasms or estrogen receptor-negative tumours, called non-luminal subtypes. The luminal subtypes are associated with low/medium grade and good/intermediate prognosis, whereas the non-luminal are associated with higher grade and poorer prognosis.

The cohort used to test the prognosis associated with molecular subtype of cancer was provided by the Carlos Haya Regional and the Virgen de la Victoria University Hospitals in Málaga, Spain. The data are part of a clinical study of a group of 165 breast cancer patients diagnosed between 2008 and 2013. The data set consists of patient gene expression (molecular variables encoded by the MHC) and cancer subtype. The division between the two groups according to estrogen receptor positivity was taken into account as classes. The most common molecular subtype in the dataset was luminal with 135 samples (82%), whereas 30 samples were non-luminal (18%).

4   Data Augmentation Methods

A simple but effective method to start testing DA is noise addition, technique based on a random modification of the original instances. To apply it, we did a random selection of samples and modified a maximum of 25% of the features [13]. Equation 1 mathematically describes the process of obtaining a new feature value $\tilde{x}$ from the original one x, where min_V and max_V are the actual limits of each feature. An oversampling noise addition method ("Noise bal") is also applied, differing from the previous method in that it performs the random selection only on the training samples belonging to the minority class.

$$\tilde{x} = \min(\text{max\_V}, \max(\text{min\_V}, x + \text{RND}(-0.2, 0.2)))   \tag{1}$$

SMOTE is a classic technique specifically designed for unbalanced data sets [3]. To generate synthetic minority class data and balance the classes, SMOTE uses a k-nearest neighbour algorithm on the minority class instead of random sampling with replacement. SMOTE performs an interpolation of the selected instance and its nearest neighbours, creating new instances for the minority class that are located in the region between the sample and its neighbours.

Currently, the generation of images has shown impressive success through the application of GAN models [5]. GANs attempt to learn the distribution of the original dataset in order to generate new samples from the learned distribution. The standard GAN model has a structure divided into two networks (Generator and Discriminator) trained simultaneously, so that they can learn from each other. In this context, the goal of the discriminator (D) is to distinguish whether a sample comes from the set of real data or is a generated sample. On the other hand, the generator (G) produces as output a distribution assigned to the space of real samples with the purpose of presenting similar features.

Variations in network architecture, loss function or the inclusion of additional information have been proposed from the basic GAN model. Specifically, since a supervised task is performed in the study, the models considered are the Conditional GAN (CGAN) [10], the Auxiliary Classifier GAN (ACGAN) [14] and the Generative-Classifier GAN (ModCGAN) [11]. The CGAN model is a simple variant of the vanilla GAN model in which the information about the label of samples y is taken into account in D and G networks. The CGAN cost function (Eq. 2) contains two parts identified with the two networks involved in the competitive process. One related to the detection of samples which are in the real distribution, and the other involved in detecting those samples generated by G.

$$\min_{G} \max_{D} \mathbb{E}_{x \sim p_{data}(x)}[\log D(x|y)] + \mathbb{E}_{z \sim p_z(z)}[\log(1 - D(G(z|y)))] \qquad (2)$$

The ACGAN model was also applied [14]. Like CGAN, ACGAN takes both the latent space and the class label as input to G. The main difference is that D receives only the sample as input, without the class label. The discriminator process still predicts whether the given sample is real or false, but also the class label of the sample. The architecture of D includes a single neural network model with two outputs: one output to distinguish the origin of the sample (real/false) and another to obtain the probability that the sample belongs to each class.

Finally, ModCGAN model was also considered [11]. This method is similar to ACGAN, but instead of integrating the classifier within the discriminator, uses externally a so-called generative classifier (GC). This GC is used to label the synthetic samples produced by the generator and discard them if they are of insufficient quality. A ModCGAN with 'Balanced Multiclass' (_BM) was also considered due to the unbalanced nature of the problem. This GAN-based modification uses two independent models, where each model is trained with an unbalanced set corresponding to each class. The purpose is to allow each generator to focus on one of the classes of the problem, always taking into account its differences from samples from the other class.

5   Experiments and Results

In the experimental procedure followed, a Principal Component Analysis (PCA) was first performed on the original dataset to obtain the component score vectors and transform the dataset from binary to continuous variables. After the PCA transformation, a stratified cross-validation procedure was performed. Due to the small number of non-luminal samples in the dataset, a 60% split was used for training and 40% for testing. The data generation process used the training set to generate the desired number of synthetic samples, except for SMOTE. The main generative models were tested with different levels of DA percentages. The synthetic samples generated were added to the training data. Prediction was performed on the test set to obtain the evaluation metrics, which were averaged on a cross-validation scheme that was repeated 10 times with different seeds.

Table 1. Test results obtained for the binary problem with the different DA methods using a RF and a SVM system as classifiers.

Method	Perc	Accuracy	Sensitivity	Specificity	F_1 score	AUC
RF						
Original	None	.8121 ±.002	.3083 ±.012	.9241 ±.004	.3652 ±.009	.7643 ±.005
CGAN	750	.8341 ±.003	.4133 ±.019	.9276 ±.004	.4624 ±.015	.7930 ±.005
ACGAN	500	.8335 ±.003	.3917 ±.014	.9317 ±.004	.4465 ±.012	.7851 ±.006
ModCGAN	750	.8364 ±.002	.4117 ±.016	.9307 ±.003	.4672 ±.011	.7857 ±.005
ModCGAN_BM	750	.8370 ±.003	.4500 ±.016	.9230 ±.004	.4903 ±.012	.8015 ±.006
NOISE	50	.8302 ±.003	.3250 ±.014	**.9424 ±.003**	.3978 ±.014	.7674 ±.005
NOISE Bal	1000	**.8433 ±.002**	**.6500 ±.014**	.8863 ±.003	**.5988 ±.007**	**.8107 ±.005**
SMOTE	None	.8380 ±.002	.5825 ±.016	.8948 ±.003	.5597 ±.009	.8040 ±.004
SVM						
Original	None	.8177 ±.002	.2358 ±.021	**.9470 ±.004**	.2723 ±.019	.7920 ±.005
CGAN	500	.8423 ±.002	.5092 ±.019	.9163 ±.004	.5291 ±.012	.8225 ±.005
ACGAN	500	.8452 ±.002	.4767 ±.017	.9270 ±.003	.5125 ±.012	.8301 ±.004
ModCGAN	200	.8427 ±.003	.4917 ±.012	.9207 ±.005	.5204 ±.009	.8296 ±.004
ModCGAN_BM	750	.8488 ±.002	.5383 ±.013	.9180 ±.003	.5535 ±.010	.8358 ±.005
NOISE	1000	.8188 ±.003	.2483 ±.013	.9456 ±.003	.3187 ±.014	.7592 ±.006
NOISE Bal	750	**.8490 ±.002**	**.7262 ±.015**	.8763 ±.003	**.6360 ±.008**	**.8405 ±.006**
SMOTE	None	.8321 ±.003	.6600 ±.015	.8704 ±.005	.5877 ±.008	.7978 ±.007

Table 1 shows the test results obtained for the different methods and DA models applied, using Random Forest (RF) and Support Vector Machines (SVM) as classifier models. In the table, "Original" indicates that no DA method is applied, serving as a reference level. The second column ('Perc.') refers to the DA percentage applied, noting that SMOTE only produces balanced classes. The remaining columns show the values (± 'between-validation performance' SE) obtained for each of the test metrics: accuracy, sensitivity, specificity, F_1 score and Area Under Curve ROC (AUC). The good/interm. prognosis subtype (majority) is considered the negative class to measure sensitivity and specificity.

The results show an improvement in prediction performance for most evaluation metrics with the Noise Bal method for both classifier models, compared to the performance obtained with the non-augmented data set. Specifically, for the RF system, augmentation using the Noise Bal strategy with 1000% DA achieves an improvement of 3.0% in accuracy, 4.6% in AUC and 23.4% in F_1 score with

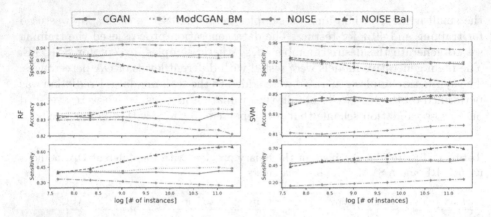

Fig. 1. Comparison of the accuracy, specificity and sensitivity obtained using different DA methods, in terms of the logarithm of the number of instances generated. Code lines: blue solid, CGAN; orange dots, ModCGAN with balancing; green dashes and dots, noise; red dashes, Noise Bal. (Color figure online)

respect to the reference levels with the original set with no DA. On the other hand, using the Noise Bal method with 750% and the SVM classifier, similar values are obtained. With regard to the GAN-based deep generative models, the evaluation metrics also indicate an improvement in the performance of the classifiers compared to the reference set. Among them, the ModCGAN model with the sample balancing modification stands out, obtaining the highest values for the different metrics, except specificity. Accuracy and AUC values obtained with ModCGAN_BM are similar to those obtained with Noise Bal, although the latter obtains a better value for sensitivity, resulting in higher values for the rest of the metrics, especially F_1 score.

The influence of the size of the generated dataset on the prediction results can be seen in Fig. 1. This presents values for accuracy, specificity and sensitivity obtained with CGAN, ModCGAN_BM, Noise and Noise Bal, against the number of instances on a logarithmic scale on the abscissa axis. The CGAN and ModCGAN_BM results show stability in the values as the number of generated samples increases, with a slightly gain in sensitivity. On the other hand, the Noise Bal strategy shows a significant negative correlation for the gain in specificity and a positive correlation for sensitivity and accuracy. This fact explains the higher performance achieved by the Noise Bal technique in F_1 score.

After analysing the performance obtained by applying DA, it is necessary to analyse the quality of the synthetic generated data. A PCA was carried out to visualise the configuration of the samples in a two-dimensional space (PC1 vs. PC2). This enables comparison of the distribution of synthetic samples and original samples, analysis of the performance of DA methods, and explanation of the obtained classification performance. Figure 2 shows the distribution of synthetic samples created using the Noise Bal method and ModCGAN (with BM modification) model. These analyses reflect the inner workings of each DA method and model in sample generation. The Noise Bal method generates samples only for the minority (non-luminal) class, so it can be seen how the samples generated are around the original samples modified with noise, although since these are generated from component scores transformed data, the noise in this case generates more variability, i.e. samples further away.

As representative of the deep generative models, ModCGAN_BM has also generated majority class samples. It is important to note that the synthetic samples are adapted to the real distribution of the samples. When there are samples that can be considered as outliers, the Noise Bal method generates samples around them. However, the deep generative models, due to their internal discriminative operation, consider these outliers as data generated by the generator, which prevents the model from generating samples close to, or simply influenced by, these samples.

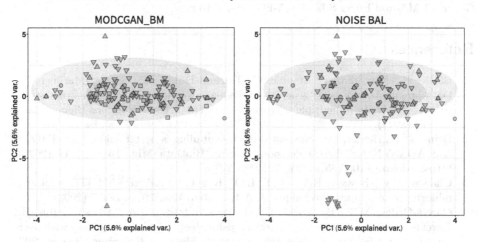

Fig. 2. PCA plot with the original samples and the samples generated by the Noise Bal and ModCGAN_BM methods. Colour codes: green circle, luminal; red triangle, non-luminal; blue square, synthetic luminal; purple triangle, synthetic non-luminal. (Color figure online)

6   Conclusion and Future Work

This paper presents the application of DA techniques as a way to improve the prognosis prediction for breast cancer associated with molecular subtype using a dataset of MHC gene expression profiles. A binary problem is addressed by performing a classification according to whether the subtype is associated with a good/intermediate prognosis (luminal) or a poor prognosis (non-luminal). The overall results obtained suggest and confirm that DA is quite effective when small and unbalanced data are used, leading to a high increase in predictive performance. Among the DA methods applied in the study, the Noise Bal method showed the best performance, leading to an increase in accuracy of 3.0%, with particular attention to the increase in sensitivity, resulting in an improvement of $4.5 - 5\%$ in AUC and $23 - 36\%$ in F_1 score. This resulted in an F1 score of 64%, compared to 30% reference level for the original data, representing a major advance in the ability to predict early treatments in patients who may have a poor prognosis and require more aggressive therapies.

In future works, we would like to extend the present approach to a multiclass problem (prediction of each subtype) and we aim to utilise generative AI and Transformer-based GAN models to enhance the quality of generated data.

Acknowledgements. The authors acknowledge the support from the the the Ministerio de Ciencia e Innovación (MICINN) under project PID2020-116898RB-I00, from the Universidad de Málaga through grant UMA20-FEDERJA-045, and from the Fundación General UMA and Pfizer S.L. (UMA-FGUMA-Pfizer).

References

1. Barile, B., Marzullo, A., Stamile, C., Durand-Dubief, F., Sappey-Marinier, D.: Data augmentation using generative adversarial neural networks on brain structural connectivity in multiple sclerosis. Comput. Methods Programs Biomed. **206**, 106113 (2021)
2. Beinecke, J., Heider, D.: Gaussian noise up-sampling is better suited than SMOTE and ADASYN for clinical decision making. BioData Min. **14**(1), 1–11 (2021). https://doi.org/10.1186/s13040-021-00283-6
3. Chawla, N.V., Bowyer, K.W., Hall, L.O., Kegelmeyer, W.P.: SMOTE: synthetic minority over-sampling technique. J. Artif. Intell. Res. **16**, 321–357 (2002)
4. García-Ordás, M.T., Benavides, C., Benítez-Andrades, J.A., Alaiz-Moretón, H., García-Rodríguez, I.: Diabetes detection using deep learning techniques with over-sampling and feature augmentation. Comput. Methods Programs Biomed. **202**, 105968 (2021)
5. Goodfellow, I., et al.: Generative adversarial nets. In: Advances in Neural Information Processing Systems, pp. 2672–2680 (2014)
6. Guttà, C., Morhard, C., Rehm, M.: Applying a GAN-based classifier to improve transcriptome-based prognostication in breast cancer. PLoS Comput. Biol. **19**(4), e1011035 (2023)
7. He, K., et al.: Transformers in medical image analysis. Intell. Med. **3**(1), 59–78 (2023)
8. Marouf, M., Machart, P., Bansal, V., Kilian, C., Magruder, D.S., et al.: Realistic in silico generation and augmentation of single-cell RNA-seq data using generative adversarial networks. Nat. Commun. **11**(1), 1–12 (2020)
9. Martin, J.D., Cabral, H., Stylianopoulos, T., Jain, R.K.: Improving cancer immunotherapy using nanomedicines: progress, opportunities and challenges. Nat. Rev. Clin. Oncol. **17**(4), 251–266 (2020)
10. Mirza, M., Osindero, S.: Conditional generative adversarial nets. arXiv preprint arXiv:1411.1784 (2014)

11. Moreno-Barea, F.J., Jerez, J.M., Franco, L.: Improving classification accuracy using data augmentation on small data sets. Expert Syst. Appl. **161**, 113696 (2020)
12. Moreno-Barea, F.J., Jerez, J.M., Franco, L.: GAN-based data augmentation for prediction improvement using gene expression data in cancer. In: Groen, D., de Mulatier, C., Paszynski, M., Krzhizhanovskaya, V.V., Dongarra, J.J., Sloot, P.M.A. (eds.) Computational Science – ICCS 2022: 22nd International Conference, London, UK, June 21–23, 2022, Proceedings, Part III, pp. 28–42. Springer, Cham (2022). https://doi.org/10.1007/978-3-031-08757-8_3
13. Moreno-Barea, F.J., Strazzera, F., Jerez, J.M., Urda, D., Franco, L.: Forward noise adjustment scheme for data augmentation. In: IEEE Symposium Series on Computational Intelligence, pp. 728–734 (2018)
14. Odena, A., Olah, C., Shlens, J.: Conditional image synthesis with auxiliary classifier GANs. In: International Conference on Machine Learning, pp. 2642–2651 (2017)
15. Perou, C.M., et al.: Molecular portraits of human breast tumours. Nature **406**(6797), 747–752 (2000)
16. Wei, K., Li, T., Huang, F., Chen, J., He, Z.: Cancer classification with data augmentation based on generative adversarial networks. Front. Comp. Sci. **16**, 1–11 (2022)
17. Zhang, P., Kamel Boulos, M.N.: Generative AI in medicine and healthcare: Promises, opportunities and challenges. Future Internet **15**(9), 286 (2023)

Threshold Optimization in Constructing Comparative Network Models: A Case Study on Enhancing Laparoscopic Surgical Skill Assessment with Edge Betweenness

Saiteja Malisetty[1]([✉]) [iD], Elham Rastegari[2] [iD], Ka-Chun Siu[3] [iD],
and Hesham H. Ali[1] [iD]

[1] University of Nebraska at Omaha, Omaha, NE 68182, USA
{smalisetty,hali}@unomaha.edu
[2] Creighton University, Omaha, NE 68178, USA
elhamrastegari@creighton.edu
[3] University of Nebraska Medical Center, Omaha, NE 68198, USA
kcsiu@unmc.edu

Abstract. Accurate and robust assessment of non-traditional approaches used for training students and professionals in improving laparoscopic surgical skills has been attracting many research studies recently. Such assessment is particularly critical with the recent advances related to virtual environments and AI tools in addressing the need to expand the education and training in the medical domains. Network models and population analysis methods have been identified as excellent approaches in providing the much-needed assessment. This study aims at further advancing the surgical skill assessment by introducing a comparative approach to threshold optimization in analyzing the network models. While the majority of network methods often on arbitrary or hard thresholds for network construction and analysis, this research explores the efficacy of network-based parameters for identifying key elements and clusters in extracting useful information from the constructed networks. We report the positive impact of using network structural parameters, such as edge betweenness and modularity, to conduct robust analysis of the assessment networks. In this work, we employ electromyography (EMG) data and the NASA Task Load Index (NASA-TLX) scores for comprehensive skill evaluation. We present a case study that highlights the advantage of selecting thresholds based on the highest edge betweenness associated with the obtained assessment networks. This proposed approach method proved to be more effective in identifying participants who exhibit significant learning progression, aligning their muscle activation patterns closely with top performers. We demonstrate that optimizing thresholds through edge betweenness offers a more accurate visualization and assessment of skill acquisition in laparoscopic surgery training.

Keywords: Network Models · Threshold Optimization · Training Assessment · Modularity · Edge Betweenness · Clustering Analysis

© The Author(s), under exclusive license to Springer Nature Switzerland AG 2024
L. Franco et al. (Eds.): ICCS 2024, LNCS 14837, pp. 28–42, 2024.
https://doi.org/10.1007/978-3-031-63778-0_3

1   Introduction

Laparoscopic surgery represents one of the most significant advancements in the medical field, offering patients minimally invasive options with faster recovery times [1, 2]. However, the transition from traditional open surgery to laparoscopic techniques presents a steep learning curve for surgical trainees. The inherent complexity of these procedures, combined with the limited tactile feedback and reliance on two-dimensional video feeds, demands a high degree of cognitive and psychomotor skills [1–3].

The assessment of laparoscopic surgical training effectiveness becomes a major concern, not only for ensuring the competency of future surgeons but also for advancing patient safety and care quality. Traditional methods for evaluating surgical skills, primarily based on direct observation and performance metrics such as task completion time or error rates, have shown limitations in capturing the depth and breadth of a trainee's learning progression [4–6]. These assessments often fail to consider the nuanced interactions between cognitive load, psychomotor abilities, and technical proficiency that define a surgeon's capability. Moreover, the subjective nature of these evaluations can introduce variability and bias, further complicating the assessment process.

Network models have emerged as a powerful tool in surgical education research, providing a novel framework to analyze and visualize the complex interactions between various aspects of surgical performance [7–9]. By constructing networks from data captured during training sessions, such as electromyography (EMG) signals, researchers can visualize and analyze the intricate relationships that exist within the learning environment. Network analysis facilitates the construction of a visual and analytical framework, where nodes (representing individual trainees) and edges (denoting relationships or similarities between them) collectively unveil the dynamics of skill acquisition, knowledge transfer, and performance improvement. By employing network models, this study seeks not merely to evaluate surgical skills in isolation but to understand how these skills evolve, intersect, and amplify within a cohort of learners. This approach is particularly well-suited to address the nuanced demands of laparoscopic surgery training, where the cognitive load, dexterity, and precision play pivotal roles [8,9]. Through network analysis, we aim to identify meaningful clusters of learners who exhibit similar patterns of skill development or face comparable challenges, thereby offering insights into the effectiveness of training interventions.

However, the effectiveness of network analysis hinges on the selection of appropriate thresholds for defining meaningful connections within the network. The problem, therefore, lies in identifying a method for threshold optimization that accurately reflects the complex dynamics of surgical skill development. Traditional approaches to threshold selection, which often rely on arbitrary or fixed values, may oversimplify the network, omitting critical information or, conversely, clutter the network with irrelevant connections. This study seeks to address this gap by exploring the potential of network-based parameters, such as edge betweenness [10–12] and modularity [12–14], in enhancing the granularity and accuracy of network models for surgical skill assessment. These parameters offer a more nuanced and data-driven approach to threshold selection, potentially enhancing the visualization and assessment of laparoscopic surgical skills.

By optimizing thresholds based on network properties, researchers can ensure that the resulting models are both robust and sensitive to the subtleties of skill development. This approach not only promises to improve the accuracy and relevance of laparoscopic skill assessments but also contributes to the broader field of surgical education by providing a more sophisticated understanding of how skills are acquired and refined over time. Furthermore, the study explores how the integration of NASA-TLX scores can enrich the network analysis, offering insights into the subjective dimensions of learning and performance in laparoscopic surgery.

2   Methodology

2.1   Needle Passing Task

The needle passing (NP) task is a fundamental component of laparoscopic surgery training, designed to simulate the precision and dexterity required for suturing in a minimally invasive surgical environment [8]. In this task, participants are required to manipulate a virtual needle through a series of designated points (Fig. 1), mimicking the challenges faced during actual surgical procedures. This task was selected for its relevance to core surgical skills and its ability to provide measurable outcomes of psychomotor performance. The study enrolled eighteen participants, ensuring a broad spectrum of experiences by including both medical and non-medical students in equal measure. Criteria for selection included no prior experience with the specific training simulator and no recent injuries that could affect the ability to perform the tasks.

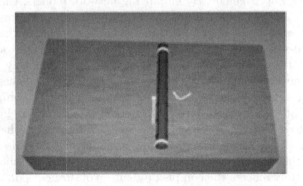

Fig. 1. Needle Passing Task

2.2 EMG Data Collection and Pre-processing

Data collection centered around the use of electromyography (EMG), a technique that records the electrical activity produced by muscles to quantify muscle effort and fatigue levels [15]. EMG data were collected using surface electrodes placed on target muscle groups involved in the task, such as the biceps brachii for arm movements and the flexor carpi radialis for wrist actions. This approach allowed for a detailed analysis of the physiological aspects of task performance, providing insights into the muscular demands of laparoscopic procedures.

For the preprocessing of electromyography (EMG) data, we utilized MAT-LAB due to its advanced signal processing capabilities. The processing of EMG data involved filtering and normalization steps to ensure that measurements could be accurately compared across participants. The signals were band-pass filtered to eliminate noise and then smoothed using a root-mean-square (RMS) technique [16], which provides a measure of the muscle's electrical activity over time. Normalization was performed against a maximal voluntary contraction (MVC) [17] for each muscle group, allowing for the comparison of muscle activation levels relative to the participant's maximum capacity. This rigorous data collection and processing methodology facilitated a nuanced understanding of the physical demands of the needle passing task, laying the foundation for subsequent analyses of skill acquisition and progression.

We collected data at three key times: before training started (session-0), after one week of training (session-1), and four weeks after the first training session (session-2). However, we are mainly focusing on the networks formed during the first and second sessions, leaving out the initial baseline session. This decision helps us highlight how participants' performance changes over time, showing how training affects their laparoscopic surgical skills.

2.3 NASA-TLX Scores

The NASA Task Load Index (NASA-TLX) is a subjective workload assessment tool used to evaluate the perceived workload experienced by individuals while performing tasks [18–20]. The NASA-TLX scores, encompassing dimensions such as mental demand, physical demand, temporal demand, performance, effort, and frustration, offer insights into the subjective experiences of participants as they engage in the needle passing task. The integration of NASA Task Load Index (NASA-TLX) scores into our network model serves to enrich the analysis by providing a subjective measure of workload associated with task performance. By correlating these scores with network properties, we can explore how subjective perceptions of task difficulty and workload relate to objective measures of skill acquisition and learning progression within the network. This holistic approach, combining objective network metrics with subjective workload assessments, provides a more comprehensive framework for understanding the multifaceted nature of surgical skill learning and development.

2.4   Nodes and Edges

In constructing the network for assessing laparoscopic surgical skills, nodes represent the participants who engage in the needle passing task. Each participant is thus represented as a node in the network, capturing their individual involvement in the skill acquisition process. Edges in the network correspond to the connections between participants, reflecting the relationships inferred from the correlation analysis of EMG data. In our study, the assessment of participants' performance similarities was conducted through Pearson's pairwise correlation coefficient (ρ), which quantifies the linear relationship between the EMG activation patterns of different participants. We have selected Pearson's correlation coefficient (ρ) for our analysis because it is ideally suited for data that adheres to a normal distribution. Pearson's correlation is sensitive to linear relationships between variables, making it a robust choice for quantifying the degree of association between the EMG activation patterns of participants, which were verified to be normally distributed. This statistical method enables a precise measurement of the linear inter-dependencies central to understanding the dynamics of surgical skill development.

This correlation coefficient ranges from 0, indicating no linear correlation, to 1, signaling an exceptionally strong linear relationship between the muscle activation patterns of two participants. Through the computation of (ρ) for every possible pair of participants, we constructed a Correlation Matrix (CM), encapsulating the strength of pairwise linear associations among the participants' EMG data. These correlations are used to establish the connections (edges) between nodes in the network, with stronger correlations yielding more meaningful edges. The resulting network structure encapsulates the interrelationships between participants' performances, offering a holistic view of skill acquisition and progression. To discern substantial correlations indicative of similar performance or learning patterns, we established a correlation threshold, k, as our criterion for similarity. This threshold is pivotal for identifying meaningful connections in the context of surgical skill acquisition, reflecting a deliberate focus on the most substantively similar EMG activation profiles. The decision matrix, henceforth referred to as the Adjacency Matrix (AM) as shown in Eq. 1, delineates the adjacency matrix for our network graph, based on the following criterion:

$$\text{AM}(i,j) = \begin{cases} 1, & \text{if } \rho(P_i, P_j) \geq k \\ 0, & \text{for other cases} \end{cases} \tag{1}$$

where P_i and P_j represent the EMG activation patterns of participants i and j respectively. A value of 1 in $AM[i,j]$ signifies a substantial correlation between the participants i and j, hence establishing a link in the network graph, while a 0 denotes the absence of a link between the nodes. The threshold k for defining significant correlations was determined by a statistical method that considers the distribution of correlation values across all participant pairs. We aimed to establish k at a level that ensures the correlations above it are not only statistically significant but also practically relevant to the skills being assessed. The exact process of calculating k and its role in shaping the network structure is further elaborated in the subsequent section on edge betweenness and modularity for threshold selection, where we discuss how this threshold optimizes the balance between capturing meaningful skill relationships and maintaining network coherence. This methodological step ensures that our network analysis focuses on the most pertinent relationships, shedding light on the dynamics of skill acquisition among participants engaging in the needle passing task.

2.5 Edge Betweenness and Modularity for Threshold Selection

Edge betweenness, a measure of centrality, identifies the most meaningful connections within the network by quantifying the number of shortest paths passing through an edge (Eq. 2) [11]. By selecting a threshold at the point where edge betweenness peaks, we ensure that the network encapsulates the most crucial interactions and learning patterns among participants, thereby preserving the integrity and complexity of the skill acquisition network. This selection criterion is vital for identifying the pivotal connections that facilitate learning progression and skill transfer among trainees.

$$C_b(e) = \sum_{s \neq t} \frac{\sigma_{st}(e)}{\sigma_{st}} \tag{2}$$

where:

- $C_b(e)$ is the edge betweenness centrality of edge e.
- σ_{st} is the total number of shortest paths from node s to node t.
- $\sigma_{st}(e)$ is the number of those paths passing through edge e.
- The sum is calculated over all pairs of nodes (s, t), where $s \neq t$.

Modularity, on the other hand, assesses the strength of division of a network into modules or communities as shown in Eq. 3 [12]. High modularity indicates that the network is effectively partitioned into clusters with dense connections internally and fewer connections between clusters. By validating our edge betweenness-based threshold selection with modularity, we ensure that while maintaining crucial learning pathways, the network also accurately reflects distinct groups or communities of skill progression, facilitating a nuanced understanding of how different skills or learning strategies cluster together among participants.

$$Q = \frac{1}{2m} \sum_{ij} \left[A_{ij} - \frac{k_i k_j}{2m} \right] \delta(c_i, c_j) \tag{3}$$

where:

- A_{ij} represents the edge weight between nodes i and j; for unweighted networks, $A_{ij} = 1$ if there is an edge between i and j, and $A_{ij} = 0$ otherwise.

- k_i and k_j are the sum of the weights of the edges attached to nodes i and j, respectively.
- m is the sum of all edge weights in the network.
- $\delta(c_i, c_j)$ is a delta function that equals 1 if nodes i and j are in the same community, and 0 otherwise.
- The sum runs over all pairs of nodes.

In our study, we plotted the mean edge betweenness of the network at various thresholds (Fig. 2) and selected the threshold where the mean edge betweenness

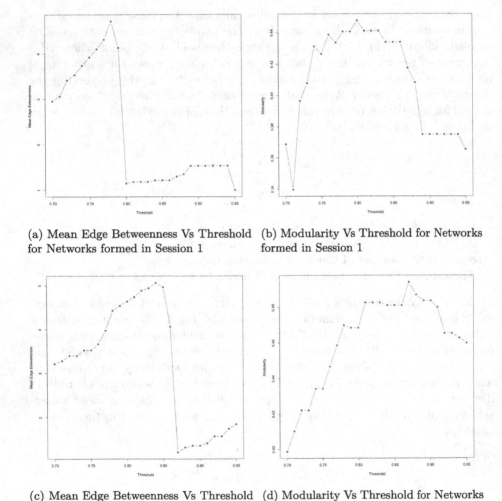

(a) Mean Edge Betweenness Vs Threshold for Networks formed in Session 1

(b) Modularity Vs Threshold for Networks formed in Session 1

(c) Mean Edge Betweenness Vs Threshold for Networks formed in Session 2

(d) Modularity Vs Threshold for Networks formed in Session 2

Fig. 2. Comparative Analysis of Edge Betweenness and Modularity across Different Thresholds for Sessions 1 and 2.

is maximized. Concurrently, we analyzed how modularity varied across these thresholds.

It was observed that at the threshold where modularity reached its maximum, there was a significant decrease in mean edge betweenness. This suggests that optimizing for maximum modularity alone might not yield the most insightful network structure but instead, selecting a threshold based on peak mean edge betweenness, while ensuring modularity does not significantly deviate from its optimum, offers a more balanced approach.

To elucidate these findings, Fig. 2a represents the mean edge betweenness plotted against various thresholds for networks created in Session 1, providing insights into the optimal threshold selection for this cohort. Similarly, Fig. 2b depicts the modularity plotted against various thresholds for networks developed in Session 1, highlighting the relationship between modularity and network structure at different thresholds. For networks created in Session 2, Fig. 2c illustrates the mean edge betweenness plotted against various thresholds, offering a comparative perspective on the impact of session variability on network properties. Lastly, Fig. 2d represents the modularity plotted against various thresholds for networks created in Session 2, further underlining the nuances of modularity optimization across different sessions. These visual representations substantiate our approach by demonstrating the variability and interplay between mean edge betweenness and modularity across sessions, reinforcing the argument for a balanced threshold selection. This methodological insight underscores the complexity of network analysis, advocating for a strategy that harmonizes both mean edge betweenness and modularity.

3 Results

In the analysis of network models constructed from the electromyography (EMG) data, three distinct thresholds were applied: the commonly utilized hard threshold of 0.70 (Fig. 3a and Fig. 3b), one where the mean edge betweenness was maximized (Fig. 3c and Fig. 3d) and another where modularity was at its highest (Fig. 3e and Fig. 3f). This tripartite-threshold approach facilitated a comparative analysis of network structures and their ability to capture skill progression among participants. Notably, when thresholds were set to maximize mean edge betweenness, the network models yielded significant insights into participant skill improvement, particularly for subjects 4 and 7, who were highlighted in green. This observation suggests that selecting a threshold based on edge betweenness effectively identifies learners who have made notable advancements in their surgical skills.

The color coding of nodes within these networks provided an intuitive and visually compelling representation of participant performance and progression. In our models, nodes representing participants were color-coded based on their performance and improvement over the course of the training sessions. Participants who were identified as superior performers, based on their consistently high levels of skill demonstration across tasks, were color-coded in yellow. This color selection was intended to visually highlight these individuals as bench-marks of excellence within the network, making it easier to identify the core group of participants who have mastered the essential skills required for laparo-scopic surgery. In contrast, participants who showed significant improvement in their performance throughout the training period were color-coded in green. The green nodes indicate individuals who, while not initially among the top perform-ers, made notable strides in their skill development, effectively narrowing the gap with the superior performers. This color distinction serves to spotlight the dynamic nature of skill acquisition, showcasing the potential for progress with targeted training and practice. The remaining participants, whose performance did not exhibit substantial improvement or who remained consistent without reaching the level of superior performers, were not highlighted with a specific color. This default representation underscores the variability in learning rates and outcomes among individuals, emphasizing the personalized nature of skill development in laparoscopic surgery.

Moreover, the networks optimized for edge betweenness revealed a compelling structural organization, segregating into two clusters. One of these clusters dis-tinctly grouped the best performers, marked in yellow, alongside subjects 4 and 7, thereby indicating their convergence towards higher skill levels by the end of the training sessions. This emergent clustering phenomenon was less pro-nounced in networks optimized for modularity, underscoring the advantage of edge betweenness-guided threshold selection in discerning the nuances of skill acquisition and performance improvement. The rest of the subjects, represented in grey, formed the second cluster, suggesting a delineation between those who demonstrated significant improvement and those whose performance remained relatively static or improved at a slower rate.

Furthermore, our analysis identified a significant concentration of top per-formers within the first cluster, distinguished by their higher variability in EMG data. This grouping suggests that the variability observed might be indicative of more effective or advanced learning strategies being employed by these par-ticipants. In contrast, the second cluster, characterized by less EMG variability, lacked a significant presence of top performers. This observation implies that while stability in muscle activation patterns can signify proficiency, the dynamic variability in EMG responses among top performers may reflect ongoing opti-mization and refinement of surgical techniques.

These insights, drawn from the comparative analysis of EMG data and net-work model clustering, underscore the multifaceted nature of surgical skill devel-opment. The presence of distinct clusters, delineated by EMG variability and the concentration of top performers, offers valuable perspectives on the learning tra-jectories of surgical trainees. It suggests that both the exploration of diverse muscle activation strategies and the stabilization of efficient patterns play criti-cal roles in the acquisition of laparoscopic surgical skills.

By leveraging the detailed data provided by EMG analysis and the struc-tural insights from network models, this study contributes to a deeper under-standing of how surgical trainees progress and differentiate in their skill levels.

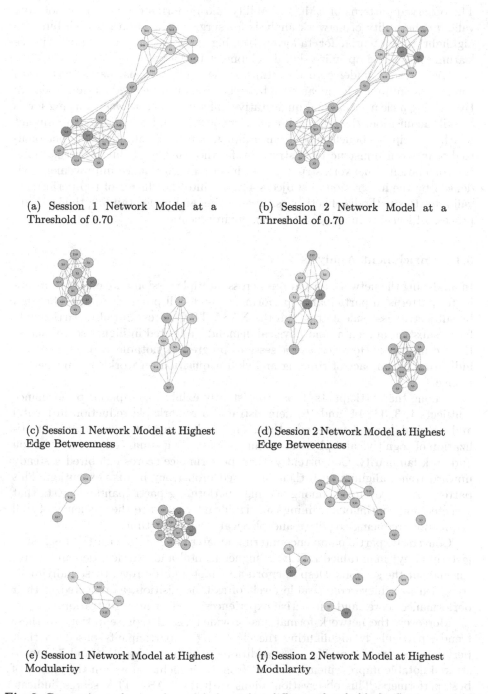

(a) Session 1 Network Model at a Threshold of 0.70

(b) Session 2 Network Model at a Threshold of 0.70

(c) Session 1 Network Model at Highest Edge Betweenness

(d) Session 2 Network Model at Highest Edge Betweenness

(e) Session 1 Network Model at Highest Modularity

(f) Session 2 Network Model at Highest Modularity

Fig. 3. Comparison of network models formed at different thresholds for Session 1 and Session 2. Nodes colored in yellow are the best performers. Nodes colored in green are the participants who have improved their performance moving from session 1 to session 2. The rest of the participant nodes are colored in grey. (Color figure online)

The observed patterns of EMG variability and performance clustering not only validate the utility of network analysis for surgical education research but also highlight the potential for tailoring training interventions to support diverse learning needs and optimize skill development pathways.

These results underscore the utility of network analysis, particularly with thresholds optimized for mean edge betweenness, in surgical education research. By offering a clear visual and quantitative differentiation between varying levels of skill acquisition, the approach enables educators and researchers to pinpoint which participants benefit most from training, potentially guiding more personalized or intensified instructional strategies for those lagging behind. Furthermore, the correlation of network structures with actual performance improvement evidenced by the integration of subjects 4 and 7 into the cluster of top performers validates the method's effectiveness in assessing and understanding the intricate process of learning in surgical training environments.

3.1   Enrichment Analysis

In analyzing the network formations across multiple sessions, we observed meaningful patterns in participants' performance and skill progression, corroborated by subjective assessments through the NASA TLX scores. Initially, participants faced substantial mental and physical demands, reflected in higher scores across these dimensions. However, as the sessions progressed, notable changes emerged, indicating the impact of training and skill acquisition on workload and performance.

Among the participants, those consistently exhibiting superior performance (subjects 1, 3, 11, 14, and 16) demonstrated a remarkable reduction in mental and physical demand scores over time. This decline suggests a more efficient utilization of cognitive and physical resources, indicating enhanced skill acquisition and task familiarity. Concurrently, their performance scores exhibited a steady upward trend, aligning with their improved proficiency in task execution. This pattern of improvement among the high-performing participants suggests that targeted and continuous training can significantly enhance the efficiency of skill acquisition, reducing cognitive and physical strain over time.

Conversely, participants encountering greater challenges with the task (subjects in Grey) maintained relatively higher mental and physical demand scores throughout the sessions. Despite efforts to adapt and improve, these individuals struggled to achieve comparable levels of skill acquisition, as reflected in their performance scores and subjective experiences of effort and frustration.

Moreover, the network formations provided visual representations of these trends, particularly highlighting the clustering of participants based on their performance levels. Notably, after additional training, subjects 4 and 7 demonstrated notable improvements, evident from their inclusion within the cluster of best performers. This observation aligns with the NASA TLX scores, indicating reduced mental and physical demands, alongside enhanced performance and decreased feelings of effort and frustration.

Overall, the integration of subjective assessments through NASA TLX scores provided valuable insights into the nuanced dynamics of skill acquisition and performance progression. These findings underscore the importance of considering both objective performance metrics and subjective experiences in evaluating surgical training outcomes, paving the way for more comprehensive and tailored approaches to skill development in laparoscopic surgery.

4 Discussion

The selection of an appropriate threshold for constructing network models from electromyography (EMG) data significantly influences the interpretation and assessment of surgical skill development. Traditional methods of applying hard thresholds, although simple and straightforward, often fail to account for the complexity and variability inherent to the process of acquiring surgical skills. This study underscores the importance of selecting an optimal threshold to accurately identify subtle patterns of skill acquisition and progression among participants. Utilizing a threshold that is excessively high risks ignoring crucial yet weaker correlations, while a threshold that is too low may introduce irrelevant connections into the network, obscuring significant patterns of improvement.

Optimizing threshold selection through edge betweenness offers several advantages. Firstly, it enhances the sensitivity of detecting performance changes over time, enabling the identification of participants who demonstrate significant advancements in their skills. This method's adaptability is particularly beneficial in the dynamic context of learning and skill acquisition, as it can accommodate a wide range of correlation strengths and complexities within the network. Moreover, by concentrating on the most impactful interactions, edge betweenness optimization ensures a focused analysis on the relationships that most accurately reflect skill progression, thereby improving the overall accuracy and relevance of assessments.

Analysis of the EMG data revealed that the superior performers (yellow nodes) and those who improved significantly (green nodes) displayed distinct patterns of muscle activation. Specifically, one group exhibited more variability in EMG muscle values, suggesting a more dynamic and adaptive approach to task execution, while the other group showed less variability, indicating a possibly more consistent but less flexible technique. These findings, visualized through the network models and the color-coded nodes, offer valuable insights into the multifaceted nature of skill acquisition in laparoscopic surgery. By correlating EMG data patterns with performance levels and improvements, our study highlights the potential of using network analysis and physiological data to inform and enhance surgical training methodologies.

The incorporation of subjective workload assessments, such as the NASA Task Load Index (NASA-TLX), alongside network model analyses introduces a more comprehensive approach to evaluating surgical training. This combination acknowledges that acquiring surgical skills is not purely a physical endeavor but also encompasses cognitive, psychological, and emotional facets. Incorporating subjective workload measures provides a holistic evaluation of a trainee's experience, integrating both physical performance and cognitive aspects of surgical skills learning. Such a dual approach enables the provision of more personalized feedback and training interventions, enhancing the effectiveness of training programs. Furthermore, understanding the interplay between subjective workload and performance offers insights into how stress and cognitive load affect skill development, which is crucial for designing optimized training environments and protocols. This comprehensive understanding of skill acquisition, grounded in both objective and subjective assessments, paves the way for more effective training methodologies and improved surgical education outcomes.

Beyond the confines of laparoscopic surgery and even the broader medical field, investigating the applicability of our methodology across various disciplines could uncover its potential to revolutionize professional training and skill assessment on a much wider scale. Furthermore, integrating this innovative approach with simulation-based training technologies could open new frontiers in surgical education. This integration promises to facilitate real-time feedback and adjustments, significantly enhancing the training process by providing trainees with immediate insights into their performance and areas for improvement.

5   Conclusion

With the emergence of various models for training and education, innovative assessment models need to be developed. This is particularly critical in domains where there is shortage of workforce and where assessment is essential before adopting new training practices. This is certainly the case for the medical domain, especially for training medical students and professionals to master surgical skills. This study introduced a novel methodology for further advancing how network models and population analysis can be used for the assessment of surgical skills training. We showed that the quality and robustness of using network-based approaches for assessment can be enhanced by how thresholds are selected in the clustering analysis of the constructed networks. We propose the use of soft thresholds based on well-defined network parameters such as modularity and edge betweenness in assessing the performance of trainees during training sessions of skill acquisition in laparoscopic surgery. By integrating electromyography (EMG) data with subjective workload assessments, we developed a comprehensive framework that offers a deeper insight into the complex dynamics of surgical skill development. Our findings highlight the value in using network models for the assessment process and the significance of conducting the associated clustering analysis using optimizing threshold selection based on edge betweenness.

We introduce an advanced analytical framework that not only surpasses conventional evaluation methods but also supports the development of personalized training programs. These programs can be tailored based on individual progress and specific needs identified through the network analysis, thereby enhancing the effectiveness of surgical training. Furthermore, by combining EMG data with

NASA-TLX scores, our study underscores the importance of considering both physiological and psychological factors in the training assessment. This holistic approach paves the way for more precise and effective training methodologies in laparoscopic surgery.

References

1. Lee, W.J., Chan, C.P., Wang, B.Y.: Recent advances in laparoscopic surgery. Asian J. Endosc. Surg. **6**(1), 1–8 (2013)
2. Cuschieri, A.: Laparoscopic surgery: current status, issues and future developments. Surgeon **3**(3), 125–138 (2005)
3. Aggarwal, R., Moorthy, K., Darzi, A.: Laparoscopic skills training and assessment. J. Br. Surg. **91**(12), 1549–1558 (2004)
4. Gumbs, A.A., Hogle, N.J., Fowler, D.L.: Evaluation of resident laparoscopic performance using global operative assessment of laparoscopic skills. J. Am. Coll. Surg. **204**(2), 308–313 (2007)
5. Oropesa, I., et al.: Methods and tools for objective assessment of psychomotor skills in laparoscopic surgery. J. Surg. Res. **171**(1), e81–e95 (2011)
6. Aggarwal, R., et al.: An evaluation of the feasibility, validity, and reliability of laparoscopic skills assessment in the operating room. Ann. Surg. **245**(6), 992 (2007)
7. Rastegari, E., Orn, D., Zahiri, M., Nelson, C., Ali, H., Siu, K.C.: Assessing laparoscopic surgical skills using similarity network models: a pilot study. Surg. Innov. **28**(5), 600–610 (2021)
8. Malisetty, S., Rastegari, E., Siu, K.C., Ali, H.H.: Exploring the impact of hand dominance on laparoscopic surgical skills development using network models. J. Clin. Med. **13**(4), 1150 (2024)
9. Malisetty, S., Ali, H.H., Rastegari, E., Siu, K.C.: An innovative comparative analysis approach for the assessment of laparoscopic surgical skills. Surgeries **4**(1), 46–57 (2023)
10. Goh, K.I., Oh, E., Kahng, B., Kim, D.: Betweenness centrality correlation in social networks. Phys. Rev. E **67**(1), 017101 (2003)
11. Gago, S., Hurajová, J.C., Madaras, T.: Betweenness Centrality in Graphs. Quantitative Graph Theory: Mathematical Foundations and Applications. Discrete Mathematics and Its Applications, pp. 233–257. Chapman and Hall/CRC, New York (2014)
12. Yoon, J., Blumer, A., Lee, K.: An algorithm for modularity analysis of directed and weighted biological networks based on edge-betweenness centrality. Bioinformatics **22**(24), 3106–3108 (2006)
13. Newman, M.E.: Modularity and community structure in networks. Proc. Natl. Acad. Sci. **103**(23), 8577–8582 (2006)
14. Shang, R., Bai, J., Jiao, L., Jin, C.: Community detection based on modularity and an improved genetic algorithm. Phys. A **392**(5), 1215–1231 (2013)
15. Potvin, J.R., Bent, L.R.: A validation of techniques using surface EMG signals from dynamic contractions to quantify muscle fatigue during repetitive tasks. J. Electromyogr. Kinesiol. **7**(2), 131–139 (1997)
16. Kannan, K., Kanna, B.R., Aravindan, C.: Root mean square filter for noisy images based on hyper graph model. Image Vis. Comput. **28**(9), 1329–1338 (2010)

17. Ekstrom, R.A., Soderberg, G.L., Donatelli, R.A.: Normalization procedures using maximum voluntary isometric contractions for the serratus anterior and trapezius muscles during surface EMG analysis. J. Electromyogr. Kinesiol. 15(4), 418–428 (2005)
18. Hart, S.G.: NASA-task load index (NASA-TLX); 20 years later. In: Proceedings of the Human Factors and Ergonomics Society Annual Meeting, vol. 50, no. 9, pp. 904–908. Sage Publications, Los Angeles, October 2006
19. Zheng, B., Jiang, X., Tien, G., Meneghetti, A., Panton, O.N.M., Atkins, M.S.: Workload assessment of surgeons: correlation between NASA TLX and blinks. Surg. Endosc. 26, 2746–2750 (2012)
20. Yurko, Y.Y., Scerbo, M.W., Prabhu, A.S., Acker, C.E., Stefanidis, D.: Higher mental workload is associated with poorer laparoscopic performance as measured by the NASA-TLX tool. Simul. Healthc. 5(5), 267–271 (2010)

Graph Vertex Embeddings: Distance, Regularization and Community Detection

Radosław Nowak[1,2](✉) [ID], Adam Małkowski[1,2,3] [ID], Daniel Cieślak[1,3] [ID],
Piotr Sokół[1] [ID], and Paweł Wawrzyński[1] [ID]

[1] IDEAS NCBR, Warsaw, Poland
radoslawa.nowak@ideas-ncbr.pl
[2] Polish Academy of Sciences, Warsaw, Poland
[3] Gdańsk University of Technology, Gdańsk, Poland
https://pg.edu.pl/en,https://ideas-ncbr.pl,https://pan.pl/en/

Abstract. Graph embeddings have emerged as a powerful tool for representing complex network structures in a low-dimensional space, enabling efficient methods that employ the metric structure in the embedding space as a proxy for the topological structure of the data. In this paper, we explore several aspects that affect the quality of a vertex embedding of graph-structured data. To this effect, we first present a family of flexible distance functions that faithfully capture the topological distance between different vertices. Secondly, we analyze vertex embeddings as resulting from a fitted transformation of the distance matrix rather than as a direct result of optimization. Finally, we evaluate the effectiveness of our proposed embedding constructions by performing community detection on a host of benchmark datasets. The reported results are competitive with classical algorithms that operate on the entire graph while benefiting from a substantially reduced computational complexity due to the reduced dimensionality of the representations.

Keywords: Graphs · Embeddings · Graph drawing · Community detection

1 Introduction

Low-dimensional metric embeddings of non-metric data play a crucial role in various domains of computer science, e.g.: **a)** machine learning, where probabilistic generative models are constructed to capture variability through low dimensional factors [26,53]; **b)** natural language processing, where symbolic/text data is represented vectorially in order to facilitate learning of statistical dependencies [50]; **c)** information retrieval, where embeddings allow for efficient search [6]; **d)** data visualization, where complex, high-dimensional data is represented in a two- or three-dimensional space [8]. Crucially, low-dimensional embeddings of data mitigate the computational and statistical challenges collectively referred to as the *curse of dimensionality*.

Graph embeddings present a particularly interesting potential application since the data is inherently non-metric and high-dimensional while

L. Franco et al. (Eds.): ICCS 2024, LNCS 14837, pp. 43–57, 2024.
https://doi.org/10.1007/978-3-031-63778-0_4

simultaneously being of tremendous practical interest, due to the ubiquity of graphs in various domains, e.g.: social networks, biological networks, energy grids, and knowledge graphs [24,38]. Addressing the challenges posed by 'graph problems' requires faithful and compact representations of the data, i.e. they should retain information about important structural properties and allow flexible, computationally efficient use for downstream processing/tasks.

Graph embeddings have been widely studied; past approaches include spectral embeddings, graph kernels, multi-dimensional scaling (MDS), locally linear embedding (LLE), and Laplacian eigenmaps (LE) [2,56]. Recently, graph neural networks (GNNs) have emerged as an empirically successful model class, which offers improved scalability and more flexible processing of the embeddings [56].

In this work, we introduce a novel method for representing graph-structured data in low-dimensional metric spaces, which combines the efficiency of optimization-based embeddings and the expressiveness of neural network-based approximate to accurately reflect the topological distances within the graph. By framing the embedding of vertices as an optimization problem, we parametrize the embedding using a small neural network, to regularize the resulting representation. Furthermore, our formulation can accommodate various distance functions, which allows us to adapt the geometry of the embedding space to better reflect the structure of the original graph. The resulting embeddings offer a compact yet faithful representation, which combined with off-the-shelf clustering algorithms allow us to effectively identify communities within the graph. Our approach not only ensures a more favorable computational and statistical scaling, comparable to that of Graph Neural Networks (GNNs) but also provides a highly expressive representation of the graph structure, capturing the intricate relationships and distances within the data.

2   Related Work

The body of research on embeddings is extensive and has a long history. Consequently, we begin by recalling foundational results before providing a concise overview of the methodologies presently in use. For a comprehensive review see [56,60].

Historically, the graph embedding problem has been initially studied in the context of dimensionality reduction and data visualization, while preserving important structural properties of the data. Seminal examples of such methods include PCA, graph kernels [9], Laplacian eigenmaps [4], Isomap [49], LLE [43], maximum variance unfolding [55], t-SNE [32], LargeVis [48], UMAP [34], and the latent variable models (LVM).

[2] introduced *minimum distortion embeddings* as a unifying framework that subsumes all of the previously mentioned methods, except t-SNE and LVM. In their formulation, low-dimensional embeddings are constructed by minimizing the distortion of pairwise distances between data points in the original and the embedded spaces. Independently, [10,21,23,31] have shown that an m-dimensional embedding of graph with n vertices incurs a distortion of order

$\mathcal{O}(\log n)$, where m is $\mathcal{O}(\log^2 n)$. Remarkably, the landmark results of Johnson, Lindenstrauss, and Bourgain [31] not only provide the mathematical foundation of low-dimensional representations of data but also have led to the development of fast, randomized algorithms.

Currently prevalent algorithmic approaches present a variety of design options. An initial consideration requires the specification of a geometry for the embedding space; potential choices include hyperbolic [28,40,44], spherical [2], and vector embeddings. The latter are compatible with a wide range of machine-learning algorithms and have proven versatile in disparate domains [30,35], and model classes, e.g.: Node2vec [19]; graph neural networks [17,51]; Isomap [49]; M-NMF [54]. Among these, we distinguish graph neural networks (GNNs) [17,36,51] due to their wide-spread adoption, and ability to directly learn embeddings from graph-structured data. GNNs are predominantly trained in a supervised or semi-supervised fashion [27]; optimizing a node classification loss, potentially augmented by auxiliary terms, e.g. reconstruction error or noise contrastive estimate [11,13,52]. The resulting algorithms have proven to be efficient, and empirically successful when applied to tasks such as graph regression, node classification, and link prediction [15].

Comparatively, the use of GNNs for community detection has been relatively underexplored [46], despite the pivotal role that communities play in understanding the structure of biology, social, and economic networks [57]. In this context, the applicability of GNNs is curtailed by their reliance on labeled data, which is often unavailable or requires cost-intensive curation.

Classically, clustering algorithms have been used to address the community detection problem, e.g.: Girvan-Newman [18], Louvain [7], and spectral clustering [41] algorithms. More recent approaches, e.g. [22,46], use GNN's to construct an embedding of the data, which is then post-processed using an off-the-shelf clustering algorithm, such as mean shift [29], DBSCAN [16], HDBSCAN [33], Birch [59], OPTICS [3], AffinityPropagation [47], AgglomerativeClustering [39].

Our proposed approach improves on GNN approaches by offering a lightweight neural network model, that can be optimized efficiently without label supervision.

Recent studies citing Agrawal's work underscore its relevance across diverse biological research areas. For example, research highlighting the pitfalls of extreme dimensionality reduction in single-cell genomics– [12] leverages Agrawal's insights to critique conventional visualization techniques, advocating for targeted embedding strategies for more meaningful biological analysis. Similarly, an integrated single-cell dataset study of the hypothalamic paraventricular nucleus (PVN) [5] applies Agrawal's principles to achieve a nuanced molecular and functional classification, revealing the complexity of neuroendocrine regulation. Additionally, advancements in brain-wide neuronal activity recording through blazed oblique plane microscopy, as detailed in a study from Nature [20], demonstrate the utility of Agrawal's work in bridging cellular and macroscale understanding. This particular study challenges the prevailing view that higher resolution is invariably superior, illustrating the impact of Agrawal's contributions to enhancing our understanding of complex biological systems.

3   Problem Statement

We consider a graph given by a set of vertices, $V = \{v_1, \ldots, v_n\}$, where $|V| = n$ is the order of the graph. Given a fixed, we represent the shortest path length between pairs of vertices as the matrix $D \in \mathbb{R}_+^{n \times n}$. These shortest distances may be based on unitary distances between adjacent vertices but also on any positive distances between adjacent vertices. In the present study, we focus exclusively on undirected graphs; therefore, D is symmetric. In the case of disconnected graphs, we assume the topological distance between vertices in different subgraphs is equal to the maximum distance between vertices in the same subgraph plus one.

The problem is to assign an embedding, $e_i \in \mathbb{R}^m$, to each vertex $v_i \in V$, in such a way that the approximate equality

$$d(e_i, e_j) \cong D_{i,j} \tag{1}$$

holds for each $i, j \in \{1, \ldots, n\}$, where $d(\cdot, \cdot)$ is a certain distance in $\mathbb{R}^m$, e.g., the Euclidean distance.

4   Method

4.1   Embeddings as Solution to Optimization Problem

Following [2], we obtain the embeddings e_i (1) as a solution to an optimization problem which minimizes the average discrepancy between $d(e_i, e_j)$ and $D_{i,j}$:

$$[e_1, \ldots, e_n] = \mathrm{argmin}_{[e_1, \ldots, e_n]} \frac{2}{n(n-1)} \sum_{i=1}^{n} \sum_{j=1}^{i-1} L(d(e_i, e_j), D_{i,j}). \tag{2}$$

These discrepancies, denoted above by $L(\cdot, \cdot)$, may be absolute

$$L(d(e_i, e_j), D_{i,j}) = (d(e_i, e_j) - D_{i,j})^2 \tag{3}$$

or relative

$$L(d(e_i, e_j), D_{i,j}) = (d(e_i, e_j)/D_{i,j} - 1)^2. \tag{4}$$

4.2   Regularized Embeddings

We notice that we can identify a vertex by a vector of distances to all vertices in the graph, i.e., a row or column in the D matrix. This is because such a vector contains only one 0, and its position identifies the vertex of the question. Moreover, if two columns in the D matrix are similar, their respective vertices are in similar distance to other vertices and are usually close. Therefore, their respective embeddings should also be close. Therefore, we consider the embeddings as results of continuous transformations of the columns in the D matrix. This continuity is a form of regularization that prevents the optimization process (2) from getting stuck in a poor local minimum.

We consider the above transformation in the form of the deep neural network,

$$e_j = f(D_{\cdot,j}; \theta), \tag{5}$$

where f denotes the network and θ represents its weights. Then, to compute the embeddings, we need to optimize the weights θ of the network:

$$\theta = \operatorname{argmin}_\theta \frac{2}{n(n-1)} \sum_{i=1}^{n} \sum_{j=1}^{i-1} L(d(f(D_{\cdot,i}; \theta), f(D_{\cdot,j}; \theta)), D_{i,j}). \tag{6}$$

4.3 Distance Functions

Fig. 1. Simple square graph

Let us consider the graph in Fig. 1 to motivate the need for flexible distance functions in the embedding space. The topological distance for adjacent vertices equals 1, and for diagonally-opposite vertices equals 2. Under an ℓ_2 distance, there is no possible arrangement of the embeddings e_i in $\mathbb{R}^m$ that preserves $D_{i,j}$. If the distance between the adjacent vertex embeddings equals 1, the distances between both pairs of opposite vertex embeddings cannot equal 2 *at the same time*.

Let us consider a generalized distance formula

$$d(e_i, e_j) = \|e_i - e_j\|^\kappa, \tag{7}$$

where $\kappa \in [0, 2]$. For $\kappa = 1$, (7) represents the Euclidean distance. Treating κ as a parameter, we optimize it jointly, along with the vertex embeddings of each graph.

Coming back to the graph in Fig. 1, let us consider $m = 2$ and the most natural embeddings for the vertices: $e_A = [0, 0]^T$, $e_B = [0, 1]^T$, $e_C = [1, 1]^T$, $e_D = [1, 0]^T$. It is seen that for $\kappa = 2$, we have $d(e_i, e_j) = 1$ for each pair of adjacent vertices and $d(e_i, e_j) = 2$ for each pair of the opposite vertices. This exactly reflects the topological distances.

For an arbitrary graph, variable κ may improve the fitness of embeddings measured by (3) and by (4). Additionally, optimizing κ to minimize the discrepancy allows the model to adapt to the specific structure of the graph, which may be beneficial for the quality of the embeddings.

5 Experiments

Our experiments are centered on three key areas, which we have identified as particularly compelling during the development of our method. These encompass evaluating the quality of graph embeddings determined by the loss function, generating graphs with embeddings produced by our method, and identifying communities within graphs using our approach. The challenge of community detection can be assessed through the *modularity* of the detected communities and by comparing them to ground truth communities. Since we did not restrict our benchmarks to sets of graphs exclusively comprised of community structures, most of our measurements are based on the *modularity* metric.

To verify the neural network approach, we used a multilayer perceptron composed of four hidden layers with: 2048, 1024, 512, 256 neurons. Each hidden layer uses the ReLU activation function.

5.1 Datasets

For graph analysis and modularity measurement, we employed *TUDataset* [37], specifically opting for one graph set from each thematic category within the dataset. The selected graph sets are as follows: *MUTAG (small molecules)*, *ENZYMES (Bioinformatics)*, *Cuneiform (Computer vision)*, *IMDB-BINARY (Social networks)*, and *SYNTHETIC (Synthetic)*. These sets comprise graphs of relatively small size and similar characteristics.

Additionally, we utilized the *CORA* dataset [45], which features a single, large graph. For evaluating the community detection task, we employed two specific graphs representing real-world communities: the *Zachary Karate Club* [58] and *American Football* [18]. To visualize how different parameters affect final embeddings, we employed *American Revolution* graph [1].

5.2 Graph Analysis

This section is dedicated to presenting the results regarding the performance of each analyzed method in preserving the actual topological distances between graph nodes within the embedding space. We introduce two metrics, RMSE and RMRSE (depending on the utilized loss function for generating embeddings), defined for a single graph as follows:

$$RMSE = \sqrt{\frac{2}{n(n-1)} \sum_{i=1}^{n} \sum_{j=1}^{i-1} (d(e_i, e_j) - D_{i,j})^2} \qquad (8)$$

$$RMRSE = \sqrt{\frac{2}{n(n-1)} \sum_{i=1}^{n} \sum_{j=1}^{i-1} (d(e_i, e_j)/D_{i,j} - 1)^2}. \qquad (9)$$

To analyze both the RMSE and RMRSE measures, we compared two approaches: a predefined $\kappa = 1$ with a value of 1.0 (Table 1) and an automatically learned κ (Table 2). From both tables, we observe that the loss generally decreases as the dimensionality of embeddings increases. This observation is expected, as embeddings in higher-dimensional vector spaces can compress a graph's nodes to preserve topological distances for more complex graphs. However, for each dataset, there is a specific limit embedding dimension above which its further increase does not lead to any loss gain.

As seen in Table 1, the embeddings optimized directly provide us with worse results than those produced by the fitted neural network. This is especially visible for small embedding dimensions. The regularization introduced by the neural network prevents the optimization of the embeddings from getting stuck in local minima. For larger dimensions, the neural network does not introduce any improvement. The only exception is the CORE graph, for which the network could not learn a well-fitting transformation.

Comparing Table 1 with Table 2, we note that the automatically learned κ parameter yields better results, usually by a large margin.

In the Table 3 we present values of the automatically learned κ parameter. We see the κ usually increases with m. This relation reflects the increasing difficulty in preserving the topological distances of vertices in embedding space of a decreasing dimension. Notably, for large m, the optimal κ is usually well above 1.

Method	Dataset	m = 2	m = 3	m = 5	m = 10	m = 15	m = 30	m = 50
RMSE Direct	MUTAG	0.38 ± 0.23	0.24 ± 0.04	0.24 ± 0.03	0.24 ± 0.03	0.24 ± 0.03	0.24 ± 0.03	0.24 ± 0.03
	Cuneiform	0.58 ± 0.16	0.33 ± 0.13	0.17 ± 0.07	0.11 ± 0.02	0.11 ± 0.02	0.11 ± 0.02	0.11 ± 0.02
	SYNTHETIC	1.32 ± 0.04	0.94 ± 0.03	0.60 ± 0.01	0.44 ± 0.00	0.43 ± 0.00	0.43 ± 0.00	0.43 ± 0.00
	ENZYMES	0.53 ± 0.71	0.30 ± 0.44	0.22 ± 0.28	0.20 ± 0.15	0.20 ± 0.11	0.20 ± 0.10	0.19 ± 0.10
	IMDB-BINARY	0.38 ± 0.09	0.25 ± 0.06	0.15 ± 0.04	0.09 ± 0.03	0.07 ± 0.03	0.07 ± 0.04	0.07 ± 0.04
	CORA	3.30	2.47	1.80	1.17	0.90	0.57	0.45
RMSE Neural	MUTAG	0.29 ± 0.05	0.25 ± 0.04	0.25 ± 0.03	0.25 ± 0.04	0.25 ± 0.04	0.25 ± 0.03	0.25 ± 0.03
	Cuneiform	0.55 ± 0.14	0.33 ± 0.12	0.18 ± 0.07	0.12 ± 0.02	0.12 ± 0.02	0.12 ± 0.02	0.11 ± 0.02
	SYNTHETIC	1.20 ± 0.05	0.88 ± 0.03	0.60 ± 0.00	0.44 ± 0.00	0.44 ± 0.00	0.44 ± 0.00	0.44 ± 0.00
	ENZYMES	0.34 ± 0.22	0.26 ± 0.14	0.24 ± 0.11	0.23 ± 0.10	0.23 ± 0.10	0.23 ± 0.10	0.24 ± 0.12
	IMDB-BINARY	0.38 ± 0.09	0.26 ± 0.06	0.16 ± 0.04	0.09 ± 0.03	0.08 ± 0.04	0.08 ± 0.04	0.08 ± 0.04
	CORA	2.28	1.82	1.53	1.35	1.40	1.45	1.45
RMRSE Direct	MUTAG	0.12 ± 0.03	0.10 ± 0.01	0.10 ± 0.01	0.10 ± 0.01	0.10 ± 0.01	0.10 ± 0.01	0.10 ± 0.01
	Cuneiform	0.24 ± 0.04	0.14 ± 0.05	0.09 ± 0.03	0.07 ± 0.01	0.07 ± 0.01	0.07 ± 0.01	0.07 ± 0.01
	SYNTHETIC	0.38 ± 0.01	0.29 ± 0.01	0.21 ± 0.00	0.17 ± 0.00	0.17 ± 0.00	0.17 ± 0.00	0.17 ± 0.00
	ENZYMES	0.15 ± 0.05	0.10 ± 0.04	0.09 ± 0.03	0.09 ± 0.03	0.09 ± 0.03	0.09 ± 0.03	0.09 ± 0.03
	IMDB-BINARY	0.29 ± 0.04	0.20 ± 0.04	0.12 ± 0.03	0.08 ± 0.02	0.06 ± 0.03	0.06 ± 0.03	0.06 ± 0.03
	CORA	0.44	0.34	0.23	0.12	0.09	0.08	0.08
RMRSE Neural	MUTAG	0.11 ± 0.01	0.10 ± 0.00	0.10 ± 0.01	0.10 ± 0.01	0.10 ± 0.01	0.10 ± 0.01	0.10 ± 0.01
	Cuneiform	0.24 ± 0.04	0.14 ± 0.05	0.09 ± 0.03	0.07 ± 0.01	0.07 ± 0.01	0.07 ± 0.01	0.07 ± 0.01
	SYNTHETIC	0.37 ± 0.02	0.28 ± 0.02	0.21 ± 0.01	0.18 ± 0.00	0.18 ± 0.00	0.18 ± 0.00	0.18 ± 0.00
	ENZYMES	0.13 ± 0.04	0.10 ± 0.04	0.09 ± 0.03	0.09 ± 0.03	0.09 ± 0.03	0.09 ± 0.03	0.09 ± 0.04
	IMDB-BINARY	0.29 ± 0.04	0.20 ± 0.04	0.13 ± 0.03	0.08 ± 0.03	0.07 ± 0.03	0.07 ± 0.03	0.07 ± 0.03
	CORA	0.30	0.23	0.17	0.13	0.13	0.13	0.12

Table 1. Mean loss function results for $\kappa = 1.0$. Methods: RMSE = absolute loss, RMRSE = relative loss, Direct = embeddings optimized directly, Neural = embeddings regularized with a neural network

Method	Dataset	m = 2	m = 3	m = 5	m = 10	m = 15	m = 30	m = 50
RMSE Direct	MUTAG	0.35 ± 0.19	0.21 ± 0.04	0.15 ± 0.03	0.04 ± 0.03	0.02 ± 0.01	0.02 ± 0.01	0.02 ± 0.00
	Cuneiform	0.50 ± 0.13	0.30 ± 0.11	0.16 ± 0.08	0.05 ± 0.02	0.03 ± 0.02	0.01 ± 0.00	0.01 ± 0.00
	SYNTHETIC	1.13 ± 0.04	0.87 ± 0.03	0.60 ± 0.01	0.39 ± 0.00	0.29 ± 0.00	0.24 ± 0.00	0.24 ± 0.00
	ENZYMES	0.39 ± 0.33	0.25 ± 0.18	0.17 ± 0.10	0.10 ± 0.07	0.08 ± 0.05	0.06 ± 0.04	0.06 ± 0.04
	IMDB-BINARY	0.23 ± 0.10	0.17 ± 0.07	0.12 ± 0.05	0.06 ± 0.04	0.03 ± 0.03	0.01 ± 0.02	0.02 ± 0.02
	CORA	3.25	2.56	1.76	0.93	0.69	0.48	0.39
RMSE Neural	MUTAG	0.30 ± 0.06	0.22 ± 0.04	0.20 ± 0.03	0.20 ± 0.04	0.20 ± 0.04	0.20 ± 0.04	0.20 ± 0.04
	Cuneiform	0.42 ± 0.08	0.29 ± 0.10	0.17 ± 0.07	0.09 ± 0.02	0.08 ± 0.01	0.08 ± 0.02	0.08 ± 0.02
	SYNTHETIC	0.79 ± 0.01	0.71 ± 0.01	0.58 ± 0.00	0.40 ± 0.01	0.37 ± 0.01	0.36 ± 0.01	0.36 ± 0.01
	ENZYMES	0.32 ± 0.17	0.26 ± 0.12	0.23 ± 0.10	0.22 ± 0.10	0.22 ± 0.09	0.22 ± 0.10	0.22 ± 0.10
	IMDB-BINARY	0.20 ± 0.07	0.17 ± 0.06	0.12 ± 0.05	0.08 ± 0.04	0.07 ± 0.04	0.06 ± 0.03	0.06 ± 0.03
	CORA	1.10	1.02	1.01	0.98	0.92	1.95	0.96
RMRSE Direct	MUTAG	0.10 ± 0.03	0.07 ± 0.01	0.05 ± 0.01	0.01 ± 0.01	0.00 ± 0.00	0.01 ± 0.00	0.01 ± 0.00
	Cuneiform	0.22 ± 0.03	0.14 ± 0.05	0.07 ± 0.04	0.02 ± 0.01	0.01 ± 0.01	0.00 ± 0.00	0.00 ± 0.00
	SYNTHETIC	0.36 ± 0.01	0.28 ± 0.01	0.21 ± 0.00	0.12 ± 0.00	0.09 ± 0.00	0.08 ± 0.00	0.08 ± 0.00
	ENZYMES	0.14 ± 0.05	0.09 ± 0.04	0.06 ± 0.03	0.04 ± 0.02	0.03 ± 0.02	0.03 ± 0.02	0.03 ± 0.02
	IMDB-BINARY	0.16 ± 0.06	0.14 ± 0.05	0.10 ± 0.04	0.04 ± 0.03	0.02 ± 0.02	0.01 ± 0.01	0.01 ± 0.02
	CORA	0.42	0.29	0.19	0.11	0.09	0.07	0.06
RMRSE Neural	MUTAG	0.09 ± 0.01	0.07 ± 0.01	0.06 ± 0.01	0.05 ± 0.01	0.05 ± 0.01	0.05 ± 0.01	0.05 ± 0.01
	Cuneiform	0.22 ± 0.03	0.14 ± 0.05	0.07 ± 0.03	0.03 ± 0.01	0.03 ± 0.01	0.03 ± 0.01	0.03 ± 0.01
	SYNTHETIC	0.32 ± 0.01	0.27 ± 0.01	0.21 ± 0.01	0.13 ± 0.01	0.11 ± 0.00	0.10 ± 0.00	0.10 ± 0.00
	ENZYMES	0.13 ± 0.04	0.09 ± 0.03	0.07 ± 0.02	0.06 ± 0.02	0.07 ± 0.02	0.07 ± 0.03	0.07 ± 0.03
	IMDB-BINARY	0.17 ± 0.05	0.14 ± 0.05	0.10 ± 0.04	0.05 ± 0.03	0.04 ± 0.02	0.04 ± 0.02	0.04 ± 0.02
	CORA	0.20	0.17	0.15	0.14	0.14	0.14	0.12

Table 2. Mean loss function results for κ optimized along with embeddings. Method: see Table 1

5.3 Graph Drawing

Figure 2 illustrates vertex embeddings for different κ and loss types (absolute or relative). We selected the American Revolution [1] graph for visualization purposes, as the graph perfectly shows local node communities being represented in the embeddings' space. It is seen that minimization of the relative loss puts more emphasis on the local graph structure, while minimization of the absolute loss focuses more on its global structure. Also, with higher κ comes a tendency to compress local clusters of vertices.

5.4 Community Detection in Graphs

Methodology. Our study aimed to identify communities within graphs using a diverse array of unsupervised clustering algorithms applied to the embeddings of graph nodes generated through our methods. The clustering algorithms employed encompassed MeanShift [29], DBSCAN [16], HDBSCAN [33], Birch [59], OPTICS [3], AffinityPropagation [47] and AgglomerativeClustering [39]

Method	Dataset	m = 2	m = 3	m = 5	m = 10	m = 15	m = 30	m = 50
RMSE Direct	MUTAG	1.02 ± 0.06	1.13 ± 0.04	1.33 ± 0.09	1.81 ± 0.16	1.89 ± 0.07	1.88 ± 0.07	1.85 ± 0.08
	Cuneiform	0.62 ± 0.11	0.81 ± 0.14	1.06 ± 0.11	1.25 ± 0.02	1.34 ± 0.04	1.45 ± 0.11	1.46 ± 0.11
	SYNTHETIC	0.70 ± 0.01	0.72 ± 0.01	0.84 ± 0.01	1.41 ± 0.00	1.81 ± 0.01	1.95 ± 0.02	1.95 ± 0.02
	ENZYMES	1.02 ± 0.12	1.08 ± 0.07	1.19 ± 0.11	1.47 ± 0.22	1.66 ± 0.22	1.74 ± 0.19	1.70 ± 0.19
	IMDB-BINARY	0.39 ± 0.16	0.54 ± 0.24	0.75 ± 0.32	1.06 ± 0.45	1.32 ± 0.41	1.50 ± 0.29	1.48 ± 0.24
	CORA	1.24	1.22	1.20	1.18	1.16	1.24	1.27
RMSE Neural	MUTAG	1.03 ± 0.03	1.11 ± 0.04	1.16 ± 0.06	1.14 ± 0.06	1.16 ± 0.06	1.16 ± 0.06	1.17 ± 0.05
	Cuneiform	0.57 ± 0.16	0.81 ± 0.14	1.05 ± 0.11	1.18 ± 0.03	1.19 ± 0.04	1.20 ± 0.04	1.21 ± 0.04
	SYNTHETIC	0.38 ± 0.01	0.55 ± 0.01	0.84 ± 0.00	1.24 ± 0.01	1.31 ± 0.02	1.35 ± 0.01	1.35 ± 0.01
	ENZYMES	0.95 ± 0.10	1.01 ± 0.08	1.06 ± 0.10	1.07 ± 0.10	1.07 ± 0.10	1.07 ± 0.10	1.07 ± 0.10
	IMDB-BINARY	0.38 ± 0.16	0.54 ± 0.23	0.75 ± 0.29	0.99 ± 0.35	1.05 ± 0.33	1.09 ± 0.30	1.10 ± 0.29
	CORA	1.10	1.02	1.01	0.98	0.92	1.95	0.96
RMRSE Direct	MUTAG	1.10 ± 0.03	1.19 ± 0.03	1.40 ± 0.07	1.87 ± 0.12	1.94 ± 0.06	1.91 ± 0.05	1.84 ± 0.05
	Cuneiform	0.76 ± 0.08	0.97 ± 0.08	1.13 ± 0.06	1.26 ± 0.03	1.35 ± 0.05	1.45 ± 0.11	1.46 ± 0.11
	SYNTHETIC	0.69 ± 0.01	0.81 ± 0.01	1.11 ± 0.00	1.59 ± 0.00	1.90 ± 0.01	1.96 ± 0.00	1.91 ± 0.01
	ENZYMES	1.00 ± 0.08	1.09 ± 0.05	1.26 ± 0.11	1.63 ± 0.20	1.80 ± 0.18	1.83 ± 0.15	1.76 ± 0.13
	IMDB-BINARY	0.39 ± 0.16	0.56 ± 0.24	0.82 ± 0.35	1.15 ± 0.48	1.37 ± 0.43	1.52 ± 0.29	1.49 ± 0.25
	CORA	1.01	1.06	1.11	1.11	1.14	1.31	1.50
RMRSE Neural	MUTAG	1.11 ± 0.02	1.19 ± 0.03	1.31 ± 0.05	1.33 ± 0.04	1.33 ± 0.05	1.33 ± 0.04	1.35 ± 0.04
	Cuneiform	0.76 ± 0.09	0.97 ± 0.08	1.12 ± 0.06	1.22 ± 0.04	1.22 ± 0.04	1.23 ± 0.04	1.23 ± 0.04
	SYNTHETIC	0.58 ± 0.01	0.81 ± 0.02	1.11 ± 0.01	1.50 ± 0.01	1.58 ± 0.02	1.62 ± 0.01	1.61 ± 0.02
	ENZYMES	0.97 ± 0.09	1.08 ± 0.08	1.19 ± 0.12	1.22 ± 0.14	1.22 ± 0.13	1.22 ± 0.12	1.24 ± 0.13
	IMDB-BINARY	0.39 ± 0.15	0.57 ± 0.23	0.82 ± 0.32	1.12 ± 0.41	1.19 ± 0.38	1.22 ± 0.36	1.23 ± 0.36
	CORA	0.46	0.58	0.71	0.79	0.79	0.86	0.91

Table 3. Mean learned κ

We assessed the modularity score for all graphs, as this metric does not necessitate a graph with defined communities. For the Zachary Karate Club and American Football graphs, we additionally examined the ARS (adjusted rand index) and NMI (normalized mutual information) scores, which are valuable metrics for assessing how effectively an algorithm detects known communities.

We compared the results of our method with five widely used community detection algorithms: greedy modularity communities [14], Louvain communities [7], Kernighan Lin bisection [25], Girvan Newman [18], and asynchronous label propagation algorithm (asyn LPA) [42].

Results. Table 4 presents the Zachary Karate Club graph results. Our method outperforms other community detection algorithms in terms of ARS and NMI scores. Interestingly, the modularity score in this case does not align with the performance of the best method. Table 5 exhibits similar results but for the American Football graph. Our method also performs admirably, although slightly worse than the Girvan Newman method.

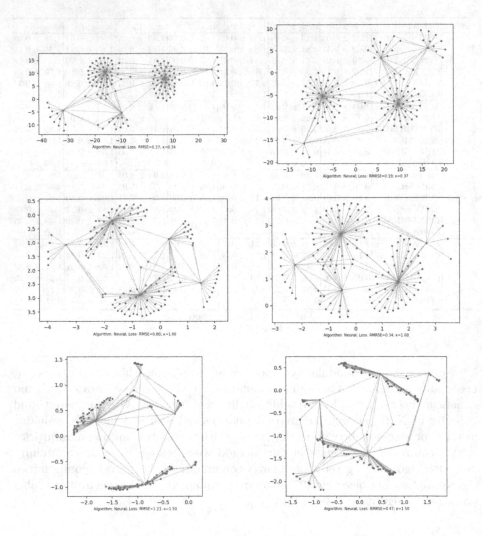

Fig. 2. American Revolution [1] graph visual representation ($m = 2$). *Left:* absolute error minimized. *Right:* relative error minimized. *Top:* $\kappa = auto \cong 0.4$, *Middle:* $\kappa = 1$, *Bottom:* $\kappa = 1.5$.

The modularity results for the TUDataset are detailed in Table 7. We compared these results with other algorithms, as shown in Table 6. Our method achieves competitive modularity scores, but not the highest ones. Notably, the method of designating the embedding does not impact the achieved modularity.

Table 4. Community detection in Zachary Karate Club

Embed. method	Error	κ	m	Clustering algorithm	ARS	NMI	Modularity
–	–	–	–	Greedy Modularity Communities	0.57	0.56	0.38
–	–	–	–	Louvain communities	0.51	0.60	**0.42**
–	–	–	–	Kernighan Lin Bisection	0.77	0.68	0.37
–	–	–	–	Girvan Newman	0.77	0.73	0.36
–	–	–	–	Asyn Lpa Communities	0.66	0.65	0.38
Direct RMSE	0.45	0.72	2	MeanShift	0.88	0.84	0.37
Direct RMRSE	0.19	0.73	5	MeanShift	0.88	0.84	0.37
Direct RMSE	0.54	1.00	2	MeanShift	0.88	0.84	0.37
Direct RMRSE	0.26	1.00	2	AffinityPropagation	0.88	0.84	0.37
Direct RMSE	0.76	1.50	2	MeanShift	0.88	0.84	0.37
Direct RMRSE	0.37	1.50	2	MeanShift	0.88	0.84	0.37
Neural RMSE	0.37	0.81	3	MeanShift	0.88	0.84	0.37
Neural RMRSE	0.19	0.93	3	MeanShift	0.88	0.84	0.37
Neural RMSE	0.23	1.00	50	AffinityPropagation	**1.00**	**1.00**	0.36
Neural RMRSE	0.20	1.00	3	MeanShift	0.88	0.84	0.37
Neural RMSE	0.74	1.50	2	MeanShift	0.88	0.84	0.37
Neural RMRSE	0.08	1.50	50	AffinityPropagation	0.88	0.84	0.37

Table 5. Community detection in American Football

Embed. method	Error	κ	m	Clustering algorithm	ARS	NMI	Modularity
–	–	–	–	Greedy Modularity Communities	0.47	0.70	0.55
–	–	–	–	Louvain Communities	0.81	0.89	**0.60**
–	–	–	–	Kernighan Lin Bisection	0.14	0.38	0.35
–	–	–	–	Girvan Newman	**0.92**	**0.94**	0.36
–	–	–	–	Asyn Lpa Communities	0.75	0.87	0.58
Direct RMSE	0.46	0.54	10	AffinityPropagation	0.78	0.86	0.58
Direct RMRSE	0.24	0.51	5	HDBSCAN	0.89	0.92	0.58
Direct RMSE	0.36	1.00	15	AffinityPropagation	0.86	0.92	0.58
Direct RMRSE	0.18	1.00	15	AffinityPropagation	0.86	0.92	0.58
Direct RMSE	0.51	1.50	5	MeanShift	0.83	0.89	0.51
Direct RMRSE	0.13	1.50	50	AffinityPropagation	0.78	0.87	0.54
Neural RMSE	0.32	1.38	15	AffinityPropagation	0.83	0.90	0.57
Neural RMRSE	0.14	1.49	15	AffinityPropagation	0.80	0.88	0.54
Neural RMSE	0.36	1.00	15	AffinityPropagation	0.86	0.91	0.58
Neural RMRSE	0.18	1.00	15	AffinityPropagation	0.87	0.92	**0.60**
Neural RMSE	0.33	1.50	10	AffinityPropagation	0.81	0.90	0.55
Neural RMRSE	0.14	1.50	15	HDBSCAN	0.79	0.87	0.50

Table 6. Mean modularity – graph community detection algorithms

Dataset	GMC	LC	GN	KLB	ALC
MUTAG	**0.46 ± 0.06**	**0.46 ± 0.06**	**0.46 ± 0.06**	0.34 ± 0.05	0.41 ± 0.05
Cuneiform	**0.53 ± 0.03**	**0.53 ± 0.03**	0.47 ± 0.07	0.30 ± 0.07	0.47 ± 0.10
SYNTHETIC	**0.48 ± 0.00**	0.47 ± 0.01	0.44 ± 0.00	0.31 ± 0.01	0.35 ± 0.03
ENZYMES	0.57 ± 0.11	**0.58 ± 0.12**	**0.58 ± 0.13**	0.40 ± 0.08	0.54 ± 0.12
IMDB-BINARY	**0.30 ± 0.16**	**0.30 ± 0.16**	0.27 ± 0.16	0.20 ± 0.12	0.25 ± 0.17
CORA	0.81	**0.82**	0.81	0.41	0.5

		$\kappa = 1$			$\kappa = $ auto			
Method	Dataset	Clustering algorithm	m	Modularity	Dataset	Clustering algorithm	m	Modularity
RMSE Direct	MUTAG	AffinityPropagation	50	0.43 ± 0.05	MUTAG	AffinityPropagation	5	0.44 ± 0.05
	Cuneiform	AffinityPropagation	2	0.44 ± 0.12	Cuneiform	AffinityPropagation	10	0.52 ± 0.07
	SYNTHETIC	AffinityPropagation	5	0.39 ± 0.02	SYNTHETIC	AffinityPropagation	5	0.39 ± 0.01
	ENZYMES	AffinityPropagation	10	0.54 ± 0.13	ENZYMES	AffinityPropagation	2	0.54 ± 0.13
	IMDB-BINARY	MeanShift	2	0.26 ± 0.16	IMDB-BINARY	HDBSCAN	3	0.25 ± 0.17
	CORA	AffinityPropagation	30	0.54	CORA	AffinityPropagation	50	0.57
RMSE Neural	MUTAG	AffinityPropagation	2	0.44 ± 0.05	MUTAG	AffinityPropagation	3	0.44 ± 0.05
	Cuneiform	AffinityPropagation	2	0.52 ± 0.07	Cuneiform	AffinityPropagation	50	0.52 ± 0.05
	SYNTHETIC	AffinityPropagation	5	0.40 ± 0.01	SYNTHETIC	AffinityPropagation	5	0.41 ± 0.00
	ENZYMES	AffinityPropagation	2	0.55 ± 0.12	ENZYMES	AffinityPropagation	2	0.55 ± 0.12
	IMDB-BINARY	MeanShift	2	0.26 ± 0.16	IMDB-BINARY	AffinityPropagation	2	0.26 ± 0.17
	CORA	AffinityPropagation	50	0.49	CORA	AffinityPropagation	15	0.58
RMRSE Direct	MUTAG	AffinityPropagation	5	0.44 ± 0.05	MUTAG	AffinityPropagation	3	0.44 ± 0.05
	Cuneiform	AffinityPropagation	10	0.53 ± 0.03	Cuneiform	AffinityPropagation	5	0.51 ± 0.09
	SYNTHETIC	AffinityPropagation	10	0.41 ± 0.00	SYNTHETIC	AffinityPropagation	5	0.40 ± 0.02
	ENZYMES	AffinityPropagation	5	0.55 ± 0.13	ENZYMES	AffinityPropagation	2	0.55 ± 0.12
	IMDB-BINARY	HDBSCAN	3	0.25 ± 0.17	IMDB-BINARY	AffinityPropagation	2	0.26 ± 0.18
	CORA	AffinityPropagation	30	0.64	CORA	AffinityPropagation	30	0.64
RMRSE Neural	MUTAG	AffinityPropagation	15	0.44 ± 0.05	MUTAG	AffinityPropagation	3	0.44 ± 0.05
	Cuneiform	AffinityPropagation	30	0.53 ± 0.04	Cuneiform	AffinityPropagation	5	0.52 ± 0.06
	SYNTHETIC	AffinityPropagation	50	0.41 ± 0.01	SYNTHETIC	AffinityPropagation	5	0.40 ± 0.01
	ENZYMES	AffinityPropagation	10	0.55 ± 0.13	ENZYMES	AffinityPropagation	2	0.55 ± 0.12
	IMDB-BINARY	AffinityPropagation	5	0.24 ± 0.18	IMDB-BINARY	AffinityPropagation	2	0.26 ± 0.18
	CORA	AffinityPropagation	15	0.58	CORA	AffinityPropagation	50	0.59

Table 7. Best mean modularity in different datasets

6   Conclusions

In this paper, we introduced a regularization method for graph vertex embeddings that preserves distances in the graph. This method uses a neural network to transform a column of the distance matrix into the embedding. In our experimental study, this regularization significantly improved the embeddings, especially when their dimension was low.

We also introduced a generalized measure of distance between the embeddings. With our proposed measure, the error of distance preservation by the embeddings was reduced by a large margin.

Finally, we performed a study on community detection in graphs, in which we compared results obtained by combining graph embeddings and clustering methods for numerical data with community detection algorithms dedicated to graphs. The analyzed combination achieved competitive results, although it yielded the best results only in the case of Zachary Karate Club.

References

1. American revolution network dataset – KONECT, October 2017. http://konect.cc/networks/brunson_revolution
2. Agrawal, A., Ali, A., Boyd, S.: Minimum-distortion embedding. Found. Trends® Mach. Learn. **14**(3), 211–378 (2021)
3. Ankerst, M., Breunig, M., Kröger, P., Sander, J.: Optics: ordering points to identify the clustering structure. SIGMOD Rec. **28**, 49–60 (1999)
4. Belkin, M., Niyogi, P.: Laplacian eigenmaps and spectral techniques for embedding and clustering. In: NIPS, vol. 14. MIT Press (2001)
5. Berkhout, J.B., Poormoghadam, D., Yi, C., Kalsbeek, A., Meijer, O.C., Mahfouz, A.: An integrated single-cell RNA-seq atlas of the mouse hypothalamic paraventricular nucleus links transcriptomic and functional types. J. Neuroendocrinol. **36**, e13367 (2024)
6. Beygelzimer, A., Kakade, S., Langford, J.: Cover trees for nearest neighbor. In: ICML, pp. 97–104. ACM Press (2006)
7. Blondel, V.D., Guillaume, J.L., Lambiotte, R., Lefebvre, E.: Fast unfolding of communities in large networks. J. Stat. Mech: Theory Exp. **2008**(10), P10008 (2008)
8. Böhm, J.N., Berens, P., Kobak, D.: Attraction-repulsion spectrum in neighbor embeddings. J. Mach. Learn. Res. **23**(1), 4118–4149 (2022)
9. Borgwardt, K., Ghisu, E., Llinares-López, F., O'Bray, L., Rieck, B.: Graph Kernels: state-of-the-art and future challenges. Found. Trends® Mach. Learn. **13**(5-6), 531–712 (2020)
10. Bourgain, J.: On Lipschitz embedding of finite metric spaces in Hilbert space. Israel J. Math. **52**, 46–52 (1985)
11. Chamberlain, B.P., Clough, J., Deisenroth, M.P.: Neural embeddings of graphs in hyperbolic space. In: ICLR (2017)
12. Chari, T., Pachter, L.: The specious art of single-cell genomics. PLoS Comput. Biol. **19**, e1011288 (2023)
13. Chen, Z., Li, L., Bruna, J.: Supervised community detection with line graph neural networks. In: ICLR (2018)
14. Clauset, A., Newman, M.E., Moore, C.: Finding community structure in very large networks. Phys. Rev. E **70**(6), 066111 (2004)
15. Dwivedi, V.P., Joshi, C.K., Luu, A.T., Laurent, T., Bengio, Y., Bresson, X.: Benchmarking graph neural networks. J. Mach. Learn. Res. **24**(43), 1–48 (2023)
16. Ester, M., Kriegel, H.P., Sander, J., Xu, X.: A density-based algorithm for discovering clusters in large spatial databases with noise. In: International Conference on Knowledge Discovery and Data Mining, pp. 226–231 (1996)
17. Gilmer, J., Schoenholz, S.S., Riley, P.F., Vinyals, O., Dahl, G.E.: Neural message passing for quantum chemistry. In: ICML, pp. 1263–1272 (2017)
18. Girvan, M., Newman, M.E.: Community structure in social and biological networks. Proc. Natl. Acad. Sci. **99**(12), 7821–7826 (2002)

19. Grover, A., Leskovec, J.: node2vec: scalable feature learning for networks. In: International Conference on Knowledge Discovery & Data Mining, pp. 855–864 (2016)
20. Hoffmann, M., Henninger, J., Veith, J., Richter, L., Judkewitz, B.: Blazed oblique plane microscopy reveals scale-invariant inference of brain-wide population activity. Nat. Commun. **14**, 1–11 (2023)
21. Indyk, P.: Algorithmic applications of low-distortion geometric embeddings. In: IEEE Symposium on Foundations of Computer Science, pp. 10–33 (2001)
22. Ji, P., Zhang, T., Li, H., Salzmann, M., Reid, I.: Deep subspace clustering networks. In: NIPS, vol. 30 (2017)
23. Johnson, W., Lindenstrauss, J.: Extensions of Lipschitz maps into a Hilbert space. Contemp. Math. **26**, 189–206 (1984)
24. Jumper, J., et al.: Highly accurate protein structure prediction with AlphaFold. Nature **596**(7873), 583–589 (2021)
25. Kernighan, B.W., Lin, S.: An efficient heuristic procedure for partitioning graphs. Bell Syst. Tech. J. **49**(2), 291–307 (1970)
26. Kipf, T.N., Welling, M.: Variational graph auto-encoders (2016). arXiv:1611.07308
27. Kipf, T.N., Welling, M.: Semi-supervised classification with graph convolutional networks (2017). arXiv:1609.02907
28. Klimovskaia, A., Lopez-Paz, D., Bottou, L., Nickel, M.: Poincaré maps for analyzing complex hierarchies in single-cell data. Nat. Commun. **11**(1), 2966 (2020)
29. Koohpayegani, S.A., Tejankar, A., Pirsiavash, H.: Mean shift for self-supervised learning. In: ICCV, pp. 10326–10335, October 2021
30. Le, Q., Mikolov, T.: Distributed representations of sentences and documents. In: ICML, vol. 4, May 2014
31. Linial, N., London, E., Rabinovich, Y.: The geometry of graphs and some of its algorithmic applications. Combinatorica **15**(2), 215–245 (1995)
32. van der Maaten, L., Hinton, G.: Visualizing data using t-SNE. J. Mach. Learn. Res. **9**, 2579–2605 (2008)
33. Malzer, C., Baum, M.: A hybrid approach to hierarchical density-based cluster selection. In: IEEE International Conference on Multisensor Fusion and Integration for Intelligent Systems, September 2020
34. McInnes, L., Healy, J., Melville, J.: UMAP: uniform manifold approximation and projection for dimension reduction (2020)
35. Mikolov, T., Sutskever, I., Chen, K., Corrado, G.S., Dean, J.: Distributed representations of words and phrases and their compositionality. In: Burges, C.J., Bottou, L., Welling, M., Ghahramani, Z., Weinberger, K.Q. (eds.) NIPS (2013)
36. Morris, C., et al.: Future directions in foundations of graph machine learning (2024). arXiv:2402.02287
37. Morris, C., Kriege, N.M., Bause, F., Kersting, K., Mutzel, P., Neumann, M.: TUDataset: a collection of benchmark datasets for learning with graphs. In: ICML 2020 Workshop on Graph Representation Learning and Beyond (GRL+ 2020) (2020). www.graphlearning.io
38. Morselli Gysi, D., et al.: Network medicine framework for identifying drug-repurposing opportunities for COVID-19. Proc. Natl. Acad. Sci. **118**(19), e2025581118 (2021)
39. Müllner, D.: Modern hierarchical, agglomerative clustering algorithms (2011). https://arxiv.org/abs/1109.2378
40. Nickel, M., Kiela, D.: Poincaré embeddings for learning hierarchical representations. In: Advances in Neural Information Processing Systems, vol. 30. Curran Associates, Inc. (2017)

41. Pothen, A., Simon, H.D., Liou, K.P.: Partitioning sparse matrices with eigenvectors of graphs. SIAM J. Matrix Anal. Appl. **11**(3), 430–452 (1990)
42. Raghavan, U.N., Albert, R., Kumara, S.: Near linear time algorithm to detect community structures in large-scale networks. Phys. Rev. E **76**(3), 036106 (2007)
43. Roweis, S.T., Saul, L.K.: Nonlinear dimensionality reduction by locally linear embedding. Science **290**, 2323–2326 (2000)
44. Sala, F., De Sa, C., Gu, A., Ré, C.: Representation tradeoffs for hyperbolic embeddings. In: ICML, pp. 4460–4469. PMLR (2018)
45. Sen, P., Namata, G., Bilgic, M., Getoor, L., Gallagher, B., Eliassi-Rad, T.: Collective classification in network data articles. AI Mag. **29**, 93–106 (2008)
46. Su, X., et al.: A comprehensive survey on community detection with deep learning. IEEE Trans. Neural Netw. Learn. Syst. **35**, 4682–4702 (2024)
47. Sun, L., Liu, R., Xu, J., Zhang, S., Tian, Y.: An affinity propagation clustering method using hybrid kernel function with LLE. IEEE Access **PP**, 1 (2018)
48. Tang, J., Liu, J., Zhang, M., Mei, Q.: Visualization large-scale and high-dimensional data (2016). http://arxiv.org/abs/1602.00370
49. Tenenbaum, J.B., de Silva, V., Langford, J.C.: A global geometric framework for nonlinear dimensionality reduction. Science **290**, 2319–2323 (2000)
50. Vaswani, A., et al.: Attention is all you need. In: NIPS, vol. 30 (2017)
51. Veličković, P., Cucurull, G., Casanova, A., Romero, A., Liò, P., Bengio, Y.: Graph attention networks. In: ICLR (2018)
52. Veličković, P., Fedus, W., Hamilton, W.L., Liò, P., Bengio, Y., Hjelm, R.D.: Deep graph infomax (2018). arXvi:1809.10341
53. Vincent, P., Larochelle, H., Lajoie, I., Bengio, Y., Manzagol, P.A., Bottou, L.: Stacked denoising autoencoders: learning useful representations in a deep network with a local denoising criterion. J. Mach. Learn. Res. **11**(12), 3371–3408 (2010)
54. Wang, X., Cui, P., Wang, J., Pei, J., Zhu, W., Yang, S.: Community preserving network embedding. In: AAAI Conference on Artificial Intelligence, vol. 31 (2017)
55. Weinberger, K.Q., Lawrence, S.K.: Unsupervised learning of image manifolds by semidefinite programming. Int. J. Comput. Vision **70**, 77 (2006)
56. Wu, L., Cui, P., Pei, J., Zhao, L. (eds.): Graph Neural Networks: Foundations, Frontiers, and Applications. Springer, Singapore (2022)
57. Yang, J., Leskovec, J.: Defining and evaluating network communities based on ground-truth. In: ACM SIGKDD Workshop on Mining Data Semantics, pp. 1–8. ACM (2012)
58. Zachary, W.: An information flow model for conflict and fission in small groups. J. Anthropol. Res. **33**, 452–473 (1977)
59. Zhang, T., Ramakrishnan, R., Livny, M.: BIRCH: an efficient data clustering method for very large databases. SIGMOD Rec. **25**, 103–114 (1996)
60. Zhang, Y.J., Yang, K.C., Radicchi, F.: Systematic comparison of graph embedding methods in practical tasks. Phys. Rev. E **104**(4), 044315 (2021)

A Robust Network Model for Studying Microbiomes in Precision Agriculture Applications

Suyeon Kim, Ishwor Thapa, and Hesham H. Ali[✉]

College of Information Science and Technology, University of Nebraska at Omaha, Omaha, NE 68182, USA
{suyeonkim,ithapa,hali}@unomaha.edu

Abstract. Recent rapid advancements in high-throughput sequencing technologies have made it possible for researchers to explore the microbial universe in high degrees of depth that were not possible even few years ago. Microbial communities occupy numerous environments everywhere and significantly impact the health of the living organisms in their environments. With the availability of microbiome data, advanced computational tools are critical to conduct the needed analysis and allow researchers to extract meaningful knowledge leading to actionable decisions. However, despite many attempts to develop tools to analyze the heterogeneous datasets associated with various microbiomes, such attempts lack the sophistication and robustness needed to efficiently analyze these complex heterogeneous datasets and produce accurate results. In addition, almost all current methods employ heuristic concepts that do not guarantee the robustness and reproducibility needed to provide the biomedical community with trusted analysis that lead to precise data-driven decisions. In this study, we present a network model that attempts to overcome these challenges by utilizing graph-theoretic concepts and employing multiple computational methods with the goal of conducting robust analysis and produce accurate results. To test the proposed model, we performed the analysis on plant microbiome datasets to obtain distinctive functional modules based on key microbial interrelationships in a given host environment. Our findings establish a framework for a new understanding of the association between functional modules based on microbial community structure.

Keywords: Precision agriculture · plant microbiomes · robust analysis · network models · graph algorithms

1 Introduction

Numerous studies have shown that the health of all living organism is underpinned by the many roles of microbiome in their environments. New technologies have revolutionized the way we understand these little microbes, encompassing

L. Franco et al. (Eds.): ICCS 2024, LNCS 14837, pp. 58–71, 2024.
https://doi.org/10.1007/978-3-031-63778-0_5

interactions between the microbes, that perform vital functions, both individually and as a group. For the human host, many clinical conditions interact with microbes that impact developmental conditions for early childhood, quality of life in aging, and everything in between. Besides human microbiome research, understanding the impact of plant microbiomes on plant growth, health, and productivity have also gained significant attention recently. This is again due to the advancement in high-throughput technologies and the consequent availability of the data required to study such impact. Studying the impact of microbial composition and their properties in plants represent an opportunity to develop and improve proposed methodologies, considering that plant environments are relatively more accessible and easier to manage, as compared to those associated with humans. All living organisms in our ecosystems continue to evolve together. Such co-evolution is influenced by the intra-relationship among microbes and inter-relationship with their surrounding environment. Plants also host their own microbial communities, and they are influenced by a wide range of ecological interactions such as symbiotic, competitive, neutral, and mutualistic relationships [6, 17].

Several previous studies have identified microbial interactions that form important microbiomes and are essential for promoting plant growth and modulating disease outbreak [3, 4]. Considering the complexity of the nature of microbiome data, researchers have attempted to model and study the relationship of plant phenotype with their microbiome using various computational and statistical methods [15]. In accordance with a rapid advancement in technological capabilities, machine learning-based and deep learning-based methods have been applied in recent studies to study the impact of the microbiome on plant growth [7, 11].

Co-occurrence network-based analysis represents one of the powerful analyses. This approach explores these significant co-occurrence patterns of microorganisms. It has become a widely adopted method in ecological studies [2]. Such co-occurrence patterns are significant in the understanding of microbial community structure and are utilized for the potential prediction of species interactions in associated environments [5]. To measure these patterns, correlation coefficients or mutual information measures are often used to identify significant microbial abundance relationships. For example, an association network inference tool (CoNet) provides multiple types of network inference methods, and the ensemble approaches to network inference were proposed to increase the network accuracy [8]. Current computational and bioinformatics tools, including CoNet, are required to have their own parameters and, in many cases, specific thresholds to solve an each computational biology problem. However, it is a major challenge in Biomedical Informatics to find the optimal parameters for each case study. In order to minimize the accuracy concerns of the obtained results, it is critical to have a robust approach that limits the impact of the imperfections associated with heuristic steps and/or randomly selected thresholds.

In this study, we focus on addressing two research questions: 1) How to capture the distinctive and impactful microbial interactions in the co-expression networks from raw microbial abundance data in a given environment. 2) How to develop a precise and robust approach for mining accurate microbial associations with relation to functional modules. To address these two issues, we present a new computational pipeline as the central component of a robust analysis in order to have a better understanding of the significant associations among functional modules and host plants based on microbial community structure. In addition, multiple computational methods or algorithms will be employed to model and solve such complex problems to avoid or limit the impact of selecting a computational approach on the outcome of the study. We argue that this will potentially lead to more trustworthy results that are less influenced by the characteristics as well as limitations associated with each computational tool.

2   Methods

2.1   Overview of Workflow

This pipeline takes in microbial abundance data together with phenotypical properties and returns 1) Highly associated bacteria in a given environment; 2) Operational taxonomic unit pairs (OTUs-pairs hereafter) specific to samples from higher and lower percentile of a phenotypic property (in this study 'total biomass'); and 3) Functional modules enriched in these groupings. The entire workflow of the proposed method is shown in Fig. 1. It consists of four components: Identification of microbial association networks using multiple statistical methods, discovery of bacteria of interest based on phenotypical characteristics, functional enrichment analysis, and comparative analysis.

2.2   Data Description

In this study, we employed bacterial abundance data and associated metadata obtained from collaborators at the University of Nantes. The dataset comprised relative abundance of bacterial OTUs at the genus level across *Medicago truncatula* plant samples, along with metadata detailing 16 phenotypical parameters related to plant growth, such as estimated quantity of nitrogen and total biomass. Genus names were standardized using the NCBI taxonomy database [9]. Additionally, we acquired two sets of functional reference information for enrichment analysis based on the genus list derived from the abundance data. The first reference was generated using Phylogenetic Investigation of Communities by Reconstruction of Unobserved States (PICRUSt) bioinformatics software, providing counts of KEGG Orthologs (KOs) for referenced taxa [13]. The second reference contained information on pathway modules, representing functional units of gene sets in metabolic pathways, retrieved using KEGG REST API in a Unix environment. This experimental dataset was employed as a case study to test the proposed methodology. All steps of the computational approach are designed and implemented to address similar scenarios in various applications. Hence, we suggest that our approach can be extended by incorporating additional datasets from diverse domains.

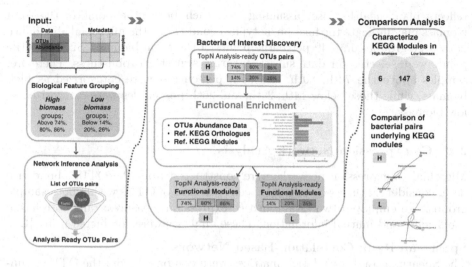

Fig. 1. Overview of the pipeline. The pipeline consists of four main components: 1) Construction of co-expression OTUs networks in a given environment; 2) Identification bacteria of interest; 3) Functional modules enrichment analysis; 4) Comparison analysis.

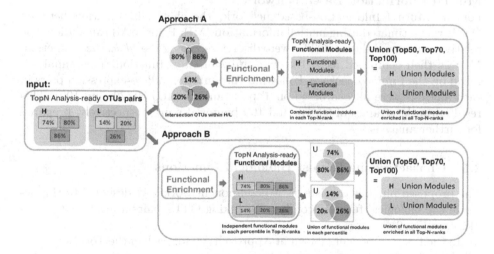

Fig. 2. Detection of distinctive OTUs pairs through Approaches A and B.

2.3 Biological Feature Grouping

Given that the input attributes of each plant included the total biomass, we observed differences in the total biomass across plant samples, presuming it

reflects plant health. Also, assuming that each percentile point of the total
biomass may indicate the host phenotypical conditions, the microbial abundance
data were grouped into 49 percentile points of the total biomass. Six percentile
points were selected for data subset, ensuring statistical robustness. Therefore,
samples above the 74th, 80th, and 86th percentiles were categorized as high
biomass, while those below 14th, 20th, and 26th percentiles were categorized as
low biomass.

2.4   Co-expression Network Analysis

Microbial co-expression networks were constructed from the OTUs abundance
data, considering the association between each pair of OTUs within both sample
groups. We employed two established co-expression measures: rank correlation
coefficients and mutual information. These analyses were conducted using R.

Spearman Rank Correlation-Based Network
The Spearman rank correlation matrices were computed from the OTUs abundance data within each sample group, utilizing the *'corr.test'* function in the
psych package with Bonferroni correction applied. The correlation coefficient
rank values, ranging from −1 to 1, were sorted from positive to negative for
subsequent analysis.

Mutual Information-Based Network
For the Mutual Information-based network, the shared information between
OTUs was estimated using mutual information (MI). Prior to MI calculation, the
relative abundance data of OTUs were discretized using the *'discretize'* function.
MI was then computed using the *'mutinformation()'* function. These analyses
relied on the *infotheo* package. Following computation, all co-expression pairs in
each analysis were filtered based on Top-N-ranking (Top 50, 70, and 100). This
resulted in dataframes of 50, 70, and 100 bacterial pairs, respectively, prepared
for further analysis.

2.5   Characterizing Robust Biological Functions

The proposed pipeline consists of two component workflows designed to characterize the biological functions of Top-N-ranking OTUs pairs using KEGG modules (see Fig. 2).

In the first component (shown in Approach A), for each of the Top-N-ranking
OTUs pairs, we identify an intersection of microbial associations across high
biomass groups (74%, 80%, and 86%) and low biomass groups (14%, 20%, and
26%), respectively. The microbial communities based on these common microbial
associations were considered bacteria of interest for functional enrichment analysis. The union of functional modules enriched in all Top-N-ranking defined as
the high and low biomass functional features. In the second component (shown
in Approach B), beginning with each Top-N-ranking OTUs pairs, we analyzed
each of the OTUs pairs in the individual percentile group of the high and low
biomass. Subsequently, the high and low biomass functional features sets at each

Top-N ranking were obtained by the union of functional modules enriched in all three percentiles of the high and low biomass groups.

2.6   Comparison of OTUs Pairs Underlying KEGG Modules

For each co-expression method, we compared the union modules from the high and low biomass groups and identified distinctive functional modules in each group. The intersection of these unique functional modules derived from two co-expression methods are used for further analysis. The OTUs annotated with these common distinctive functional modules are labeled as driver OTUs. OTUs associated with the driver OTUs were captured from both high and low biomass groups and were represented as networks. To visualize the networks, R package *igraph* was used. Next, we compared the dynamic change of relationships between driver OTUs in the high and low biomass group networks.

2.7   Bacterial Functional Enrichment Analysis

To identify functional pathways that are over-represented in any given OTUs list of interest, we performed an enrichment analysis. A fisher's exact test (*'phyper'* function in R environment) was used to identify pathways enriched in the OTUs list of interest from Top-N-ranking OTUs pairs. Table 1 illustrates a contingency table with a KEGG module (M00001 Glycolysis) as an example. In this case, the 40 OTUs of interest have been analyzed. A total count of known KEGG modules and the total count of background OTUs are also required for this comparison.

Table 1. An example contingency table for enrichment analysis.

	Present	Absent	Total
Present in KEGG module	40	0	40 (OTUs of interest)
Absent in KEGG module	320	1	321
Total OTUs in KEGG module	360	1	361 (Total OTUs)

For all enriched functional modules in high and low biomass groups, we identified the overlapping of these modules using a Venn diagram. The *'venn.diagram()'* function in R was implemented for this analysis.

3   Results and Discussion

3.1   A Comparison of Networks in High and Low Biomass Groups

We collected the Jaccard similarity index (J.I) for comparing Top-N-rank analysis-ready OTUs pairs at each percentile point (see Table 2). The Jaccard similarity index was computed based on the edges, and Table 2 presents a summary of the similarity between networks within and between high and low biomass groups. Notably, within high biomass (74th, 80th, and 86th percentiles) and low biomass groups (14th, 20th, and 26th percentiles), we observed high degrees of similarity. Despite the relatively homogeneous microbial abundance data, distinct similarity patterns emerged between high and low biomass groups of OTU pairs in the Top-N rankings. These findings suggest that the thresholds for high or low biomass groups minimally affect the overall results, indicating a high level of robustness in our approach.

Table 2. Jaccard similarity index $(J.I)$ for comparison of network in high and low biomass groups.

Network comparison within high biomass group			
Range of high (H) biomass groups	*J.I* in Top50	*J.I* in Top70	*J.I* in Top100
74th and 80th percentile	0.79	0.75	0.83
80th and 86th percentile	0.56	0.59	0.60
74th and 86th percentile	0.52	0.51	0.57
Network comparison within low biomass group			
Range of low (L) biomass groups	*J.I* in Top50	*J.I* in Top70	*J.I* in Top100
14th and 20th percentile	0.75	0.73	0.74
20th and 26th percentile	0.61	0.63	0.60
14th and 26th percentile	0.64	0.71	0.67
Network comparison between groups			
Range of H and L biomass group	*J.I* in Top50	*J.I* in Top70	*J.I* in Top100
74th and 26th percentile	0.45	0.41	0.38
80th and 20th percentile	0.47	0.41	0.42
86th and 14th percentile	0.39	0.33	0.29

3.2 Co-expression Networks Analysis

Table 3 provides an overview of the obtained networks, including their size parameters and densities. It reports the total count of vertices and edges present in our co-expression networks. The edges in these networks are undirected and represent associations between vertices, which signify bacterial communities. Edge density reflects the ratio of actual edges in the network to the total possible edges, calculated as n(n − 1)/2. Here, we present the network descriptions of Spearman rank correlation-based networks in Top 50, 70, and 100.

3.3 Characterization of Distinctive KEGG Modules in High and Low Biomass Groups

We begin with each of the Top-N-ranking OTUs pairs derived from co-expression networks to implement our proposed methods (Approach A and B). The two

Table 3. Summary of the spearman correlation-based networks at different percentile of the total biomass (V = Vertices, E = Edges, D = Edge density).

	Top50			Top70			Top100		
Percentile	V	E	D	V	E	D	V	E	D
74th	54	50	0.03	63	70	0.04	76	100	0.04
80th	52	50	0.04	64	70	0.03	75	100	0.04
86th	55	50	0.03	72	70	0.03	79	100	0.03
14th	53	50	0.04	69	70	0.03	88	100	0.03
20th	51	50	0.04	65	70	0.03	80	100	0.03
26th	53	50	0.04	61	70	0.04	78	100	0.03

approaches are independent of one another, each having its own benefits and limitations. The Approach A consists of common OTUs pairs from different percentiles followed by the union of functional modules enriched in Top-N-ranks. On the other hand, the Approach B includes functional modules enriched in all percentiles with Top-N-ranks.

Through each approach, we first compared the number of distinctive significant functional modules underlying both high and low biomass groups shown in a Venn diagram (see Fig. 3). Figure 3 effectively illustrates the number of unique functional modules for both high (blue) and low (pink) biomass groups obtained from the usage of two approaches (Approach A and B) and two co-expression analyses. The first row of Fig. 3 is the result of the comparison between Spearman analysis (left) and mutual information (right) through the Approach A.

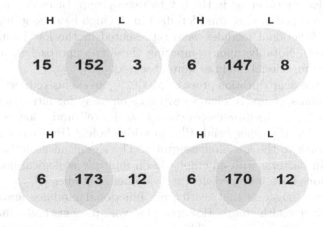

Fig. 3. KEGG modules are identified in high (H) and low (L) biomass groups. First row is the Venn diagram results in Spearman analysis on the left and mutual information on the right from Approach A. Second row has the same order of information as the first row, but from Approach B. (Color figure online)

In a similar manner, the second row depicts the comparison result from the Approach B.

Table 4. List of distinctive KEGG functional modules in High (H) and Low (L) Biomass groups.

Approach A	
High	M00159 V-type ATPase, prokaryotes
	M00529 Denitrification, nitrate $\Rightarrow$ nitrogen
	M00567 Methanogenesis, CO_2 $\Rightarrow$ methane
	M00736 Nocardicin A biosynthesis, L-pHPG + arginine + serine $\Rightarrow$ nocardicin A
	M00804 Complete nitrification, comammox, ammonia $\Rightarrow$ nitrite $\Rightarrow$ nitrate
Low	M00091 Phosphatidylcholine (PC) biosynthesis, PE $\Rightarrow$ PC
	M00745 Imipenem resistance, repression of porin OprD
Approach B	
High	M00754 Nisin resistance, phage shock protein homolog LiaH
Low	M00152 Cytochrome bc1 complex
	M00555 Betaine biosynthesis, choline $\Rightarrow$ betaine
	M00564 Helicobacter pylori pathogenicity signature, cagA pathogenicity island
	M00721 Cationic antimicrobial peptide (CAMP) resistance, arnBCADTEF operon

Incorporating both approaches serves the dual purpose of ensuring highly confident results and retaining significant information. Approach A reveals five distinctive functional modules in the high biomass group, whereas only two distinctive modules are observed in the low biomass group. Likewise, in Approach B, only one functional module was identified in the high biomass group, whereas four exclusive functional modules were represented in the low biomass group (refer to Table 4). Notably, upon comparing these two approaches, no overlapping distinctive functional modules were observed.

In Table 4, the denitrification process (M00529), an exclusive functional module of high biomass group, is a known primary pathway for nitrogen (NO) production by bacteria, which promotes energy to the cell under low oxygen conditions [12, 19]. On the other hand, Phosphatidylcholine (PC) biosynthesis is a pathway exclusive to the low biomass group. The significance of phosphatidylcholine (PC) in bacteria with an emphasis on multiple ecological microbe-host plant interactions have been revealed in a number of studies [1].

This evidence suggests that distinctive functional modules are associated with specific plant health groups. However, this work presents preliminary results characterizing functional modules of high and low biomass groups. The biological association between these groups may be more complex than initially observed. It's possible that these functional modules may occasionally contribute to opposite health conditions.

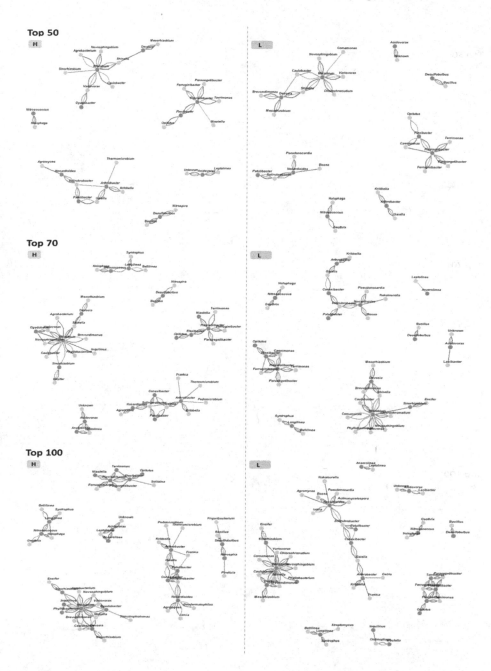

Fig. 4. Comparison of dynamics of OTUs pairs underlying the denitrification process (M00529) in high (H) and low (L) biomass networks. Driver OTUs nodes are red and their neighbor nodes are yellow. Edges are colored by percentile of samples: 74th and 14th percentile (red), 80th and 20th percentile (green), and 86th and 26th percentile (turquoise). (Color figure online)

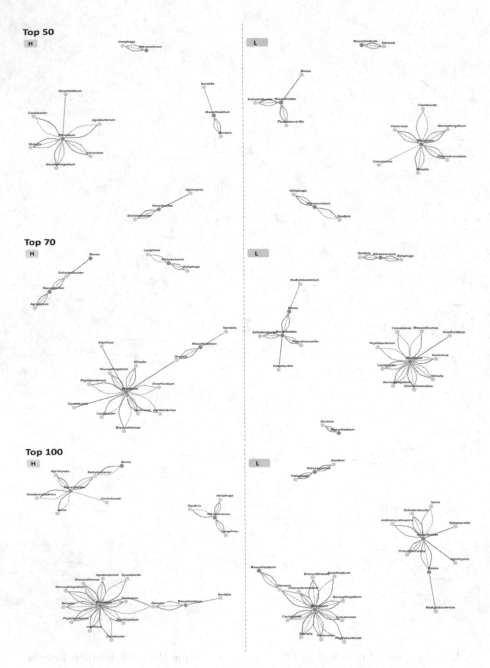

Fig. 5. Comparison of dynamics of OTUs pairs underlying the dhosphatidylcholine (PC) biosynthesis (M00091) in high biomass (H) and low biomass (L) networks.

3.4   Comparison of Dynamics of OTUs Pairs Underlying KEGG Modules

To comprehend the biological significance of identifying distinctive functional modules in OTU pairs, we examined how the dynamics of association between driver OTUs, derived from functional modules, varied in high and low biomass networks. Specifically, we compared the representation of association dynamics between driver OTUs involved in the denitrification process (M00529) across Top-N-ranking high and low biomass groups (see Fig. 4). We hypothesized that these driver OTUs' relationships might serve as key OTU pairs within functional modules related to host environmental conditions. The differing representation of association dynamics between driver OTUs across Top-N-ranking networks in high and low biomass groups is illustrated in different rows of Fig. 4. We posited that the relationship between driver OTUs could serve as crucial pairs within functional modules associated with host environmental conditions. The representation of this relationship differs across Top-N-ranking networks between high and low biomass group networks, as depicted in different rows of Fig. 5.

In addition, the microbiome community among *Flexibacter*, *Flavisolibacter*, *Opitutus*, and *Terrimonas* are commonly represented in both high and low biomass networks across Top-N-ranks. However, these microbes are consistently associated with *Niastella* only in a high biomass group network. Moreover, each of following pairs, the group of *Longilinea* and *Nitrosococcus*, the group of *Acidovorax* and *Anaerolinea*, the group of *Desulfobulbus* and *Nitrospira*, have shown strong correlations in high biomass group networks whereas the relationship of these pairs are not represented in low biomass group networks. In particular, the *Dyadobacter* has shown exclusive associations with *Rhizobium* in high biomass group networks. *Rhizobium* is one kind of rhizosphere bacteria, which is also known as plant beneficial microbes, and can establish a symbiotic relationship with legumes species [10]. It can fix atmospheric nitrogen to help plant growth and improve soil fertility. The *Dyadobacter* is also known as core rhizosphere bacteria [18]. Therefore, it could be inferred that these driver OTUs pairs may have the potential to facilitate plant health and disease suppression.

Figure 5 illustrates the network comparisons of the association between driver OTUs derived from phosphatidylcholine (PC) biosynthesis (M00091) in high and low biomass groups. The M00091 module was commonly enriched in both Spearman analysis and mutual information and was distinctive to low biomass groups. The relationship between *Bosea* and *Nocardioides* is depicted differently in high and low biomass group networks. This OTUs pair constantly appears in low biomass group networks, while this pair has shown an indirect relationship through *Solirubrobacter* in the high biomass group (See Fig. 5). The *Bosea* belongs to non-rhizobial endophytes (NRE), which were also detected in nodules of legume species, although it does not have a similar function like the rhizosphere bacteria group [14]. This suggests that they might cause the rhizobial infection when co-inoculated with rhizosphere microbial communities [14,16].

4   Conclusion

In this study, we introduced a new approach for modeling and analyzing heterogeneous datasets associated with microbiome. Our proposed pipeline aims to mine microbial associations in relation to functional modules in a given environment. The pipeline is designed with robustness in mind in order to achieve a high degree of accuracy and trustworthiness. Our proposed approach facilitated the identification of distinct functional modules based on key OTUs pairs in a given health condition. The result produced by this pipeline highlighted the association between distinctive functional modules and legume species on the basis of key OTUs pairs in both high and low biomass samples. The obtained findings are consistent with previous relevant results reported in the literature. There is a great potential to further develop the proposed approach to analyze microbiome in different environments in future studies. The model presented in the reported study can serve as a framework within which future modifications, in line with choosing percentiles based on given phenotypical factors with different types of microbiome data, can be easily incorporated.

References

1. Aktas, M., Wessel, M., Hacker, S., Klüsener, S., Gleichenhagen, J., Narberhaus, F.: Phosphatidylcholine biosynthesis and its significance in bacteria interacting with eukaryotic cells. Eur. J. Cell Biol. **89**(12), 888–894 (2010)
2. Barberán, A., Bates, S.T., Casamayor, E.O., Fierer, N.: Using network analysis to explore co-occurrence patterns in soil microbial communities. ISME J. **6**(2), 343–351 (2012)
3. Berendsen, R.L., Pieterse, C.M., Bakker, P.A.: The rhizosphere microbiome and plant health. Trends Plant Sci. **17**(8), 478–486 (2012)
4. Berg, G., Rybakova, D., Grube, M., Köberl, M.: The plant microbiome explored: implications for experimental botany. J. Exp. Bot. **67**(4), 995–1002 (2016)
5. Berry, D., Widder, S.: Deciphering microbial interactions and detecting keystone species with co-occurrence networks. Front. Microbiol. **5**, 219 (2014)
6. Deng, Y., Jiang, Y.H., Yang, Y., He, Z., Luo, F., Zhou, J.: Molecular ecological network analyses. BMC Bioinform. **13**(1), 113 (2012)
7. Deng, Z., Zhang, J., Li, J., Zhang, X.: Application of deep learning in plant-microbiota association analysis. Front. Genet. **12**, 697090 (2021)
8. Faust, K., Raes, J.: CoNet app: inference of biological association networks using Cytoscape. F1000Research **5**, 1519 (2016)
9. Federhen, S.: The NCBI taxonomy database. Nucleic Acids Res. **40**(D1), D136–D143 (2012)
10. Gage, D.J.: Infection and invasion of roots by symbiotic, nitrogen-fixing rhizobia during nodulation of temperate legumes. Microbiol. Mol. Biol. Rev. **68**(2), 280–300 (2004)
11. Hernández Medina, R., et al.: Machine learning and deep learning applications in microbiome research. ISME Commun. **2**(1), 98 (2022)
12. Horchani, F., et al.: Both plant and bacterial nitrate reductases contribute to nitric oxide production in Medicago Truncatula nitrogen-fixing nodules. Plant Physiol. **155**(2), 1023–1036 (2011)

13. Langille, M.G., et al.: Predictive functional profiling of microbial communities using 16S rRNA marker gene sequences. Nat. Biotechnol. **31**(9), 814–821 (2013)

14. Leite, J., et al.: Cowpea nodules harbor non-rhizobial bacterial communities that are shaped by soil type rather than plant genotype. Front. Plant Sci. **7**, 2064 (2017)

15. Lucaciu, R., et al.: A bioinformatics guide to plant microbiome analysis. Front. Plant Sci. **10**, 1313 (2019)

16. Pandya, M., Naresh Kumar, G., Rajkumar, S.: Invasion of rhizobial infection thread by non-rhizobia for colonization of Vigna radiata root nodules. FEMS Microbiol. Lett. **348**(1), 58–65 (2013)

17. Vishwakarma, K., Kumar, N., Shandilya, C., Mohapatra, S., Bhayana, S., Varma, A.: Revisiting plant-microbe interactions and microbial consortia application for enhancing sustainable agriculture: a review. Front. Microbiol. **11**, 560406 (2020)

18. Xu, J., et al.: The structure and function of the global citrus rhizosphere microbiome. Nat. Commun. **9**(1), 1–10 (2018)

19. Zumft, W.G.: Cell biology and molecular basis of denitrification. Microbiol. Mol. Biol. Rev. **61**(4), 533–616 (1997)

A Graph-Theory Based fMRI Analysis

Luca Barillaro[1]([✉])(iD), Marianna Milano[1,5](iD), Maria Eugenia Caligiuri[4](iD),
Jelle R. Dalenberg[3,6](iD), Giuseppe Agapito[1,2](iD), Michael Biehl[7](iD),
and Mario Cannataro[1](iD)

[1] Data Analytics Research Center, Department of Medical and Surgical Sciences,
University "Magna Græcia" of Catanzaro, 88100 Catanzaro, Italy
{luca.barillaro,m.milano,agapito,cannataro}@unicz.it
[2] Department of Law, Economics and Social Sciences, University "Magna Græcia" of
Catanzaro, 88100 Catanzaro, Italy
[3] Department of Neurology, University Medical Centre Groningen, University of
Groningen, Groningen, The Netherlands
j.r.dalenberg@umcg.nl
[4] Neuroscience Research Center, University "Magna Græcia" of Catanzaro,
88100 Catanzaro, Italy
me.caligiuri@unicz.it
[5] Department of Experimental and Clinical Medicine, University Magna Græcia of
Catanzaro, 88100 Catanzaro, Italy
[6] Expertise Centre Movement Disorders Groningen, University Medical Center
Groningen, Groningen, The Netherlands
[7] Bernoulli Institute for Mathematics, Computer Science and Artificial Intelligence,
Computer Science Department, Intelligent Systems Group, University of Groningen,
Groningen, The Netherlands
m.biehl@rug.nl

Abstract. In this study, we employed a clustering approach to analyze
fMRI data from a publicly available dataset of patients with mild depression. We utilized the CONN toolbox, a widely recognized tool, to extract
functional networks from the fMRI data. Subsequently, these networks
were aligned using MULTIMAGNA++, a global multiple alignment software, to ensure consistency across individual datasets. The aligned data
was then subjected to a clustering analysis to investigate the presence
of distinct patterns. Our findings demonstrate that not only is it feasible to accurately cluster patients using this approach, but there is also
potential to uncover previously unidentified subgroups among both control subjects and those affected by the disease. These results suggest
new avenues for understanding the neurobiological underpinnings of mild
depression and for developing targeted interventions.

Keywords: Functional Magnetic Resonance Imaging · Machine
learning · Graph theory · Clustering · Global Network alignment

L. Franco et al. (Eds.): ICCS 2024, LNCS 14837, pp. 72–85, 2024.
https://doi.org/10.1007/978-3-031-63778-0_6

1 Introduction

Functional Magnetic Resonance Imaging is an MRI modality that measures oxygenation changes in cerebral blood flow and provides a proxy for the activity of the neurons in the brain. fMRI signal is sensitive to blood dynamic changes which drive the neuron firing. This relationship is known as the Blood Oxygenation Level Dependent (BOLD) effect [8].

fMRI data in general, are the subject of a growing interest since they may open the way to a deeper understanding of brain functioning and may lead to discovering hidden patterns in diseases [21].

To better understand brain functioning and pathology, it is very important to investigate how communication between brain areas changes across different tasks or due to disease. Such communication can be modeled as connections between nodes in the mathematical structure of a graph. Graph theory is a widely applied method in several fields of bioinformatics [11].

Machine learning approaches are often used to discover hidden information or patterns within data. A possible strategy is clustering, an unsupervised learning approach that aims to find groups in data [2]. Even on labeled data, as in our case, it can still unveil hidden information, e.g. a hidden group.

Here, we present a graph-based pipeline for clustering of fMRI data. To test our method, we used data from a publicly available dataset of patients with mild depression. We worked on these data with a popular tool, CONN toolbox [27], to extract networks which later we aligned through a global multiple alignment software, MULTIMAGNA++ [26] and then used this data to perform a clustering task.

The structure of this paper is the following: Sect. 2 describes all the background concepts involved in the experiments described in this paper; Sect. 3 illustrates the dataset and the proposed approach. Section 4 discusses the preliminary results, and finally, Sect. 5 concludes the paper.

2 Background

In this section, we aim to describe background concepts that are involved in the work presented in this paper. We first discuss our data type, functional magnetic resonance imaging, then we recall some machine learning and clustering concepts. In the latter, a quick recall of graph theory concepts and network alignment is made.

2.1 fMRI

Here we introduce a brief description of the functional Magnetic Resonance (fMRI) technique and its applications.

The Functional Magnetic Resonance Imaging (fMRI) is a form of the Magnetic Resonance Imaging (MRI). Whereas MRI provides information about anatomical structure, fMRI provides information about function over time. It was introduced by S. Ogawa et all in 1990 [19].

fMRI is based on the hemodynamic response [23]. When neurons in the brain are active, they consume more oxygen and cause increased regional blood flow. By using a very strong magnetic field and radio waves, fMRI scanners can detect these changes in blood oxygenation and create detailed maps of brain activity.

fMRI methods can highlight the areas of the brain that are activated during specific tasks or cognitive processes. These active areas are represented as statistical maps known as activation maps, representing the regions of the brain that show a significant increase or decrease in neural activity.

Functional Magnetic Resonance Imaging (fMRI) can be divided into two main typologies based on the experimental design:

1. Task-Based fMRI: In task-based fMRI, participants are presented with specific tasks or stimuli while their brain activity is measured using fMRI [24,25]. The goal is to identify brain regions that are selectively activated during task performance and, when appropriate, compare these across groups or conditions.
2. Resting-State fMRI: Resting-state fMRI involves measuring spontaneous brain activity in the absence of any specific task or stimulation. Participants are instructed to relax and keep their minds at rest while fMRI scans are conducted. The analysis focuses on identifying intrinsic patterns of brain connectivity, known as resting-state networks (RSNs), which represent synchronized activity between different brain regions. By doing so, insights into the brain's intrinsic organization and potential biomarkers for neurological and psychiatric disorders are possible.

These typologies of fMRI reflect different experimental designs and data analysis approaches, each offering unique insights into the functional organization of the brain.

2.2   Machine Learning

Machine learning is a branch of artificial intelligence that uses computers to learn from data and make predictions or decisions without explicit programming.

There is a distinction between supervised and unsupervised learning. In supervised learning, machine learning models are trained on labeled data to learn mappings between input and output. Popular tasks like classification and regression use this approach. Unsupervised learning, instead, involves extracting patterns or structures from unlabeled data.

Clustering [10] is part of the unsupervised learning methods and is a technique used to group similar data points based on certain features or characteristics. Clustering algorithms, such as k-means, hierarchical clustering, or DBSCAN, aim to partition data into distinct groups, or clusters, where data points within the same cluster share similarities while being dissimilar to those in other clusters.

2.3   Graph Theory

Graph theory is a powerful formalism for modeling and analyzing complex systems composed of interconnected elements. The graph is the mathematical structure used to model relations between objects. It consists of vertices or nodes, which represent the objects, and edges, which are the connections or relationships between these objects. Graph theory brings together lots of concepts and techniques, including paths, cycles, connectivity, trees, and a wide range of algorithms for analyzing and manipulating graphs.

In bioinformatics, graph theory is mainly used to represent and model biological entities and their interactions as networks [1]. Biological entities such as proteins, DNA, RNA, and metabolites can be effectively modeled as nodes, while their interactions and relationships are represented as edges in a graph [20]. Popular graph-based approaches deal, e.g., with genetic sequences, prediction of protein structures, and understanding metabolic pathways.

A graph is characterized by several topological indices, which quantify different aspects of its structure. Below are some key indices along with brief explanations:

- **Degree**: The number of edges connected to a node. This indicates the immediate number of connections that the node has with others in the graph.
- **Clustering Coefficient**: Measures how complete the neighborhood of a node is by calculating the proportion of actual connections between a node's neighbors to the total possible connections.
- **Global Efficiency**: Represents the average efficiency of parallel information transfer in the network, calculated as the average inverse of the shortest path length between each pair of nodes. This index provides insight into the node's centrality and its overall connectivity within the graph.
- **Cost**: Typically defines the ratio of existing edges to the maximum number of possible edges among nodes, reflecting the density of connections in the graph.
- **Average Path Length**: The average number of steps along the shortest paths for all possible pairs of network nodes. It offers a measure of the efficiency of information or traffic flow in the network.
- **Shortest Path**: The minimum path length between any two non-adjacent nodes, is important for understanding the most efficient route for communication between them.
- **Betweenness Centrality**: Measures the extent to which a node lies on paths between other nodes. Nodes with high betweenness may have considerable influence within a network by virtue of their control over information passing between others.

2.4   Network Alignment

Network alignment is a technique that is used in bioinformatics and network science, to compare and align two or more biological networks. These networks can represent various biological entities and interactions, such as protein-protein interactions, metabolic pathways, or gene regulatory networks.

Network alignment aims to find the correspondence between nodes (biological entities) in different networks, identifying conserved functional modules and evolutionary relationships across species or biological conditions. The result of a network alignment, allows researchers to discover, i.e., shared biological processes, and predict protein functions.

There are different typologies of network alignment. They can be distinguished between local and global.

Global alignment compares entire sequences and is suitable for identifying overall similarity, while local alignment focuses on identifying specific regions of similarity within sequences, making it ideal for detecting conserved functional elements amidst sequence variation.

Moreover, Pairwise or multiple alignments are possible: pairwise compare two sequences to identify regions of similarity. Multiple, instead, extends this concept to align three or more sequences simultaneously, enabling the identification of conserved regions and evolutionary relationships.

3   Material and Methods

3.1   Dataset Description

The dataset comprises resting-state fMRI (rs-fMRI) data from 72 individuals, including 51 patients diagnosed with mild depression and 21 control subjects. The gender distribution within this group includes 19 males (6 controls) and 53 females (15 controls). Depression severity is classified according to the International Statistical Classification of Diseases and Related Health Problems, Tenth Revision (ICD-10), referencing Chapter V: Mental and Behavioral Disorders (codes F00-F99), specifically the section on Mood [affective] disorders (codes F30-F39). While the dataset specifies four types of depression (F32.0, F32.1, F34.1, F34.9), we aggregated these into a single class for analysis. Each scanning session in the dataset consists of 100 scans with a repetition time (TR) of 2.5 s [15]. The data[1] are publicly available through OpenNeuro [4,5] and have been preprocessed using fMRIPrep, a comprehensive tool for fMRI data preparation, with a voxel size set at $2 \times 2 \times 2$ mm [9]. The dataset adheres to the Brain Imaging Data Structure (BIDS) format, which standardizes the organization of neuroimaging and behavioral data across studies by providing a consistent file and directory structure. This standardization facilitates data sharing and reusability, enhancing the potential for collaborative research and validation of findings.

3.2   Pipeline Description

We aimed to create a pipeline that can be automated for further development of this work. Since this work is composed of several steps, our resulting pipeline is

---

[1] https://openneuro.org/datasets/ds002748/versions/1.0.5.

heterogeneous. More specifically, we used a Matlab toolbox to perform analysis on the given dataset, some scripts for data manipulation, and finally another software to perform the clustering tasks.

We now describe which software is involved, while in the next subsection, we discuss our approach with technical details.

Given our dataset, the first step was to perform preprocessing and analyses to obtain a graph representation, i.e., an adjacency matrix.

Thus, to calculate resting state cross-correlations between brain regions we used the popular CONN toolbox [27] (release 21.a) running in Matlab (release 2023b)

From CONN toolbox results, we generated adjacency matrices. A single adjacency matrix contains graph parameters for all the patients, but to perform the multiple global alignments, we needed a single representation for a patient. Thus we used an R script to extract single edge lists, while later to perform clustering tasks we used a Python script to merge data for control and disease cases.

The multiple alignment phase was performed with MULTIMAGNA++ [26] due to its good performance as shown in [16]. Finally, the clustering experiments were conducted by using the popular Weka platform [12] for its friendly and easy-to-use environment. A synthetic description of the pipeline workflow is shown in Fig. 1.

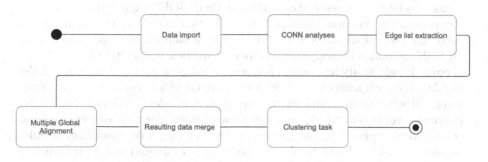

Fig. 1. Pipeline workflow

3.3 Proposed Solution

As introduced earlier, given the selected dataset, we imported it into the CONN toolbox. For this purpose, we created two different projects, one for control and the other one for disease patients. Then we performed the typical analysis pipeline available in this toolbox.

After data importing through the dedicated function for the fMRIPrep pre-processed dataset, the pipeline is the following:

- **Preprocessing**: Functional data were smoothed using spatial convolution with a Gaussian kernel of 8 mm full-width half maximum (FWHM).

– **Denoising**: functional data were denoised using a standard denoising pipeline
[17] including the regression of potential confounding effects characterized by
white matter time-series (5 CompCor noise components), CSF time-series (5
CompCor noise components), motion parameters and their first order derivatives (12 factors) [13], outlier scans (below 62 factors) [22], session effects and
their first order derivatives (2 factors), and linear trends (2 factors) within
each functional run, followed by band-pass frequency filtering of the BOLD
time-series [14] between 0.008 Hz and 0.09 Hz. CompCor [3,6] noise components within white matter and CSF were estimated by computing the average
BOLD signal as well as the largest principal components orthogonal to the
BOLD average, motion parameters, and outlier scans within each subject's
eroded segmentation masks. From the number of noise terms included in this
denoising strategy, the effective degrees of freedom of the BOLD signal after
denoising were estimated to range from 4.9 to 30.3 (average 26.2) across all
subjects [18].
– **First-level analysis**: Seed-based connectivity maps (SBC) and ROI-to-ROI
connectivity matrices (RRC) were estimated characterizing the patterns of
functional connectivity with 164 HPC-ICA networks [7] and Harvard-Oxford
atlas ROIs. Functional connectivity strength was calculated using Fisher-
transformed bi-variate correlation coefficients from a weighted general linear
model (weighted-GLM), defined separately for each pair of seed and target
areas, modeling the association between their BOLD signal time-series. To
compensate for possible transient magnetization effects at the beginning of
each run, individual scans were weighted by a step function convolved with
an SPM canonical hemodynamic response function and rectified.
– **Group-level analyses**: were performed using a General Linear Model
(GLM). For each individual, voxel a separate GLM was estimated, with first-
level connectivity measures at this voxel as dependent variables (one independent sample per subject and one measurement per task or experimental
condition, if applicable), and groups or other subject-level identifiers as independent variables. Voxel-level hypotheses were evaluated using multivariate
parametric statistics with random effects across subjects and sample covariance estimation across multiple measurements. Inferences were performed at
the level of individual clusters (groups of contiguous voxels). Cluster-level
inferences were based on parametric statistics from Gaussian Random Field
theory [28]. Results were thresholded using a combination of a cluster-forming
$p < 0.001$ voxel-level threshold and a family-wise corrected p-FDR < 0.05
cluster-size threshold [16].

Consequently, our analyses incorporate data from 164 regions of interest
(ROIs). Visual representations of the networks derived from these ROIs can
be viewed in Fig. 2 and Fig. 3.

The strength of the selected graph theory metric represents the size of the
circle while edges represent the correlation values between the nodes. The shown
networks represent the network of healthy and disease groups, respectively, generated using the clustering coefficient metric. By a visual analysis of these images,

it is notable that there is a different strength (bigger clustering coefficient) in the disease case group.

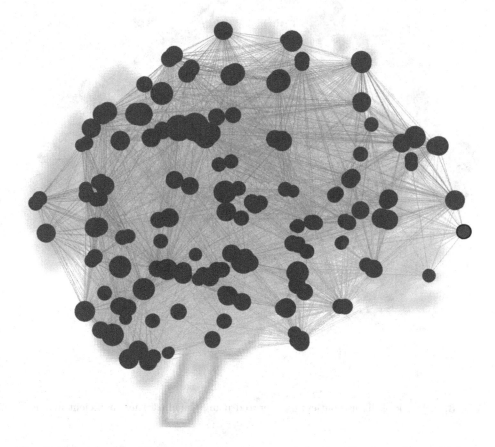

Fig. 2. Network of healthy control patients, generated using the clustering coefficient metric

Our interest was focused on graph theory results and, since CONN computes several graph network measures, we selected two popular graph metrics: global efficiency and clustering coefficient. As a result, from the group analysis results of ROI to ROI connectivity, we obtained a two adjacency matrix for all the patients of the given class.

To prepare data for multiple alignment steps, we extracted single edge lists for every participant from the complete adjacency matrix. We used R (version 4.3.2) and two packages: igraph (version 1.6.0) and R.matlab (version 3.7.0). We show our R script in Algorithm 1, for the control case:

We take the adjacency matrix generated by the CONN toolbox (in a Matlab data file, .mat) as input and extract a single edge list for every patient, saving them into a single text file.

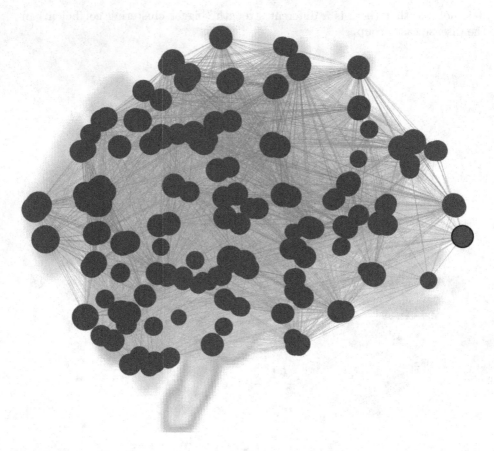

Fig. 3. Network of disease patients, generated using the clustering coefficient metric

---

Algorithm 1. R script for edge list extraction

```
library(igraph)
library(R.matlab)
x = readMat(file.choose());
m = xA
for i = 1 : 21  do
    filename < −paste("/path/edgeListName", i, ".txt", sep = "")
    a = m[, , i]
    g = graph.adjacency(a, mode = "undirected")
    g < −simplify(g)
    list = get.edgelist(g)
    write.table(list, filename, col.names = FALSE, row.names = FALSE, sep =
"tab", quote = FALSE)
end for
```

Then we performed the multiple global alignment obtaining as a result single file for a patient class that is made by ROIs as rows and patient nets on columns.

Then, we manipulated data to prepare them for the clustering phase. More specifically, we merged data from alignment into a single data structure and then transposed it to have patients on rows and ROIs on columns, ending with a comma-separated value file generation. This task was performed in Python (version 3.11.7) using two packages: Pandas (version 2.1.4) and Tkinter (in tk package, version 8.6.12) We report the Python script for that task in Algorithm 2.

Algorithm 2. Python script for multiple alignment merging

```
import pandas as pd
import tkinter as tk
from tkinter import filedialog
root = tk.Tk()
root.withdraw()
path = filedialog.askopenfilename()
df1 = pd.read − csv(path, sep =′ tab′, header = None, engine =′ c′)
root = tk.Tk()
root.withdraw()
path = filedialog.askopenfilename()
df2 = pd.read − csv(path, sep =′ tab′, header = None, engine =′ c′)
df3 = pd.concat([df1, df2], axis = 1)
df4 = df3.transpose()
df4.to − csv(′allTotalHTr.csv′, index = False, sep =′ tab′)
```

After that, we imported data into Weka (version 3.8.6) and performed several clustering tasks to check for the best results. To understand the clustering capabilities to correctly assign a patient to its group, our data are enriched with a class feature. In particular, we used it in the clustering settings, classes to clusters evaluation; using this mode, Weka first ignores the class attribute and generates the clustering. Then during the test phase, it assigns classes to the clusters, based on the majority value of the class attribute within each cluster.

We tested the clustering algorithms K-Means, Expectation Maximization, Farthest-First, and Canopy. Our interest was focused on the ability to correctly assign a patient to its group. Among all, trying different configurations, with a cluster number set to 2, K-Means resulted in better performance for our purposes, so we decided to use it for our testing. It was able to reach the lowest percentage of incorrectly clustered instances by varying the seed number, thus the clustering task was able to identify a healthy control case or a disease case.

4   Results

In this section, we present the results from our experiments.

As introduced earlier, we used the popular K-means method to perform clustering on global alignment data. We tested two scenarios: in the first, we evaluated the clustering ability to correctly assign a patient to its original group, so in this case our K value was set to 2.

In the second one, we tried something different, since our aim was to evaluate if a clustering with three group could find a significantly group of patient in a different stage of disease.

First we recall that the total number of patients (rows) is equal to the number of aligned networks, so we have a total of 72 rows in which the first 21 are control patients, and the remaining 51 are patients with disease.

We present our results, in a confusion matrix, for the first experiment in Table 1 and Table 2 where d stands for disease, c for control.

Table 1. Confusion matrix - global efficiency based network

Actual class	Predicted disease	Predicted control
Control	12	9
Disease	35	16

Table 2. Confusion matrix - clustering coefficient based network

Actual class	Predicted disease	Predicted control
Control	3	18
Disease	31	20

From these results we can understand that the clustering task over data from networks based on the clustering coefficient metric performs slightly better then the one with the global efficiency one. We have a total of 23 incorrectly assigned patients in the first case, and 28 in the second one.

The second experiment, with the K-means K value set to 3, is presented in a similar way to the first one, but our focus is to find if a third group is significant.

We present results for the first experiment in Table 3 and Table 4.

Table 3. Confusion matrix - global efficiency.

Actual class	Predicted disease	Predicted control	Third group
Control	9	7	5
Disease	21	13	17

Table 4. Confusion matrix - clustering coefficient.

Actual class	Predicted disease	Predicted control	Third group
Control	7	8	6
Disease	21	15	15

In this case, we can understand that the third group, takes, in both experiments, patients from the two groups, suggesting to be more investigated.

5   Conclusions

In conclusion, our study aimed to explore the application of machine learning techniques to graph data derived from fMRI scans. By employing a global multiple alignment method, we were able to align extracted networks and discover similarities among them. This alignment facilitated a clustering analysis, which further elucidated relationships within the data. Our methodology successfully demonstrated the capability to classify unseen data into its appropriate group based on regions of interest (ROIs) identified through multiple alignments. The results are promising and suggest that machine learning can effectively analyze fMRI graph data. However, the performance of the clustering could be enhanced by using huge data sets, and a detailed examination of preprocessing parameters to optimize the analysis.

Acknowledgement. This work was funded by the Next Generation EU - Italian NRRP, Mission 4, Component 2, Investment 1.5, call for the creation and strengthening of 'Innovation Ecosystems', building 'Territorial R&D Leaders' (Directorial Decree n. 2021/3277) - project Tech4You - Technologies for climate change adaptation and quality of life improvement, n. ECS0000009. This work reflects only the authors' views and opinions, neither the Ministry for University and Research nor the European Commission can be considered responsible for them.

References

1. Agapito, G., Guzzi, P.H., Cannataro, M.: Visualization of protein interaction networks: problems and solutions. BMC Bioinform. **14**, 1–30 (2013)
2. Agapito, G., Milano, M., Cannataro, M.: A Python clustering analysis protocol of genes expression data sets. Genes **13**(10) (2022). https://www.mdpi.com/2073-4425/13/10/1839
3. Behzadi, Y., Restom, K., Liau, J., Liu, T.T.: A component based noise correction method (CompCor) for BOLD and perfusion based fMRI. NeuroImage **37**(1), 90–101 (2007). https://doi.org/10.1016/j.neuroimage.2007.04.042. https://linkinghub.elsevier.com/retrieve/pii/S1053811907003837
4. Bezmaternykh, D.D., et al.: Brain networks connectivity in mild to moderate depression: resting state fMRI study with implications to nonpharmacological treatment. Neural Plast. **2021**, 1–15 (2021). https://doi.org/10.1155/2021/8846097. https://www.hindawi.com/journals/np/2021/8846097/

5. Bezmaternykh, D.D., Melnikov, M.E., Savelov, A.A., Petrovskii, E.D.: Resting state with closed eyes for patients with depression and healthy participants (2021). https://doi.org/10.18112/OPENNEURO.DS002748.V1.0.5. https://openneuro.org/datasets/ds002748/versions/1.0.5
6. Chai, X.J., Castañón, A.N., Öngür, D., Whitfield-Gabrieli, S.: Anticorrelations in resting state networks without global signal regression. Neuroimage **59**(2), 1420–1428 (2012). https://doi.org/10.1016/j.neuroimage.2011.08.048. https://linkinghub.elsevier.com/retrieve/pii/S1053811911009657
7. Desikan, R.S., et al.: An automated labeling system for subdividing the human cerebral cortex on MRI scans into gyral based regions of interest. Neuroimage **31**(3), 968–980 (2006). https://doi.org/10.1016/j.neuroimage.2006.01.021
8. DeYoe, E.A., Bandettini, P., Neitz, J., Miller, D., Winans, P.: Functional magnetic resonance imaging (FMRI) of the human brain. J. Neurosci. Methods **54**(2), 171–187 (1994). https://doi.org/10.1016/0165-0270(94)90191-0
9. Esteban, O., et al.: fMRIPrep: a robust preprocessing pipeline for functional MRI. Nat. Methods **16**(1), 111–116 (2019). https://doi.org/10.1038/s41592-018-0235-4. https://www.nature.com/articles/s41592-018-0235-4
10. Ezugwu, A.E., et al.: A comprehensive survey of clustering algorithms: State-of-the-art machine learning applications, taxonomy, challenges, and future research prospects. Eng. Appl. Artif. Intell. **110**, 104743 (2022). https://doi.org/10.1016/j.engappai.2022.104743. https://linkinghub.elsevier.com/retrieve/pii/S095219762200046X
11. Farahani, F.V., Karwowski, W., Lighthall, N.R.: Application of graph theory for identifying connectivity patterns in human brain networks: a systematic review. Front. Neurosci. **13** (2019). https://www.frontiersin.org/articles/10.3389/fnins.2019.00585
12. Frank, E., Hall, M.A., Witten, I.H.: The WEKA workbench. Online Appendix. In: Data Mining: Practical Machine Learning Tools and Techniques, 4th edn. Morgan Kaufmann (2016)
13. Friston, K.J., Williams, S., Howard, R., Frackowiak, R.S., Turner, R.: Movement-related effects in fMRI time-series. Magn. Reson. Med. **35**(3), 346–355 (1996). https://doi.org/10.1002/mrm.1910350312
14. Hallquist, M.N., Hwang, K., Luna, B.: The nuisance of nuisance regression: spectral misspecification in a common approach to resting-state fMRI preprocessing reintroduces noise and obscures functional connectivity. Neuroimage **82**, 208–225 (2013). https://doi.org/10.1016/j.neuroimage.2013.05.116. https://linkinghub.elsevier.com/retrieve/pii/S1053811913006265
15. ICD-10 Version:2019. https://icd.who.int/browse10/2019/en
16. Milano, M., Guzzi, P.H., Cannataro, M.: Network building and analysis in connectomics studies: a review of algorithms, databases and technologies. Netw. Model. Anal. Health Inform. Bioinform. **8**(1), 13 (2019). https://doi.org/10.1007/s13721-019-0192-6
17. Nieto-Castanon, A.: Handbook of Functional Connectivity Magnetic Resonance Imaging Methods in CONN. Hilbert Press (2020). https://doi.org/10.56441/hilbertpress.2207.6598. https://www.hilbertpress.org/link-nieto-castanon2020
18. Nieto-Castanon, A.: Preparing fMRI Data for Statistical Analysis (2022). https://doi.org/10.48550/ARXIV.2210.13564. https://arxiv.org/abs/2210.13564, publisher: [object Object] Version Number: 1
19. Ogawa, S., Lee, T.M., Kay, A.R., Tank, D.W.: Brain magnetic resonance imaging with contrast dependent on blood oxygenation. Proc. Natl. Acad. Sci. U.S.A. **87**(24), 9868–9872 (1990). https://doi.org/10.1073/pnas.87.24.9868

20. Pastrello, C., et al.: Visual data mining of biological networks: one size does not fit all. PLoS Comput. Biol. **9**(1), e1002833 (2013)
21. Poston, K.L., Eidelberg, D.: Functional brain networks and abnormal connectivity in the movement disorders. NeuroImage **62**(4), 2261–2270 (2012). https://doi. org/10.1016/j.neuroimage.2011.12.021.   https://www.sciencedirect.com/science/ article/pii/S1053811911014236
22. Power, J.D., Mitra, A., Laumann, T.O., Snyder, A.Z., Schlaggar, B.L., Petersen, S.E.: Methods to detect, characterize, and remove motion artifact in resting state fMRI. Neuroimage **84**, 320–341 (2014). https://doi.org/10.1016/j.neuroimage. 2013.08.048. https://linkinghub.elsevier.com/retrieve/pii/S1053811913009117
23. Shibasaki, H.: Human brain mapping: hemodynamic response and elec-trophysiology. Clin. Neurophysiol. **119**(4), 731–743 (2008). https://doi.org/ 10.1016/j.clinph.2007.10.026. https://www.sciencedirect.com/science/article/pii/ S1388245707006578
24. Silva, M.A., See, A.P., Essayed, W.I., Golby, A.J., Tie, Y.: Challenges and tech-niques for presurgical brain mapping with functional MRI. NeuroImage. Clinical **17**, 794–803 (2018). https://doi.org/10.1016/j.nicl.2017.12.008
25. Soares, J.F., et al.: Task-based functional MRI challenges in clinical neuroscience: choice of the best head motion correction approach in multiple sclerosis. Front. Neu-rosci. **16**, 1017211 (2022). https://doi.org/10.3389/fnins.2022.1017211. https:// www.frontiersin.org/articles/10.3389/fnins.2022.1017211/full
26. Vijayan, V., Milenkovic, T.: Multiple network alignment via MultiMAGNA+. IEEE/ACM Trans. Comput. Biol. Bioinf. **15**(5), 1669–1682 (2018). https://doi. org/10.1109/TCBB.2017.2740381
27. Whitfield-Gabrieli, S., Nieto-Castanon, A.: Conn: a functional connectivity toolbox for correlated and anticorrelated brain networks. Brain Connect. **2**(3), 125–141 (2012). https://doi.org/10.1089/brain.2012.0073
28. Worsley, K.J., Marrett, S., Neelin, P., Vandal, A.C., Friston, K.J., Evans, A.C.: A unified statistical approach for determining significant signals in images of cere-bral activation. Hum. Brain Mapp. **4**(1), 58–73 (1996). https://doi.org/10.1002/ (SICI)1097-0193(1996)4:1⟨58::AID-HBM4⟩3.0.CO;2-O

A Pipeline for the Analysis of Multilayer Brain Networks

Ilaria Lazzaro[1,3]([✉]) [iD], Marianna Milano[2,3] [iD], and Mario Cannataro[1,3] [iD]

[1] Department of Medical and Surgical Sciences, University Magna Græcia,
88100 Catanzaro, Italy
{cannataro,ilaria.lazzaro}@unicz.it
[2] Department of Experimental and Clinical Medicine, University Magna Græcia,
88100 Catanzaro, Italy
m.milano@unicz.it
[3] Data Analytics Research Center, University Magna Græcia, 88100 Catanzaro, Italy

Abstract. The formalism of multilayer networks (MLN) makes possible to model and understand the multiple relationships between entities in a system. Indeed, this representation has found its way into a wide range of disciplines, particularly in the fields of neuroscience and neuroimaging. Human brain modelling made possible the identification of the basis for the construction of morphological, structural and functional brain connectivity networks.

In this work, we propose the design and implementation of a software pipeline for the construction and analysis of multilayer brain networks. This approach aims to identify groups of strongly connected nodes within the network and to evaluate the resulting communities. We examined 10 healthy subjects and 10 patients with multiple sclerosis. We analyzed the brain MLN by applying community detection algorithm that identified recurrent communities in patients with multiple sclerosis. To assess the structure of communities within the network, we calculate modularity indices for each subject. Finally, we confirm what has already been found in the literature, i.e. a high modularity in the brain networks of diseased subjects compared to those of healthy subjects. Future developments could involve aligning these networks to identify common patterns among multiple sclerosis patients and potentially identify subgroups of patients with similar neural characteristics.

Keywords: Multilayer Network · Brain Network · Community Detection · Modularity

1 Introduction

In the last decades, network theory has undergone a profound evolution as an interdisciplinary field that combines methods and graph theories to study complex phenomena. Multilayer networks are an essential tool for understanding multiple interactions between elements of the same type, connected through

L. Franco et al. (Eds.): ICCS 2024, LNCS 14837, pp. 86–98, 2024.
https://doi.org/10.1007/978-3-031-63778-0_7

edges between the different layers or levels of a system. They are a mathematical extension [6] of traditional networks that find a place in a wide range of fields, from human and social sciences to computer science, from biology to medicine. Particularly, in the field of neuroscience and neuroimaging, multilayer networks summarise the concept of 'brain networks' to explain and model the structure and function of the brain. Starting from this concept, the intersection of network science and neuroscience has made it possible to understand precisely how multilayer networks facilitate this approach. In fact, the literature shows how multilayer modelling offers a unique opportunity to take into account the simultaneous existence of different types of relationships by including multiple network layers to explicitly represent the characteristics and functionality of the brain [5]. Specifically, we aim to bridge the gap in existing methodologies [18,23] by offering a comprehensive approach integrating information from multiple layers of brain connectivity, including structural, morphological and functional aspects.

The origins and evolution of multilayer networks have been discussed extensively [12,14], and each multilayer network model depends on how the layers and edges are implemented. In the simplest case, to describe a multilayer network one defines a set of entities as nodes, organised in different layers, whose edges connect nodes of the same layer (intra-layer) or of different layers (inter-layer). The human brain, on the other hand, is a complex system organised by a set of structural and functional relationships between its elements. It is represented as a graph that defines nodes as brain regions and edges as direct correlations between the various functionalities.

In this paper, we propose the design of a pipeline for the modelling and analysis of multilayer networks based on the functional and structural activity of the brain.

We used a public dataset provided by Jordi Casas et al. [4] on MRI images, from which we extracted 20 patients (10 healthy subjects and 10 multiple sclerosis patients). We using R programming [19] to model brain connectivity matrices in a multilayer network. We also illustrate the difference between the two groups of patients (healthy and MS patients) by computing modularity and analysing the characteristics of recurrent communities associated with particular brain regions in patients with multiple sclerosis.

The rest of paper is organised as follows: In Sect. 2, we present background of a multilayer network, in particular the approach used in brain networks. In Sect. 3, we describe the proposed dataset and pipeline modelling techniques. In Sect. 4, we discuss the data extracted from the analysis conducted on multilayer networks. We discuss the processed dataset and compare the results obtained. Finally, in the Sect. 5 section, we summarize and conclude the paper.

2 Background

In this section, we discuss the background of multilayer networks, outlining their main features and their strength in modelling complex systems.

Furthermore, we explore how this powerful paradigm has been adopted in the analysis of brain networks, highlighting the relationship between brain structure and function.

2.1 Definition of Multilayer Networks

Multilayer networks present an important architecture within network theory, involving the stratification of multiple layers of interconnections for understanding complex systems. In a multilayer network, each layer can represent a different connection or a different type of relationship between the elements of the system, allowing for a more accurate modelling of its structure and dynamics. Given a set of N nodes and L layers, a multilayer network is described as a quadruple $G = (N, L, V, E)$ where (V, E) is a graph $V \subseteq N \times L$ and $E \subseteq V \times V$ [13].

For definition, the set of nodes N represents the actors of the system, the set of layers L represent the interconnections and relationships of the system. Furthermore, entities of the graph may exist on different layers and assume different types of relationships. A distinction is made between intra-layer and iter-layer edges, the former connect nodes belonging to the same layer; inter-layer nodes, on the other hand, connect vertices of two distinct layers, as depicted in Fig. 1.

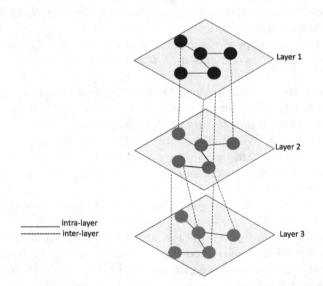

Fig. 1. Example of a multi-layer network. The figure illustrates the general model of a multilayer network, in which there is a set of nodes V, which can be connected to each other in pairs, either within the layers or across the layers themselves. Furthermore, the edges of a multilayer graph connecting nodes of the same layer are described as intra-layer (illustrated by the solid lines), and inter-layer connecting nodes of two different layers (illustrated by the dashed lines).

The definition of a multilayer network provides a more broad version, although the general model allows the specification of most systems that include the different interactions between entities; each layer may contain a complete set of nodes or a subset of them. Depending on the type of implementation and according to the network's conformation, it assumes different terminology, Kivela et al. [14] outline a general framework associating each type of network (Multiplex, Interconnected, Multilevel, Hypergraph) with each characteristic.

2.2 Brain Network Applications

The role that multilayer networks assume in the field of neuroscience and neuroimaging has become crucial, providing key insights into the possibility of modelling functional and structural connectivity between different areas of the human brain. In fact, based on the network concept, particularly multilayer networks, the human brain can be considered a complex network consisting of nodes that correspond to brain regions and edges that refer to structural or functional connections between these regions [16]. Multilayer brain network modelling approaches have been extensively employed to provide insights both into mechanisms of brain structure and function by investigating the topological characteristics of the network; but also into the diagnosis and treatment of brain diseases, as well as to deepen the understanding of cognitive and neurological processes (computational modelling). Specifically, the structural connectivity of the brain refers to the physical organisation of neural connections within the brain and their anatomical integrity and organisation [24]. Data are observed by means of diffusion tensor magnetic resonance imaging (DTI), which allows the direct mapping of neural fibres in the brain and the assessment of their integrity and organisation. This information is represented as multilayer networks, where each layer reflects a different connection mode, such as connectivity between cortical regions or connections between brain areas and white matter fibres [3].

Functional connectivity refers to patterns of synchronised neuronal activity between different brain regions during the performance of cognitive tasks or under resting conditions. This synchronisation can be measured using brain imaging techniques such as functional magnetic resonance imaging (fMRI) or electroencephalography (EEG) [10]. Functional brain networks reveal the neural circuits involved in cognitive, emotional and behavioural processes [22]. From a diagnostic point of view, through the analysis of alterations in the structure and function of brain networks, it is possible to identify biomarkers for neurodegenerative diseases such as Alzheimer's [15], Parkinson's [26] or post-traumatic stress disorder [21].

Finally, brain networks are widely used in computational brain modelling to simulate cognitive and neurological processes. Through computational models based on multilayer networks, researchers can explore the complex dynamics of the brain and test hypotheses about its organisation and functioning [25]. This holistic approach provides a comprehensive view of the complexity of the brain. In fact, the different brain regions that constitute the nodes of the network, supplemented by the multiple layers that each respectively represent the brain modalities already discussed, represent the right methodology for understanding complex neural processes.

3   Materials and Methods

This section presents the developed papelines.

3.1   Data Source

The data were obtained from a public dataset made available by the authors (https://github.com/ADaS-Lab/Multilayer-MRI/tree/main). The dataset comprises: the adjacency matrices obtained from structural, morphological and functional MRI of 142 subject samples, from which we extracted 10 healthy subjects and 10 patients with relapsing-remitting multiple sclerosis (MS). Each network has 76 label nodes, specifying each of the 76 brain regions.

3.2   Construction and Analysis of Multilayer Networks

The pipeline for modeling and analysis of multilayer networks includes 3 steps, as depicted in Fig. 2:

1. The first step consists of constructing the single-layer networks using the connectivity information extracted from the adjacency matrices. Functional matrices represent the connectivity of brain areas based on neuronal activity measured during the resting state of the brain (RS). Structural matrices reflect structural connectivity based on fractional anisotropy (FA) weighted

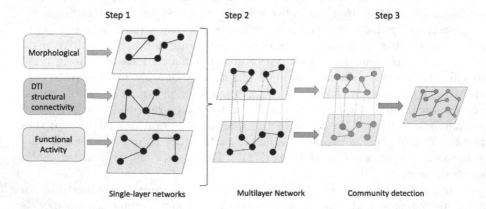

Fig. 2. Outline of workflow definition on pipeline development. Extraction of the functional, morphological and structural brain connectivity matrices and modeling of the individual network. Implementation of the multilayer network by combining the three of information, application of community detection algorithms to perform the analysis, and calculation of modularity indices to evaluate the structure of the network.

connectivity, highlighting the directionality of nerve fibres and their structural integrity. Morphological matrices provide information on the morphological association of the grey matter (GM). From these matrices, the list of arcs (edge list) is extracted for each network, thus modelling single-layer networks. In this context, these are unweighted networks, so the edgelist consists of a row with two elements, the indication of the starting node and the ending node. For each subject, three monolayer networks were generated corresponding to structural connectivity by DTI, functional activity by rs-fMRI and morphological representation.

2. Multilayer network construction. The second phase consists in modelling the multilayer network by combining the three single-layer networks. The output is a multilayer network composed of layers representing morphology and functional activity, and DTI structural connectivity encoding the intra-layer connections of the network. The nodes of the cerebral networks are all constructed by 76 labels, representing the same object in each of the different layers and representing a particular brain region.

3. Analysis of multilayer network. The third step consists in the analysis of multilayer networks, through community extraction (to extract the set of densely connected nodes) and the calculation of modularity indices (evaluating the quality of the results). Community extraction for each network was performed by applying *infomap* [7] [8], a community detection algorithm for multilayer networks. It is based on the concept of finding optimal information structures within the network, interpreting links as information flows that seek to find a balance between the compression of the network structure and the discovery of distinct communities. Modularity, on the other hand, allows the evaluation of the resulting communities and provides an indication of the modular structure of the network: a higher value indicates a greater separation and distinction between communities, while a lower value suggests a more homogeneous network structure.

The code was run in R version 4.3.0.

4   Result and Discussion

From the dataset, we extracted information from 20 subjects, 10 healthy and 10 relapsing-remitting multiple sclerosis patients. From the combination of functional, structural and morphological monolayer brain networks, we modeled a multilayer network for each patient. The characteristics of these networks are detailed in the Table 1 for MS patients, and in Table 2 for healthy subjects.

The analysis conducted on the modelled multilayer networks is based on the application of a community extraction algorithm. According to the literature, by performing a detailed evaluation of community detection algorithms, Infomap emerges as the only one that offers the optimal or near-optimal performance throughout [2]. In fact, it was observed that this algorithm revealed the presence of recurrent communities in MS patients, involved such as amygdala [20], brainstem and hippocampus [17].

Table 1. The table describes the information for each MS patients network: the number of nodes and the number of edges belonging to each layer.

Brain MLN MS patients	Nodes	Edges	Layer
P1	1	68	291
	2	74	282
	3	43	274
P2	1	54	174
	2	34	147
	3	34	147
P3	1	67	306
	2	52	101
	3	29	109
P4	1	57	247
	2	63	121
	3	31	115
P5	1	56	188
	2	67	186
	3	39	163
P6	1	56	208
	2	63	168
	3	40	180
P7	1	73	495
	2	62	261
	3	54	285
P8	1	71	337
	2	73	419
	3	29	120
P9	1	138	525
	2	70	359
	3	47	322
P10	1	74	607
	2	74	535
	3	39	195

In addition, we compared patients' brain networks with those of healthy subjects, revealing significant differences in modularity and network structure. In particular, we confirmed what is reported in the literature of greater modularity in the brain networks of patients with MS, for example Abdolalizadeh et al. showed that subjects with MS had a higher modularity and a lower overall efficiency than the controls [1]. Studies reveal that already in single-layer functional

Table 2. The table describes the information for each Healthy Subject network: the number of nodes and the number of edges belonging to each layer.

Brain MLN Healthy Subject	Nodes	Edges	Layer
S1	1	75	209
	2	75	368
	3	40	840
S2	1	72	668
	2	75	368
	3	45	220
S3	1	70	449
	2	74	691
	3	38	255
S4	1	49	360
	2	73	297
	3	36	213
S5	1	72	834
	2	59	215
	3	36	181
S6	1	75	935
	2	71	477
	3	52	379
S7	1	72	500
	2	70	835
	3	56	314
S8	1	73	517
	2	50	249
	3	63	354
S9	1	62	396
	2	64	396
	3	56	292
S10	1	75	926
	2	72	510
	3	51	301

connectivity networks there is greater modularity in MS patients than healthy subjects [9]. For example, Gamboa et al. found that the resting brain networks of MS patients showed a greater modularity of the network, i.e. a decrease in functional integration between distinct functional modules [11];

Table 3. The table presents the number of communities identified in each brain multilayer network (MLN) for both MS patients and healthy patients. The rows represent the MLNs, with each cell indicating the number of communities detected within the respective MLN for both patient groups.

	Brain MLN	Community Detection Results
M	N 1.	4
S	N 2.	2
	N 3.	9
P	N 4.	5
A	N 5.	6
T	N 6.	10
I	N 7.	7
E	N 8.	1
N	N 9.	8
T	N 10.	1
S		
	N 1.	5
	N 2.	3
H	N 3.	2
E	N 4.	7
A	N 5.	9
L	N 6.	1
T	N 7.	2
H	N 8.	10
Y	N 9.	4
	N 10.	2

Table 4. The table describes the values of the modularity indices calculated for healthy volunteers and MS patients.

Healthy Subjects	Multiple sclerosis patients
0.208	0.237
0.166	0.256
0.111	0.362
0.305	0.437
0.343	0.593
0.104	0.5790
0.117	0.479
0.454	0.134
0.184	0.376
0.114	0.114

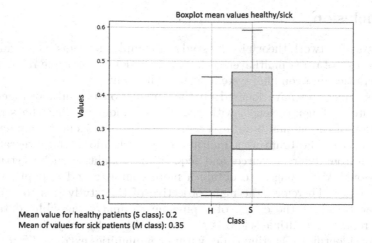

Fig. 3. Boxplot of the mean of the extracted modularity indices. Healthy subjects are shown in pink and MS patients in light blue. (Color figure online)

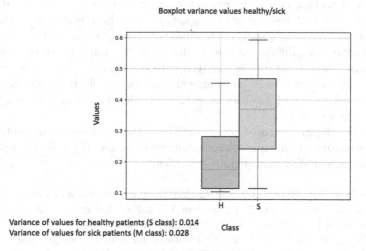

Fig. 4. Boxplot of the variance of the extracted modularity indices. Healthy subjects are shown in pink and MS patients in light blue. (Color figure online)

The following tables provide detailed information on the conducted analysis Table 3, Table 4. These tables offer a comprehensive description of the communities extracted using the applied algorithm, as well as the results of the modularity index analysis between the brain networks of patients with multiple sclerosis (MS) and those of healthy subjects.

We statistically examined the mean and variance of the modularity indices in the two groups to confirm the significance of the observed differences. Fig. 3 and Fig. 4 depict box plots illustrating the statistical analysis conducted on the modularity values in the two groups.

5   Conclusion

The success of network theory in the study of complex systems has highlighted the advantages of using multilayer networks to model the multiple relationships between entities in a complex system such as the brain.

In this paper, we have focused on the application of multilayer networks in the context of neuroscience, with the aim of understanding the structure and function of the human brain in a more detailed and accurate manner. The representation of the brain as a multilayer network allows the integration of information from different sources and acquisition modalities, such as structural and functional MRI images, to obtain a more complete and in-depth view of brain dynamics. Therefore, the main objective of this study was to construct a workflow based on the design of a pipeline to analyse a public dataset of relapsing-remitting multiple sclerosis patients and healthy subjects.

We have identified recurring nodes within communities extracted from multi-layered brain networks of patients with multiple sclerosis, allowing us to observe the presence of similar neural structures among patients. This suggests that community detection is able to extract a significant region involved in multiple sclerosis and that specific brain regions tend to be involved in a similar or related way in the pathogenesis of the disease. Moreover, the statistical analysis on the extracted modularity indices underscores the importance of modularity as a measure for understanding the structure of communities within networks, especially neural networks in pathological conditions, offering valuable insights into the development of analysis strategies needed to completely understand disease characteristics.

Future developments could involve aligning these networks to identify common patterns among multiple sclerosis patients and identify possible subgroups of patients with similar neural characteristics.

Acknowledgement. This work was funded by the Next Generation EU - Italian NRRP, Mission 4, Component 2, Investment 1.5, call for the creation and strengthening of 'Innovation Ecosystems', building 'Territorial R&D Leaders' (Directorial Decree n. 2021/3277) - project Tech4You - Technologies for climate change adaptation and quality of life improvement, n. ECS0000009. This work reflects only the authors' views and opinions, neither the Ministry for University and Research nor the European Commission can be considered responsible for them.

References

1. Abdolalizadeh, A., Ohadi, M.A.D., Ershadi, A.S.B., Aarabi, M.H.: Graph theoretical approach to brain remodeling in multiple sclerosis. Netw. Neurosci. **7**(1), 148–159 (2023)
2. Aldecoa, R., Marín, I.: Exploring the limits of community detection strategies in complex networks. Sci. Rep. **3**(1), 2216 (2013)
3. Bastiani, M., Roebroeck, A.: Unraveling the multiscale structural organization and connectivity of the human brain: the role of diffusion MRI. Front. Neuroanat. **9**, 77 (2015)

4. Casas-Roma, J., et al.: Applying multilayer analysis to morphological, structural, and functional brain networks to identify relevant dysfunction patterns. Netw. Neurosci. **6**(3), 916–933 (2022)
5. De Domenico, M.: Multilayer modeling and analysis of human brain networks. Giga Sci. **6**(5), gix004 (2017)
6. De Domenico, M., et al.: Mathematical formulation of multilayer networks. Phys. Rev. X **3**(4), 041022 (2013)
7. Edler, D., Holmgren, A., Rosvall, M.: The MapEquation software package (2023). https://mapequation.org
8. Edler, D., Holmgren, A., Rosvall, M.: The MapEquation software package (Apr 2023), https://mapequation.org
9. Eqlimi, E., et al.: Modular organization of resting state functional networks in patients with multiple sclerosis. In: Proceedings of the 37th Annual International Conference of the IEEE Engineering in Medicine and Biology Society (EMBC 2015), Milan, Italy, pp. 25–29 (2015)
10. Fox, M., Greicius, M.: Clinical applications of resting state functional connectivity. Front. Syst. Neurosci. **4**, 1443 (2010)
11. Gamboa, O., et al.: Working memory performance of early MS patients correlates inversely with modularity increases in resting state functional connectivity networks. Neuroimage **94**, 385–395 (2014). https://doi.org/10.1016/j.neuroimage.2013.12.008. https://www.sciencedirect.com/science/article/pii/S10538119
12. Hammoud, Z., Kramer, F.: Multilayer networks: aspects, implementations, and application in biomedicine. Big Data Anal. **5**(1), 2 (2020)
13. Interdonato, R., Magnani, M., Perna, D., Tagarelli, A., Vega, D.: Multilayer network simplification: approaches, models and methods. Comput. Sci. Rev. **36**, 100246 (2020). https://doi.org/10.1016/j.cosrev.2020.100246. https://www.sciencedirect.com/science/article/pii/S1574013719301923
14. Kivelä, M., Arenas, A., Barthelemy, M., Gleeson, J.P., Moreno, Y., Porter, M.A.: Multilayer networks. J. Complex Netw. **2**(3), 203–271 (2014). https://doi.org/10.1093/comnet/cnu016
15. Mahbub, N.I., Hasan, M.I., Rahman, M.H., Naznin, F., Islam, M.Z., Moni, M.A.: Identifying molecular signatures and pathways shared between Alzheimer's and Huntington's disorders: a bioinformatics and systems biology approach. Inform. Med. Unlocked **30**, 100888 (2022)
16. Mandke, K., et al.: Comparing multilayer brain networks between groups: introducing graph metrics and recommendations. NeuroImage **166**, 371–384 (2018). https://doi.org/10.1016/j.neuroimage.2017.11.016. https://www.sciencedirect.com/science/article/pii/S1053811917309230
17. Meijboom, R., et al.: Patterns of brain atrophy in recently-diagnosed relapsing-remitting multiple sclerosis. PLoS One **18**(7), e0288967 (2023)
18. Muldoon, S.F., Bassett, D.S.: Network and multilayer network approaches to understanding human brain dynamics. Philos. Sci. **83**(5), 710–720 (2016). https://doi.org/10.1086/687857
19. R Core Team: R: A Language and Environment for Statistical Computing. R Foundation for Statistical Computing, Vienna, Austria (2021). https://www.R-project.org/
20. von Schwanenflug, N., et al.: Increased flexibility of brain dynamics in patients with multiple sclerosis. Brain Commun. **5**(3), fcad143 (2023). https://doi.org/10.1093/braincomms/fcad143

21. Suo, X., et al.: Multilayer network analysis of dynamic network reconfiguration in adults with posttraumatic stress disorder. Biol. Psychiatry Cogn. Neurosci. Neuroimaging **8**(4), 452–461 (2023)
22. Ting, C.M., Samdin, S.B., Tang, M., Ombao, H.: Detecting dynamic community structure in functional brain networks across individuals: a multilayer approach. IEEE Trans. Med. Imaging **40**(2), 468–480 (2020)
23. Vaiana, M., Muldoon, S.F.: Multilayer brain networks. J. Nonlinear Sci. **30**(5), 2147–2169 (2020)
24. Wierzbiński, M., Falcó-Roget, J., Crimi, A.: Community detection in brain connectome using quantum annealer devices. bioRxiv (2022)
25. Williamson, B.J., De Domenico, M., Kadis, D.S.: Multilayer connector hub mapping reveals key brain regions supporting expressive language. Brain Connect. **11**(1), 45–55 (2021)
26. Xu, J., et al.: Identification of key genes and signaling pathways associated with dementia with lewy bodies and parkinson's disease dementia using bioinformatics. Front. Neurol. **14**, 1029370 (2023)

Numerical Algorithms and Computer Arithmetic for Computational Science

Modified CORDIC Algorithm for Givens Rotator

Pawel Poczekajlo[2], Leonid Moroz[3], Ewa Deelman[1], and Pawel Gepner[3(✉)]

[1] USC Information Sciences Institute, University of Southern California, Marina del Rey, CA 90292, USA
deelman@isi.edu
[2] Faculty of Electronics and Computer Science, Koszalin University of Technology, Koszalin, Poland
pawel.poczekajlo@tu.koszalin.pl
[3] Faculty of Mechanical and Industrial Engineering, Warsaw University of Technology, Warsaw, Poland
{leonid.moroz,pawel.gepner}@pw.edu.pl

Abstract. The article presents a modified CORDIC algorithm for implementing a Givens rotator. The CORDIC algorithm is an iterative method for computing trigonometric functions and rotating vectors without using complex calculations. The authors propose two modifications for improving the classical CORDIC algorithm: completing iterations with one-directional rotation of the vector at the final stages and choosing a scaling factor value that can be implemented with low-cost dedicated hardware utilising canonical signed digits representation. The modified algorithm is implemented in a pipeline approach using Verilog language in an Altera Cyclone V System-on-Chip FPGA. The results show that the proposed algorithm achieves higher accuracy and lower latency than the classic CORDIC algorithm.

Keywords: CORDIC algorithm · Givens rotator · FPGA

1 Introduction

With their pivotal role in numerical linear algebra, Givens rotators serve as indispensable tools across various mathematics and computer science domains. Primarily renowned for their efficacy in the QR decomposition (QRD) of matrices, where the strategic zeroing of matrix elements is a common practice [1–4], Givens rotators exhibit versatility that extends beyond traditional linear algebra applications. These rotators form the cornerstone for implementing point rotation within 2D coordinate systems [5–7], as well as in the intricate realms of 3D space, showcasing adaptability in various variants [8].

Beyond mathematics, Givens rotators find practical utility in computer vision and image processing applications. A notable example includes their integral role in estimating head direction (orientation/position) within images or frames

L. Franco et al. (Eds.): ICCS 2024, LNCS 14837, pp. 101–114, 2024.
https://doi.org/10.1007/978-3-031-63778-0_8

[9]. This feature holds significant implications for tracking positions or gazes, particularly in the dynamic landscapes of virtual reality (VR), augmented reality (AR), and mixed-reality (XR) systems. As technology advances, implementing Givens rotators emerges as a crucial element in enhancing the precision and efficacy of position tracking in these immersive environments.

While Givens rotators focus on matrix manipulations and factorisations, the coordinate rotation digital computer (CORDIC) algorithm is more geared towards trigonometric function calculations, emphasising simplicity and hardware efficiency. They are complementary in specific applications, where Givens rotations are used for matrix transformations, and CORDIC is employed for efficient angle computations, such as coordinate rotations. Currently, CORDIC remains one of the main approaches to rotation [10–14].

In this article, we propose CORDIC's modified algorithm and its embedded implementation in an FPGA. We compare it with an earlier implementation that was our reference and the starting point for our considerations.

2   Review of Algorithms and Solutions

The critical component of a Givens rotator is an orthogonal rotation matrix, and in the basic form, it is a 2×2 matrix [5,6]. The elements of this matrix assume values defined by trigonometric functions, and the orthogonality of the matrix arises from the Pythagorean trigonometric identity. This orthogonality property also allows using Givens rotators to implement orthogonal filters (orthogonal state equations) in a pipelined structure [15].

An important aspect when applying Givens rotators is their realisation to minimise computation complexity. The basic approach of software implementation of this is based on using multiplication and addition operations [3]. However, more commonly encountered implementations utilise an iterative CORDIC algorithm [16], often using an FPGA/CPLD chip [17].

In general, the CORDIC algorithm appears in two fundamental variants [18]:

- For determining the coordinates of a point after rotation according to the values of the rotation matrix (values of trigonometric functions sin() and cos()). This is a typical implementation for Givens rotators [4,19].
- For determining the values of trigonometric functions for a specified angle, usually associated with QR decomposition [1–3,20].

The mathematical basis of the CORDIC algorithm involves conditional addition/subtraction operations and bit-shifts (as division/multiplication by 2). CORDIC was invented (and is commonly used) as an algorithm based on simple fixed-point operations. But, some variants utilise floating-point arithmetic [19]. The CORDIC algorithm and its individual iterations have been modified and improved for many years [18]. The motivations for improving CORDIC are:

- Increased precision with fewer iterations.
- Reduced iteration complexity (fewer arithmetic/logical operations).

- Lower utilisation of computational unit resources (e.g., fewer ALUs elements in FPGA/CPLD).
- Higher operating frequency (higher sampling rate for iterations).

An alternative to the CORDIC algorithm is the BKM algorithm, which can also be used for implementing Givens rotators [21]. However, BKM is more complex and, therefore, rarely employed.

3 CORDIC Algorithm and Its Modified Version

The classical CORDIC algorithm takes the input vector $[x_{in}, y_{in}]^T$ and the rotation angle $\theta \in \{-Pi/2, +Pi/2\}$, then the output vector $[x_{out}, y_{out}]^T$ can be found from

$$\begin{bmatrix} x_{out} \\ y_{out} \end{bmatrix} = \begin{bmatrix} cos\theta & -sin\theta \\ sin\theta & cos\theta \end{bmatrix} \begin{bmatrix} x_{in} \\ y_{in} \end{bmatrix} \tag{1}$$

In the CORDIC method, angle θ is presented through the arctangent set of constant angles

$$\theta = \sum_{i=0}^{m} \sigma_i \arctan(2^{-i}) \tag{2}$$

where a signed basis of bi-directional rotation $\sigma_i \in \{-1, 1\}$ is used with the weight of the corresponding angle $\arctan(2^{-i})$. The main idea of the classical CORDIC method is the linear convergence of the method (only one correct bit of the result per iteration) associated with the need to implement three simultaneous iteration equations (for $x_{i+1}, y_{i+1}, z_{i+1}$) in the case of applying a pipeline structure of the computation.

$$x_{i+1} = x_i - \sigma_i 2^{-i} y_i;$$
$$y_{i+1} = y_i + \sigma_i 2^{-i} x_i;$$
$$z_{i+1} = z_i - \sigma_i \arctan(2^{-i});$$
$$\sigma_i = \text{sign}(z_i), i = (0, .., m)$$

needs $m + 1$ elementary bi-directional rotation.
Initial values:
$$z_0 = \theta,$$
$$x_0 = x_{in},$$
$$y_0 = y_{in}$$

Rotations, in this case, are called pseudo-rotations and have gain factor

$$K = \prod_{i=0}^{m} \sqrt{1 + 2^{-2i}} \tag{3}$$

and scaling factor

$$P = \frac{1}{K}$$

Rotations equations take the matrix form

$$\begin{bmatrix} x_{i+1} \\ y_{i+1} \end{bmatrix} = \begin{bmatrix} 1 & -\sigma_i 2^{-i} \\ \sigma_i 2^{-i} & 1 \end{bmatrix} \begin{bmatrix} x_i \\ y_i \end{bmatrix} \qquad (4)$$

If we complete all $m+1$ iterations according to a formula (4), we obtain:

$$\begin{bmatrix} x_{m+1} \\ y_{m+1} \end{bmatrix} = (\prod_{i=0}^{m} \begin{bmatrix} 1 & -\sigma_i 2^{-i} \\ \sigma_i 2^{-i} & 1 \end{bmatrix}) \begin{bmatrix} x_i \\ y_i \end{bmatrix} = K * \begin{bmatrix} cos\theta & -sin\theta \\ sin\theta & cos\theta \end{bmatrix} \begin{bmatrix} x_{in} \\ y_{in} \end{bmatrix} \qquad (5)$$

or

$$\begin{bmatrix} x_{m+1} \\ y_{m+1} \end{bmatrix} = \begin{bmatrix} cos\theta & -sin\theta \\ sin\theta & cos\theta \end{bmatrix} \begin{bmatrix} K * x_{in} \\ K * y_{in} \end{bmatrix} \qquad (6)$$

It follows from the last equation that the rotated vector will be modified in the Cordic procedure, so it is necessary to perform scaling operations and obtain the corrected values of the components of the output vector. To do this, we must scale the components of the output vector as follows (then the connection between vectors $[x_{m+1}, y_{m+1}]^T$ and $[x_{out}, y_{out}]^T$ is):

$$\begin{bmatrix} x_{out} \\ y_{out} \end{bmatrix} = P * \begin{bmatrix} x_{m+1} \\ y_{m+1} \end{bmatrix} = \begin{bmatrix} P * x_{m+1} \\ P * y_{m+1} \end{bmatrix} \qquad (7)$$

The Cordic Rotation Algorithm approximates rotation through iterative steps without relying on complex trigonometric computations. This makes it particularly suitable for hardware implementations or applications with generalizable computational efficiency. Algorithm 1 presents a generalised pseudocode version of the classic CORDIC rotation algorithm for eleven iterations.

Our proposed approach for improving the classic CORDIC algorithm is based on two principles:

- Completion of iterations with the help of one-directional rotation of the vector at the final stages (only the first half of iterations has a significant effect on the value of the gain factor [22], and the second half starting with $i = m/2+1$ - can be marginalised, in our paper $m = 16$).
- Calculating the scaling factor value for the first half of the iteration by selectively choosing the iteration steps used to calculate the gain factor in a way that can be implemented in inexpensive dedicated hardware, with a canonical signed digits approach. Of course, such a choice of iteration steps determines an adequate selection of the value $arctg(2^{-i})$.

The first 11 iterations of classical CORDIC are performed according to Eq. (4), but the individual iteration steps are chosen according to the following scheme $i = (1, 1, 2, 2, 4, 4, 4, 5, 6, 7, 8)$. In fact, iteration steps are selected in such a way and with full premeditation that the scaling factor P is represented as the simplest possible number in the signed canonical digit system (shifts and summations). After 11 iterations, we have the following computation results:

$$x_9, y_9, \sigma_9, z_9.$$

Algorithm 1: Cordic Rotation Algorithm

Input: $x, y, angle$
Output: x, y
1 $theta\_table = [\arctan(2^{-i})$ for i in 1 to 11]$
2 $P = 0.6072530315291345$
3 $\theta = 0.0$
4 $P2i = 1$
5 **for** $current\_angle$ **in** $theta\_table$ **do**
6 $\quad \sigma = 1$
7 $\quad$ **if** $\theta \geq angle$ **then**
8 $\quad \quad \sigma = -1$
9 $\quad \theta = \theta + \sigma \cdot current\_angle$
10 $\quad x\_temp = x - \sigma \cdot y \cdot P2i$
11 $\quad y = \sigma \cdot P2i \cdot x + y$
12 $\quad x = x\_temp$
13 $\quad P2i = P2i/2$
14 $x = x \cdot P$
15 $y = y \cdot P$
16 **return** x, y

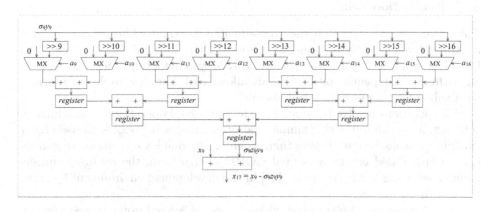

Fig. 1. The output multiplier is built on the shift-add principle

The next stage is multiplication x_9 and y_9 by scaling factor P. In this article, we use $P = 0.748065769 \approx 1 - 2^{-2} - 2^{-9} + 2^{-16}$ (approximation error 0.000004).

In the last stage, next two variables ($x9$ and $y9$) are multiplied on the residual angle $z9$. The multiplications are performed according to the principle of addition and summations (see Fig. 1):

$$x_{17} = x_9 - \sigma_9 z_9 y_9;$$
$$y_{17} = y_9 + \sigma_9 z_9 x_9;$$
$$z_9 = a_9 * 2^{-9} + a_{10} * 2^{-10} + a_{11} * 2^{-11} + a_{12} * 2^{-12} +$$
$$+ a_{13} * 2^{-13} + a_{14} * 2^{-14} + a_{15} * 2^{-15} + a_{16} * 2^{-16};$$
$$a_i \in \{0, 1\}, i = 9, ..., 16$$

Algorithm 2: Pseudocode of one iteration of the presented CORDIC algorithm

---

Input: X, Y, T, E
Output: Xi, Yi, Ti

1 **do in parallel**
2     $arctgE = arctg[E]$ //values for each iteration are tabulated

3 **do in parallel if T$\geq$0**
4     $Xi = X\text{-}(Y >> E)$
5     $Yi = Y + (X >> E)$
6     $Ti = T\text{-}arctgE$

7 **do in parallel if T$<$0**
8     $Xi = X + (Y >> E)$
9     $Yi = Y\text{-}(X >> E)$
10    $Ti = T + arctgE$

A detailed hardware implementation is given in the next section.

3.1   Realisationation

This improvement algorithm was implemented for the realisation of a rotator in Altera Cyclone V System-on-Chip (SoC) FPGA (5CSXFC6D6F31C6N). An FPGA programmable chip is an ideal hardware system for implementing iterative algorithms in a pipeline approach. This allows each iteration to do calculations with subsequent input data simultaneously.

All operations involve basic arithmetic and logical operations (e.g. addition, subtraction, bit shifts). Using simple operations makes it possible to use higher clock frequencies for data inputs than when using complex operations (e.g. multiplication). Based on the presented CORDIC algorithm, the rotator is implemented using the Verilog language [23] in the development environment Quartus Prime [24].

All elements are implemented with the use of a fixed-point representation. The precision of registers and variables is Q8.12 U2 (8 integer bits and 12 fractional bits) format for coordinates (i.e. x and y) and Q2.18 U2 (2 integer bits and 18 fractional bits) for other values (θ). All values are quantised via round (round a number to the nearest in a given representation, if it is required). The pseudocode of one iteration is presented in Algorithm 2.

The variable E is a number from the sequence, which determines the variable $arctgE$. Value of the $arctgE$ can be calculated from the equation $arctgE = arctg(2^{-E})$. Sequence for E is constant for any rotator (any angle θ). Table 1 shows values of E and $arctgE$ for each iteration.

Algorithm 3 describes a full rotator using pseudocode. The one iteration (scheme) of the presented CORDIC algorithm is also shown in Fig. 2. The scheme is generated with the use the RTL Viewer tool. Also, to test proper operation,

Table 1. Values of sequence E and $arctgE$ for each iteration

Number of iteration	Values of E	Values of $arctgE$ after quantisation to fixed-point format
1	1	0.463645935
2	1	0.463645935
3	2	0.244979858
4	2	0.244979858
5	4	0.062419891
6	4	0.062419891
7	4	0.062419891
8	5	0.031238556
9	6	0.015625
10	7	0.0078125
11	8	0.00390625

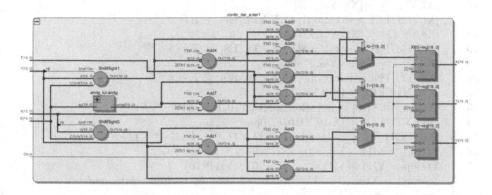

Fig. 2. The scheme of the first iteration of the presented CORDIC algorithm (from RTL Viewer of Quartus Prime), where: X[•], Y[•] - input coordinates of the rotated point; T[•] - input value of the angle θ; E[•] - input value from the sequence; Xi[•], Yi[•] - output coordinates of the rotated point; Ti[•] - output value of the angle θ.

simulations were performed in the Intel ModelSim environment (a dedicated simulator for Intel FPGA programming technology). Figure 3 shows the waveforms for the inputs of subsequent iterations and output data of the selected input data case. The values are displayed with limited precision due to the lack of space on the graph. Clock signal period is written to 20 ns.

4   Measurements

The presented CORDIC algorithm is implemented in an FPGA structure, and a measurements are made for selected input values. 100 different values each for

Algorithm 3: Pseudocode of the presented CORDIC algorithm

Input: Xin, Yin, Tin
Output: Xout, Yout

```
1  do in parallel
2  │  X1 = Xin
3  │  Y1 = Yin
4  │  T1 = Tin  //angle θ
5  │  [Xi1,Yi1,Ti1] = iteration1[X1,Y1,T1, E1]  //E1=1
6  │  [X2,Y2,T2] = [Xi1,Yi1,Ti1]
7  │  [Xi2,Yi2,Ti2] = iteration2[X2,Y2,T2, E2]  //E2=1
8  │  [X3,Y3,T3] = [Xi2,Yi2,Ti2]
9  │  [Xi3,Yi3,Ti3] = iteration3[X3,Y3,T3, E3]  //E3=2
10 │  [X4,Y4,T4] = [Xi3,Yi3,Ti3]
11 │  [Xi4,Yi4,Ti4] = iteration4[X4,Y4,T4, E4]  //E4=2
12 │  [X5,Y5,T5] = [Xi4,Yi4,Ti4]
13 │  [Xi5,Yi5,Ti5] = iteration5[X5,Y5,T5, E5]  //E5=4
14 │  [X6,Y6,T6] = [Xi5,Yi5,Ti5]
15 │  [Xi6,Yi6,Ti6] = iteration6[X6,Y6,T6, E6]  //E6=4
16 │  [X7,Y7,T7] = [Xi6,Yi6,Ti6]
17 │  [Xi7,Yi7,Ti7] = iteration7[X7,Y7,T7, E7]  //E7=4
18 │  [X8,Y8,T8] = [Xi7,Yi7,Ti7]
19 │  [Xi8,Yi8,Ti8] = iteration8[X8,Y8,T8, E8]  //E8=5
20 │  [X9,Y9,T9] = [Xi8,Yi8,Ti8]
21 │  [Xi9,Yi9,Ti9] = iteration9[X9,Y9,T9, E9]  //E9=6
22 │  [X10,Y10,T10] = [Xi9,Yi9,Ti9]
23 │  [Xi10,Yi10,Ti10] = iteration10[X10,Y10,T10, E10]  //E10=7
24 │  [X11,Y11,T11] = [Xi10,Yi10,Ti10]
25 │  [Xi11,Yi11,Ti11] = iteration11[X11,Y11,T11, E11]  //E11=8
26 │  Xtemp = Xi11−(Xi11»2)−(Xi11»9)+(Xi11»16)
27 │  Ytemp = Yi11−(Yi11»2)−(Yi11»9)+(Yi11»16)
28 │  Xtemp2 = Xtemp*Ti11  //implementation by multiple shifts and
   │     summations
29 │  Ytemp2 = Ytemp*Ti11  //implementation by multiple shifts and
   │     summations
30 │  Xout = Xtemp−Ytemp2
31 └  Yout = Ytemp+Xtemp2
```

x, y and θ are randomly selected. Table 3 and Fig. 5 illustrate the selected values. In this way, $100^3 = 1000000$ different combinations of input data are obtained. The input data is processed by the system on finite precision calculations (like a hardware structure in an FPGA chip). For comparison, the second realisations of CORDIC algorithm from [25] is implemented.

The algorithm [25] is based on the values of the trigonometric functions $sin()$ and $cos()$ for the angle of rotation θ. In successive iterations, the values of these functions approach zero (by summing/subtracting successive values of 2^{-i}). The input is also the sum and difference of the x and y coordinates. The output

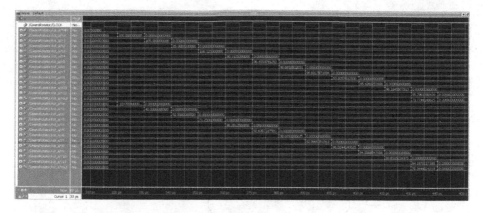

Fig. 3. Input waveforms of successive iterations and output values for the selected input data case: $Xin = 100$, $Yin = -10$, $Tin = 0.875$.

coordinates are also obtained by summing/subtracting the values of $(x + y)^{-i}$ and $(x - y)^{-i}$ (depending on the sign of $sin()$ and $cos()$ in next iterations). Thus, successive approximations of the coordinates of a point after rotation coincide with zeroing the values of the trigonometric functions of the angle of rotation.

In both realisations, the precision of the registers is the same (Q8.12 for x and y and Q2.18 for θ, $sin(\theta)$ and $cos(\theta)$). Values $sin(\theta)$ and $cos(\theta)$ are quantised via round to suitable format. To determine errors, a dedicated script is written in Scilab which for the same input data determined the output at with full precision. The scheme of the error determination for the implemented systems on Fig. 4 is presented.

Accordingly, $\{x_a(n), y_a(n)\}$ realisation are output data from our new proposed realisation of CORDIC and $\{x_b(n), y_b(n)\}$ realisation are output from referenced realisation of CORDIC from [25].

For easier comparison, from the output data of both systems, mean, max, and mean square errors are respectively determined:

$$mean_{dx} = \frac{\sum_{n=1}^{N} |dx_\bullet(n)|}{n}$$
$$mean_{dy} = \frac{\sum_{n=1}^{N} |dy_\bullet(n)|}{n}$$
$$max_{dx} = \text{MAX}\{|dx_\bullet(n)|\} \text{ for } n=1,2,...,N$$
$$max_{dy} = \text{MAX}\{|dy_\bullet(n)|\} \text{ for } n=1,2,...,N$$
$$ms_{dx} = \sqrt{\frac{\sum_{i=1}^{n} (dx_\bullet(n))^2}{N}}$$
$$ms_{dy} = \sqrt{\frac{\sum_{i=1}^{n} (dy_\bullet(n))^2}{N}}$$

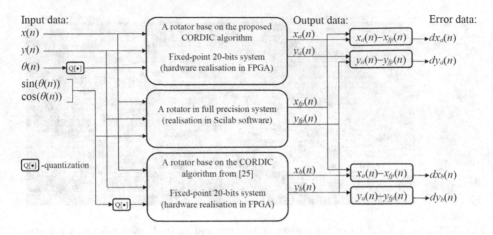

Input data:
$x(n)$
$y(n)$
$\theta(n)$ — Q[•]
$\sin(\theta(n))$
$\cos(\theta(n))$

Q[•] -quantization

A rotator base on the proposed CORDIC algorithm

Fixed-point 20-bits system (hardware realisation in FPGA)

A rotator in full precision system (realisation in Scilab software)

A rotator base on the CORDIC algorithm from [25]

Fixed-point 20-bits system (hardware realisation in FPGA)

Output data:
$x_a(n)$
$y_a(n)$

$x_{fp}(n)$
$y_{fp}(n)$

$x_b(n)$
$y_b(n)$

Error data:
$x_a(n)-x_{fp}(n)$ → $dx_a(n)$
$y_a(n)-y_{fp}(n)$ → $dy_a(n)$

$x_a(n)-x_{fp}(n)$ → $dx_b(n)$
$y_a(n)-y_{fp}(n)$ → $dy_b(n)$

Fig. 4. The scheme of error determination.

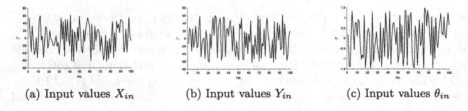

(a) Input values X_{in} (b) Input values Y_{in} (c) Input values θ_{in}

Fig. 5. Graphs of the distribution of randomly selected input values X_{in}, Y_{in}, θ_{in}

where:

$dx_\bullet(n) = x_\bullet(n) - x_{fp}(n)$ - error of x from selected CORDIC realisation of a rotator with finite precision;
$dy_\bullet(n) = y_\bullet(n) - y_{fp}(n)$ - error of y from selected CORDIC realisation of a rotator with finite precision;
$x_\bullet(n) = \{x_a(n), x_b(n)\}$;
$y_\bullet(n) = \{y_a(n), y_b(n)\}$;

Table 2. Result for both realisation of rotator with CORDIC algorithm for 100^3 different inputs

Statistical parameters	For the presented realisation of CORDIC algorithm	For CORDIC algorithm from [25]
$mean_{dx_\bullet}$	0.000404	0.000871
$mean_{dy_\bullet}$	0.000344	0.000873
$max_{dx_\bullet}$	0.002187	0.005113
$max_{dy_\bullet}$	0.001809	0.004646
$ms_{dx_\bullet}$	0.000500	0.001113
$ms_{dy_\bullet}$	0.000430	0.001099

Table 3. Values of input data X, Y and θ

No.	X_{in}	Y_{in}	θ_{in}	No.	X_{in}	Y_{in}	θ_{in}
1	54.686768	−13.691650	0.1776543	51	18.411133	38.604980	−0.8566322
2	7.2453613	27.958984	−1.3416557	52	−31.682617	25.765869	1.1263771
3	−35.985840	21.395508	−1.4022789	53	−39.565430	54.284912	−0.7227516
4	33.136475	−33.696045	0.2222099	54	23.140869	−53.439209	0.6395493
5	−2.8244629	26.485596	1.3384476	55	−10.260498	−55.881348	−1.3748894
6	−44.203369	−22.261475	0.4598045	56	21.527344	19.163818	−0.9397888
7	−32.554443	−2.5603027	−0.5633698	57	41.037842	−24.477783	1.3701820
8	2.25	33.317139	0.7503815	58	54.882324	14.881836	1.3425331
9	−53.773682	−23.946289	0.7742653	59	41.742920	−40.162109	−1.3309402
10	−8.0004883	−19.677246	0.0965271	60	30.294678	11.783936	1.1404228
11	49.181641	30.155518	0.9295235	61	−61.528076	−10.580811	0.0332298
12	60.213623	59.224609	0.8558731	62	37.803711	46.683594	1.1209145
13	−5.7312012	−47.788330	−0.3750763	63	−31.557373	−63.023438	1.2699356
14	5.9826660	−15.373535	−1.0399971	64	−29.861084	−16.056152	−0.9068604
15	−51.795898	−1.9924316	−0.8022614	65	31.992920	−53.655518	−1.3718338
16	−10.700928	1.9377441	−0.9709969	66	−8.7856445	8.2387695	−1.4570351
17	31.282471	−52.482910	1.1290512	67	17.609863	−24.588379	0.6099663
18	27.670410	22.790527	−1.4500771	68	24.364990	−42.585693	0.0637779
19	−27.226562	−0.2170410	−0.8927536	69	18.544678	60.291016	−1.3251991
20	−14.371338	−57.136475	0.3126755	70	−34.749512	−59.960205	−0.0350609
21	20.920166	−61.917969	0.4943581	71	5.9055176	51.533203	0.9105721
22	−2.0537109	12.845215	0.2130623	72	−25.404785	−52.716797	−0.3540306
23	−4.4880371	−26.485840	−1.4655914	73	−59.287842	−13.067871	−0.2065315
24	−18.754395	17.931396	−1.0358696	74	−42.360596	−19.529541	0.7719536
25	−20.282471	56.011963	1.3166656	75	8.3732910	25.858398	−1.4639931
26	−9.9829102	7.6108398	−1.1705666	76	−63.537598	−46.740723	1.4983902
27	6.9682617	60.286621	−0.4913292	77	−6.9572754	25.703125	1.3526115
28	−22.028320	54.686035	−0.1704292	78	−56.674561	53.873535	0.5973282
29	−11.200928	−51.004639	0.3290825	79	−1.0581055	10.558350	1.3801918
30	−43.733154	28.613281	−0.3980331	80	−40.010498	−13.555908	0.9557190
31	−14.684570	42.085938	−0.7365456	81	7.6435547	−50.685059	0.1027145
32	55.576416	54.473877	−1.1620560	82	44.739990	−50.711426	−0.8232880
33	−4.0180664	−57.550293	0.7624626	83	−53.906494	−2.4414062	1.0680122
34	44.377441	−42.698486	−1.3940582	84	14.691650	−62.891113	−1.4600182
35	−48.661621	56.344482	−0.2734604	85	−35.538818	44.604736	0.5777702
36	50.292236	55.258545	0.5201035	86	44.992676	−55.353760	−1.0067711
37	−8.3117676	−47.870117	−0.9015388	87	26.354736	−2.6320801	1.0023117
38	49.995850	19.341797	1.2544518	88	37.930664	41.502930	−0.2214317
39	−1.6289062	27.339600	0.4810982	89	14.147217	−4.8681641	−0.4543991
40	−20.791748	6.0939941	−1.0403938	90	49.769043	35.700439	0.3698921
41	62.570068	32.607666	−0.5035858	91	46.766357	−29.708008	0.7878532
42	−61.264404	48.277344	1.3361053	92	42.550049	−24.773682	1.2350807
43	23.336426	17.103027	−1.2712784	93	−38.079834	62.423096	0.2882652
44	−63.429443	−33.214844	0.0814323	94	−23.784912	−43.524414	0.1655083
45	17.252686	33.524170	0.6752968	95	24.986572	38.187012	1.0634651
46	4.7851562	−30.674072	0.7812843	96	−37.671631	−1.8007812	0.1593971
47	−35.454834	−48.004150	0.2856369	97	2.3264160	−57.724609	−0.0062294
48	29.522949	−6.2275391	−0.3152084	98	55.588623	−44.486084	0.7303886
49	−6.8962402	−36.974854	0.5320129	99	2.1872559	4.4125977	1.2736359
50	58.912354	−56.394775	−1.1868744	100	−6.7250977	19.757568	0.6630516

$x_{fp}(n), y_{fp}(n)$ - output data from full precision system;
$N = 100^3$ - numbrealisation samples (sets);
MAX{•} - a max function.

Table 2 shows the described statistical values for obtained results. Completed measurements and research highlight that the presented CORDIC realisation consistently yields lower errors compared to the version from [25]. The Table 2 clearly shows that the statistical parameters obtained with our algorithm are less than half of those achievable with the CORDIC from [25]. The test on a large number of input data also confirms the correct operation of the algorithm in the full data range.

At the same time, the indicated results were obtained with a smaller number of iterations (the presented algorithm requires 11, while the implementation in accordance with [25] required 16). This also means lower consumption of FPGA processor resources and lower latency (delay between the first input sample and the first output sample). All the more reason to see significant progress in the presented solution.

5   Conclusion

In summary, the implemented CORDIC algorithm within an FPGA structure underwent a comprehensive evaluation, contrasting its performance with the standard CORDIC algorithm presented in [25]. The assessment involved processing 100 varied values for x, y, and θ, generating a total of 1000000 diverse input combinations. Both FPGA based implementations employed finite precision calculations, utilizing Q8.12 precision for x and y, and Q2.18 precision for θ, $sin(\theta)$, and $cos(\theta)$.

To gauge the accuracy of the systems, a dedicated Scilab script computed output data at full precision for identical input sets. The error assessment procedure, depicted in Fig. 4, compared output data from the newly proposed CORDIC realization $(x_a(n), y_a(n))$ with the referenced implementation from [25] $(x_b(n), y_b(n))$.

For comparative analysis, mean, max, and mean square errors for output x and y components are determined (following the formulas for $mean_{dx}$, $mean_{dy}$, max_{dx}, max_{dy}, ms_{dx}, and ms_{dy}). Detailed data are presented in Table 2.

The superior accuracy demonstrated by the FPGA implementation of the CORDIC algorithm positions it as a compelling choice for real-time applications requiring precise trigonometric calculations. Beyond its performance in traditional CORDIC applications, the enhanced accuracy may particularly benefit Givens rotator implementations. The refined precision could contribute to more accurate transformations and improved overall performance in applications relying on Givens rotators, potentially leading to advancements in areas such as signal processing, numerical linear algebra, and applications like virtual reality.

Acknowledgements. This paper and the research behind it would not have been possible without the exceptional support of Graphcore Customer Engineering and Software Engineering team. We would like to express our very great appreciation to Hubert

Chrzaniuk, Krzysztof Góreczny and Grzegorz Andrejczuk for their valuable and constructive suggestions connected to testing our algorithms and developing this research work. This research was partly supported by PLGrid Infrastructure at ACK Cyfronet AGH, Krakow, Poland. This work was also partly supported by the National Science Foundation under grant #2331153.

References

1. Aslan, S., Niu, S., Saniie, J.: FPGA implementation of fast QR decomposition based on givens rotation. In: Midwest Symposium on Circuits and Systems, pp. 470–473 (2012). https://doi.org/10.1109/MWSCAS.2012.6292059
2. Zhang, J., Chow, P., Liu, H.: An efficient FPGA implementation of QR decomposition using a novel systolic array architecture based on enhanced vectoring CORDIC. In: 2014 International Conference on Field-Programmable Technology (FPT), Shanghai, China, pp. 123–130 (2014). https://doi.org/10.1109/FPT.2014.7082764
3. Omran, S., Abdul-Abbas, A.K.: Implementation QR decomposition based on triangular systolic array. Int. J. New Technol. Sci. Eng. 5(6), 150–161 (2018). ISSN 2349-0780
4. Kuan-Ting, C., Wei-Hsuan, M., Yin-Tsung, H., Kuan-Ying, C.: A low complexity, high throughput DoA estimation chip design for adaptive beamforming. Electronics 9, 641 (2020). https://doi.org/10.3390/electronics9040641
5. Dasgupta, B.: Applied Mathematical Methods. Pearson Education India (2006). ISBN 813177600X, 9788131776001
6. Golub, G.H., Van Loan, C.F.: Matrix Computations, 4th edn. The Johns Hopkins University Press, Baltimore (2013). 13: 978-1-4214-0794-4
7. Soler, T.: Active versus passive rotations. J. Surv. Eng. 144, 9 (2018). https://doi.org/10.1061/(ASCE)SU.1943-5428.0000247
8. Wittenburg, J., Lilov, L.,: Decomposition of a finite rotation into three rotations about given axes. Multibody Syst. Dyn. 9, 353–375 (2003). https://doi.org/10.1023/A:1023389218547
9. Ranganathan, A., Yang, M.-H., Ho, J.: Online sparse gaussian process regression and its applications. IEEE Trans. Image Process. 20, 391–404 (2011). https://doi.org/10.1109/TIP.2010.2066984
10. Yen, M.H., Lu, H.Y., Lin, S.Y., Lu, K.H., Chan, C.C.: A partial-givens-rotation-based symbol detector for GSM MIMO systems: algorithm and VLSI implementation. IEEE Syst. J. (2023)
11. Wang, Y.P., Wen, C.C., Kao, C.C., Huang, C.J., Liu, D.Z., Yang, C.H.: Iterative receiver with a lattice-reduction-aided MIMO detector for IEEE 802.11 ax. In: GLOBECOM 2020-2020 IEEE Global Communications Conference, pp. 1–6. IEEE (2020)
12. Chen, L., Xing, Z., Li, Y., Qiu, S.: Efficient MIMO preprocessor with sorting-relaxed QR decomposition and modified greedy LLL algorithm. IEEE Access 8, 54085–54099 (2020)
13. Yen, M.H., Lu, H.Y., Lu, K.H., Lin, S.Y., Chan, C.C.: Algorithm and VLSI architecture of a near-optimum symbol detector for QSM MIMO systems. IEEE Access (2023)
14. Hormigo-Aguilar, J., Muñoz, S.: Efficient floating-point givens rotation unit (2020)

15. Poczekajło, P.: An overview of realisation of synthesis, realization and implementation of orthogonal 3-D rotation filters and possibilities of further research and development. Int. J. Electron. Telecommun. **67**(2), 295–300 (2021). https://doi.org/10.24425/ijet.2021.135979
16. Volder, J.E.: The CORDIC trignometric computing technique. IRE Trans. Electron. Comput. **8**, 330–334 (1957)
17. Andraka, R.: A survey of CORDIC algorithm for FPGA based computers. In: Proceedings of ACM/SIGDA Sixth International Symposium on FPGAs, Monterrey CA, pp. 191–200 (1998). https://doi.org/10.1145/275107.275139
18. Nagvajara, P., Lin, Z., Nwankpa, C., Johnson, J.: State estimation using sparse givens rotation field programmable gate array. In: 2007 39th North American Power Symposium, NAPS, pp. 421–427 (2007). https://doi.org/10.1109/NAPS.2007.4402344
19. Hormigo, J., Muñoz, S.: Efficient floating-point givens rotation unit. Circ. Syst. Signal Process. **40**, 1–24 (2021). https://doi.org/10.1007/s00034-020-01580-x
20. Moroz, L.V., Nagayama S., Mykytiv, T., Kirenko, I.O., Boretskyy, T.: Simple hybrid scaling-free CORDIC solution for FPGAs. Int. J. Reconfigurable Comput. 615472:1–615472:4 (2014)
21. Bajard, J.C., Kla, S., Muller, J.M.: BKM: a new hardware algorithm for complex elementary functions. IEEE Trans. Comput. **43**, 955–963 (1994). https://doi.org/10.1109/ARITH.1993.378098
22. Timmermann, D., Hahn, H., Hosticka, B.J., Rix, B.: A new addition scheme and fast scaling factor compensation methods for cordic algorithms. VLSI J. Integr. **11**(1), 85–100 (1991)
23. IEEE Standard for Verilog Hardware Description Language, in IEEE STD 1364-2005 (Revision of IEEE STD 1364-2001), pp. 1–590, 7 April 2006. https://doi.org/10.1109/IEEESTD.2006.99495
24. Online. Intel Quartus Prime - FPGA Design Software. https://www.intel.com/content/www/us/en/products/details/fpga/development-tools/quartus-prime.htmll. Accessed 15 Jan 2024
25. Wawryn, K., Poczekajło, P., Wirski, R.: FPGA implementation of 3-D separable Gauss filter using pipeline rotation structures. In: 22nd International Conference Mixed Design of Integrated Circuits & Systems (MIXDES), Torun, Poland, pp. 589–594 (2015). https://doi.org/10.1109/MIXDES.2015.7208592

Numerical Aspects of Hyperbolic Geometry

Dorota Celińska-Kopczyńska(iD) and Eryk Kopczyński$^{(\boxtimes)}$(iD)

Institute of Informatics, University of Warsaw, Warsaw, Poland
{erykk,dot}@mimuw.edu.pl

Abstract. Hyperbolic geometry has recently found applications in social networks, machine learning and computational biology. With the increasing popularity, questions about the best representations of hyperbolic spaces arise, as each representation comes with some numerical instability. This paper compares various 2D and 3D hyperbolic geometry representations. To this end, we conduct an extensive simulational scheme based on six tests of numerical precision errors. Our comparisons include the most popular models and less-known mixed and reduced representations. According to our results, polar representation wins, although the halfplane invariant is also very successful. We complete the comparison with a brief discussion of the non-numerical advantages of various representations.

Keywords: hyperbolic geometry · numerical precision · tiling

1 Introduction

Hyperbolic geometry has recently gained interest in many fields. Notable examples include the hyperbolic random graph models of hierarchical structures [18,20], social network analysis [5], hyperbolic embeddings used in machine learning [22], as well as visualizations and video games [10,15,23].

One crucial aspect of hyperbolic geometry is its *tree-like structure*, as shown in Fig. 1. Each edge of these trees has the same length, making them grow exponentially. This tree-likeness property is crucial in the modeling hierarchical data [5,22] and game design [15]. However, this property comes with a severe numerical cost [2,7,24,26]. Since the circumference of a hyperbolic circle of radius r is exponential in r, any representation based on a fixed number b of bits will not be able to distinguish between points in a circle of radius $r = \Theta(b)$, even if the pairwise distances between these points are large.

Different communities use different representations of hyperbolic spaces. The newcomers and some experts often use the Poincaré model, most popularly used in the introductions to hyperbolic geometry. However, in visualizations [14,15,20] the Minkowski hyperboloid model seems to be commonly used for the internal representation (and converted to Poincaré for visualization purposes), and in the social network research, native polar coordinates are popular [5,13]. The users

L. Franco et al. (Eds.): ICCS 2024, LNCS 14837, pp. 115–130, 2024.
https://doi.org/10.1007/978-3-031-63778-0_9

motivate their choice of representation by factors such as ease of use, generalizability, and numerical stability. The numerical aspect needs further study; for example, the hyperboloid model and the Poincaré disk model may be better numerically depending on the computation at hand. At the moment, we know one study comparing the hyperboloid model and the Poincaré half-plane model [12]; however, this paper compares only two representations of isometries, and has been written in 2002, before the surge of interest in hyperbolic geometry.

In this paper, we compare a large number of representations of hyperbolic geometry. We primarily focus on the numerical issues. It is worth to note that, due to the exponential growth, any representation based on a fixed number of bits will introduce numerical errors, a commonly used solution to this [7,8,11, 15,17,26] is to use combinatorially generated tessellations, and represent points and isometries by a pair (t, h), where t is one of the tiles of the tessellation, and h is coordinates relative to tile t. Our research takes this into account.

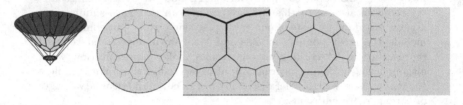

Fig. 1. The {7,3} tessellation of the hyperbolic plane in the following models: hyperboloid, Poincaré disk, upper half-plane, Beltrami-Klein disk, polar coordinates.

2   Hyperbolic Geometry and Representations

A geometry is defined by how points, lines, distances, angles, and isometries behave. A *representation* of a geometry is a method of representing points p and isometries f. We need to define and compute the following basic geometric objects and operations:

- ORIGIN, a constant C_0 representing a point which we consider the origin.
- TRANSLATE in the X direction by distance x (returns an isometry T^x).
- ROTATE around C_0 by an angle α (returns an isometry R^α).
- APPLY an isometry f to a point p.
- COMPOSE two isometries f_1 and f_2.
- INVERT an isometry f.
- DISTANCE of a given point p to the origin C_0.

Spherical Geometry. The most straightforward non-Euclidean geometry is spherical geometry. The d-dimensional sphere is $\mathbb{S}^d = \{x \in \mathbb{R}^{d+1} : g^+(x,x) = 1\}$, where g is the inner product, $g^+((x_1,\ldots,x_{d+1}),(y_1,\ldots,x_{d+1})) = x_1 y_1 + \ldots + x_{d+1}y_{d+1})$. We consider $C_0 = (0,\ldots,0,1) \in \mathbb{S}^d$ to be the origin of $\mathbb{S}^d$. Isometries of the sphere are exactly the isometries of $\mathbb{R}^{d+1}$ which map 0 to 0, i.e., orthogonal matrices in $\mathbb{R}^{(d+1)\times(d+1)}$. It is straightforward to compute the basic translations and rotations, for example, an isometry T^α which moves the origin of $\mathbb{S}^2$ by α units in the direction of the first coordinate can be written as $T^\alpha(x_1, x_2, x_3) = (x_1 \cos(\alpha) + x_3 \sin(\alpha), x_2, x_3 \cos(\alpha) - x_1 \sin \alpha)$. The distance between the points $x, y \in \mathbb{S}^{d+1}$ is the length of the shortest arc $\gamma \subseteq \mathbb{S}^{d+1}$ connecting x and y, and can be computed using the formula acos $g^+(x,y)$.

Minkowski Hyperboloid (Linear Representation). The representation above lets a person with a basic knowledge of linear algebra work with spherical geometry in an intuitive wall. We obtain hyperbolic geometry (in the Minkowski hyperboloid model) by applying the same construction in pseudo-Euclidean space, using the Minkowski inner product $g^-((x_1,\ldots,x_{d+1}),(y_1,\ldots,x_{d+1})) = x_1 y_1 + \ldots + x_d y_d - x_{d+1}y_{d+1})$. Many formulas of hyperbolic geometry are the same as the relevant formulas of spherical geometry, except that we need to change the sign in some places (due to the change of sign in g^-), and use sinh and cosh instead of sin and cos when the argument represents distance. The hyperbolic plane is $\mathbb{H}^d = \{x \in \mathbb{R}^{d+1} : x_{d+1} > 0, g^-(x,x) = -1\}$. Again, we consider $C_0 = (0,\ldots,0,1)$ to be the origin of $\mathbb{H}^d$. Isometries of $\mathbb{H}^d$ are the matrices $M \in \mathbb{R}^{(d+1)\times(d+1)}$ such that $g^-(Mx, My) = g^-(x,y)$, and $(MC_0)_{d+1} > 0$. An isometry T^α which moves the origin of $\mathbb{H}^2$ by α units in the direction of the first coordinate can be written as the Lorentz boost $T^\alpha(x_1, x_2, x_3) = (x_1 \cosh(\alpha) + x_3 \sinh(\alpha), x_2, x_3 \cosh(\alpha) + x_1 \sinh \alpha)$; however, a rotation R^α of $\mathbb{H}^2$ by α around C_0 still uses cos and sin: $R^\alpha(x_1, x_2, x_3) = (x_1 \cos(\alpha) - x_2 \sin(\alpha), x_2 \cos(\alpha) + x_1 \sin(\alpha), x_3)$. The distance between the points $x, y \in \mathbb{H}^d$ is the length of the shortest arc $\gamma \subseteq \mathbb{H}^d$ connecting x and y computed according to g^-, and can be computed using the formula acosh $g^-(x,y)$. For $x, y \in \mathbb{H}^d$, the midpoint mid(x,y) is the point in the middle of this arc. We have mid$(x,y) = \dfrac{x+y}{-\sqrt{g^-(x+y,x+y)}}$. We can use this model directly (i.e., represent points as their coordinates in $\mathbb{H}^d$ and isometries as linear transformation matrices), which we will call the *linear* representation.

Other Models of $\mathbb{H}^2$, and Polar Representation. In the Minkowski hyperboloid model, every point has three coordinates, while two are sufficient. Other common models (projections to $\mathbb{R}^2$) of $\mathbb{H}^2$ include:

- Beltrami-Klein disk model: the point $x = (x_1, x_2, x_3) \in \mathbb{H}^2$ is mapped to $P(x) = (\frac{x_1}{x_3}, \frac{x_2}{x_3})$. This maps $\mathbb{H}^2$ to the inside of the unit disk in $\mathbb{R}^2$.
- Poincaré disk model: the point $x = (x_1, x_2, x_3) \in \mathbb{H}^2$ is mapped to $K(x) = (\frac{x_1}{x_3+1}, \frac{x_2}{x_3+1})$. Again, this maps $\mathbb{H}^2$ to the inside of the unit disk in $\mathbb{R}^2$.
- Upper half-plane model: obtained from the Poincaré disk model by applying a circle inversion which maps the inside of the Poincaré disk into the upper

half-plane $\mathbb{U}^2$, which we interpret in terms of complex numbers: $\mathbb{U}^2 = \{z \in \mathbb{C} : \Im(z) > 0\}$. The center point of $\mathbb{U}^2$ is $i \in \mathbb{U}^2$.

- Native polar coordinates: in this model, we use coordinates $(\phi, r) \in \mathbb{P}$, where r is the distance from C_0 and ϕ is the angle. In other words, the coordinates (ϕ, r) in native polar coordinates correspond to the point $R^\phi(T^r(C_0)) \in \mathbb{H}^2$.

These models have natural analogs in higher dimensions. Figure 1 shows the hyperbolic plane in the models above. The pictures show the tessellation of $\mathbb{H}^2$ by regular hyperbolic heptagons; all heptagons are of the same size and shape, but these sizes and shapes had to be changed by the projection to $\mathbb{E}^2$ used.

Native polar coordinates are the first example of an alternative representation of the hyperbolic plane, that is popular in network science applications [5,13]. It is somewhat analogous to using latitude and longitude in spherical geometry. It is straightforward to compute the distance between $(\phi_1, r_1) \in \mathbb{P}$ and $(\phi_2, r_2) \in \mathbb{P}$. Indeed, $(0, r_1) \in \mathbb{P}$ corresponds to $(\sinh(r_1), 0, \cosh(r_1)) \in \mathbb{H}^2$, and $(\phi_2, r_2) \in \mathbb{P}$. corresponds to $(\cos(\phi_2)\sinh(r_2), \sin(\phi_2)\sinh(r_2), \cosh(r_2)) \in \mathbb{H}^2$; thus, the distance d satisfies the hyperbolic cosine rule

$$\cosh(d) = \cosh(r_1)\cosh(r_2) + \cos(\phi_2)\sinh(r_2)\cosh(r_2). \tag{1}$$

In general, we need to replace ϕ_2 with ϕ_1. If (ϕ_1, r_1) and (ϕ_2, r_2) are close, the following formula is numerically better [3]:

$$\cosh(d) = \cosh(r_1 - r_2) + (1 - \cos(\phi))\sinh(r_1)\sinh(r_2) \tag{2}$$

While less useful in network science applications, we also need to represent arbitrary isometries, but this is also straightforward: (ϕ, r, ψ) represents the isometry $R^\phi T^r R^\psi$. In higher dimensions, we need to replace ϕ and ψ with isometries of $\mathbb{S}^{d-1}$.

Clifford Algebras, and Mixed Representation. While using $d \times d$ matrices is a straightforward method of representing rotations of $\mathbb{R}^d$ (and thus also isometries of $\mathbb{S}^{d-1}$), it is often advantageous to use other representations. For orientation-preserving isometries of $\mathbb{R}^3$, quaternions are commonly used in computer graphics. This section explains Clifford algebras that generalize this construction.

Let $\mathbb{V}$ be $\mathbb{R}^d$ with inner product g, and let $e_1, \dots, e_d$ be the unit vectors of $\mathbb{V}$. The points of $\mathbb{V}$ can be written as $x_1 e_1 + \dots + x_d e_d$. The free algebra over $\mathbb{V}$, $T(\mathbb{V})$, is the vector space whose basis is the set of all sequences of e_i (including the empty sequence, denoted by 1). The elements of $T(\mathbb{V})$ are added and multiplied in a natural way, for example:

$$(3 + 2e_2 e_1)(1 + 2e_1) = 3 + 2e_2 e_1 + 6e_1 + 4e_2 e_1 e_1$$

Note that $\mathbb{V}$ is a subspace of $T(\mathbb{V})$ and that this multiplication is associative but not commutative. Addition is associative and commutative, and multiplication is distributive over addition.

The Clifford algebra $Cl(\mathbb{V})$ is obtained from $T(\mathbb{V})$ by identifying elements according to the following rule: for $u, v \in \mathbb{V}$, $uv + vu = 2g(u, v)$ $(\star)$. We perform all the identifications that follow from this rule and the associativity/commutativity/distributivity rules. In particular, for $\mathbb{V} = (\mathbb{R}^{d+1}, g^-)$ we have $e_i e_j = -e_j e_i$ and $e_i e_i = 1$ for $i \leq d$, and $e_i e_i = -1$ for $i = d+1$. Thus, $Cl(\mathbb{V})$ is a 2^{d+1}-dimensional space (using the rules above, we can rewrite any element of $Cl(\mathbb{V})$ using only products of e_i which are ordered and have no repeats).

For $x \in Cl(\mathbb{V})$, $\bar{x}$, called the *conjugate* of x, is defined as follows: $\bar{1} = 1$, $\bar{e_i} = -e_i$, $\overline{x+y} = \bar{x}+\bar{y}$, $\overline{xy} = \bar{y}\bar{x}$.

Let v be a non-zero vector in $\mathbb{V}$. Any vector $w \in \mathbb{V}$ can then be decomposed as $w = av + u$, where u is orthogonal to v. We have $vw\bar{v} = v(av+u)\bar{v} = a(vv\bar{v}) + vu\bar{v} = g(v,v)a\bar{v} - v\bar{v}u = -g(v,v)av + vvu = g(v,v)(u-av)$. For $\mathbb{V} = (\mathbb{R}^{d+1}, g^-)$, if $v, w \in \mathbb{H}^d$, the operation $w \mapsto vw\bar{v}$ is exactly the point reflection of $\mathbb{H}^{d-1}$ in v; and if $g^-(v,v) = 1$, it is the reflection in the hyperplane orthogonal to v.

For an $x \in Cl(\mathbb{V})$, let us denote the operation $w \mapsto xw\bar{x}$ by $M(v)$. It is easy to check that $M(xy)$ is the composition of $M(x)$ and $M(y)$. Since the basic translations and rotations of $\mathbb{H}^d$ can be obtained as compositions of two operations from the last paragraph, they can be represented as $M(x)$ for some $x \in Cl(\mathbb{V})$; furthermore, every orientation-preserving isometry of $\mathbb{H}^d$ can be obtained as a composition of basic translations and rotations, thus, also $M(x)$ for some $x \in Cl(\mathbb{V})$. It is easy to check that, since the number of basic reflections is even, we only use $Cl^{[0]}(\mathbb{V})$, which is the subspace of $Cl(\mathbb{V})$ whose base is all the products of even number of e_i's.

Therefore, for $\mathbb{H}^d$ (and $\mathbb{S}^d$), we have a representation of orientation-preserving isometries which uses only 2^d real numbers. For $d \leq 5$ this is less than $(d+1)^2$ we would have to use in the $\mathbb{R}^{(d+1)\times(d+1)}$ representation.

Reduced Representation. In the previous paragraph, we represented the points $x \in \mathbb{H}^d$ using the Minkowski hyperboloid model and the isometries using $Cl^{[0]}(\mathbb{V})$ (the *mixed* approach). Another possible representation is to represent $x \in \mathbb{H}^d$ as $y \in Cl^{[0]}(\mathbb{V})$ such that $M(y)(C_0) = x$ and $M(y)(x') = C_0$, where x' is the point such that $C_0 = \text{mid}(x, x')$ (i.e., this isometry moves C_0 to x without introducing any rotation). This y can be computed as $x''C_0$, where $x'' = \text{mid}(C_0, x)$.

Reduced representation represents every point $x \in \mathbb{H}^d$ as $y_1 e_1 e_{d+1} + y_2 e_2 e_{d+1} + \ldots + y_d e_d e_{d+1} + y_{d+1}$ (only $d+1$ out of 2^d coordinates are used). The coordinates $y = (y_1, y_2, \ldots y_{d+1})$ can be interpreted as coordinates on the Minkowski hyperboloid of the point x''. Since $x'' = \text{mid}(C_0, x)$, the distance of x'' from C_0 is half the distance of x from C_0; this will be a serious numerical advantage. We call this representation *reduced*. Another important property is that, by projecting y as in the Beltrami-Klein disk model, we obtain x in the Poincaré disk model, $K(y) = K(x'') = P(x)$.

Half-Plane Representation. Another representation of $\mathbb{H}^2$ uses the upper half-plane $\mathbb{U}$ for points and matrices $A = \begin{pmatrix} a & b \\ c & d \end{pmatrix}$, where $a, b, c, d \in \mathbb{R}$, for isometries. We apply A to $z \in \mathbb{U}^2$ as follows: $\text{APPLY}(A, z) = (az+b)/(cz+d)$. Note that, for $\alpha \in \mathbb{R}$ such that $\alpha \neq 0$, A and αA represent the same isometry. The *normalized* one is the one which has determinant $ad - bc = 1$. Therefore, the set of isometries corresponds to the set of 2×2 matrices over reals with determinant 1, which is called the *special linear group* over $\mathbb{R}$, $\mathrm{SL}(2, \mathbb{R})$.

This representation of isometries is essentially equivalent to the Clifford algebra representation up to the base change. In particular, if the Clifford algebra representation of an isometry is $k_0 + k_2 e_1 e_3 + k_1 e_2 e_3 + k_3 e_1 e_2$, then its $\mathrm{SL}(2, \mathbb{R})$ representation is $\begin{pmatrix} k_0 - k_2 & k_1 + k_3 \\ k_1 - k_3 & k_0 + k_2 \end{pmatrix}$.

Half-Space Representation. While half-plane representation works for $\mathbb{H}^2$, a similar representation exists for $\mathbb{H}^3$. We will be using quaternions $\mathbb{H}$: a four-dimensional space over reals, with the four basis vectors called 1, i, j, and k, multiplied according to rules $i^2 = j^2 = k^2 = -1$, $ij = k$, $jk = i$, $ki = j$. (To avoid confusing $\mathbb{H}$ with the hyperbolic space $\mathbb{H}^d$, note that the standard notation for quaternions, $\mathbb{H}$, has no index.) The points are represented in the Poincaré half-space model, using quaternions $x \in \mathbb{H}$ such that the j-part of x is positive, and the k-part of x is 0. The center point is $j \in \mathbb{H}$. The isometries are represented as $\mathrm{SL}(2, \mathbb{C})$, that is, 2×2 matrices over $\mathbb{C}$ with determinant 1. Applications are performed using the same formula: $\mathrm{APPLY}(A, x) = (ax + b)/(cx + d)$. Again, this representation is essentially equivalent to the Clifford algebra: $k_0 + k_3 e_1 e_2 + k_5 e_1 e_3 + k_6 e_2 e_3 + k_9 e_1 e_4 + k_{10} e_2 e_4 + k_{12} e_3 e_4 + k_{15} e_1 e_2 e_3 e_4$ is equivalent to

$$\begin{pmatrix} k_0 - k_9 + k_{15}i - k_6 i & k_3 + k_{10} - k_5 i - k_{12} i \\ k_{10} - k_3 + k_{12} i - k_5 i & k_0 + k_9 + k_6 i + k_{15} i \end{pmatrix}.$$

3   Representation Variants

In Sect. 2, we introduced the basic representations we will compare in our study (linear, mixed, reduced, half-plane/half-space, polar, and generalized polar). There are also multiple methods of dealing with numerical errors. As an example, consider a point $x = (x_1, \ldots, x_{d+1})$ in the Minkowski hyperboloid $\mathbb{H}^d$. We should have $x_{d+1}^2 = 1 + x_1^2 + \ldots + x_d^2$. However, if we apply several of representation operations to compute the point x, it may happen that this equation is not true as a result of numerical errors. We consider the following variations:

- **Invariant.** Do not do anything. Hope that numerical errors do not build up.
- **Careless.** Here we consider αx, for any $\alpha \in \mathbb{R}$ other than 0, to be a correct representation of $x \in \mathbb{H}^d$.
- **Flattened.** We normalize in another way: we multiply $x \in \mathbb{H}^d$ by $1/x_{d+1}$. This lets us conserve memory since the $d + 1$-th coordinate in our representation will always equal 1. (This is effectively the Beltrami-Klein model.)
- **Forced.** Normalize the output after every computation.
- **Weakly Forced.** Try to normalize the output after every computation, but do not do it if the norm could not be computed due to precision errors.

- **Binary.** In careless, values may easily explode and cause underflow/overflow; avoid this by making the leading coordinate in $[0.5, 2)$ range (by multiplying by powers of 2, which is presumably fast).

Similar variants can also be applied to the Clifford representation of points and isometries. For example, flattened reduced representation is effectively the Poincaré disk model. Furthermore:

- In linear representations, matrices can be *fixed* by replacing them with correct orthogonal matrices close to the current computation. HyperRogue [15] uses this method; not applying such fixes would have a visible effect of the visualization becoming visibly stretched after the user moves sufficiently far away from the starting point (interestingly, it tends to fix itself when the user moves back towards the starting point). We call non-fixed representations *linear-F*, and fixed representations *linear+F*.
- In *polar1*, we always use the basic cosine rule (1). In *polar2*, we use a better formula when the angles are close or opposite to each other (2). In 2D, we can use angles (*angles* variant), but forcing angles into $[-\pi, \pi]$ may be needed to prevent explosion (*mod* variant). In general, we have a choice of representation for $\mathbb{S}^{d-1}$. We use *forced* or *invariant* reduced representations.
- In the Clifford representation, the *gyro* variant splits the isometries into the translational part (which is flattened, making it equivalent to the Poincaré disk model) and the rotational part (for which 'invariant' is used). This fixes the problem with full flattening where rotations by 180° are flattened to infinity. This is inspired by gyrovectors [25] and is essentially doing the computation in the Poincaré disk model, a popular representation [1,4,10].

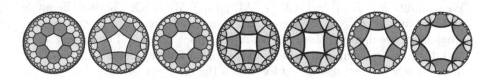

Fig. 2. Tessellations of $\mathbb{H}^2$ used in our experiments. From left to right: $\{7,3\}$, $\{5,4\}$, $\{8,3\}$, $\{4,5\}$, $\{4,6\}$, $\{6,4\}$, $\{6,6\}$.

4   Tessellations

The hyperbolic plane can be tessellated with regular p-gons of the same size, such that q of them meet in every vertex, for all p, q such that $\frac{1}{p} + \frac{1}{q} < \frac{1}{2}$. Such a tessellation has Schläfli symbol $\{p, q\}$. Figure 1 shows the $\{7,3\}$ tessellation.

Tessellations can be used to avoid numerical precision errors. With every tile t, we assign an isometry X_t that maps the central tile to t (rotation is chosen arbitrarily). If a point p is in tile t, we can write $p = X_t p_0$, where p_0 is coordinates of p relative to the center of t. We can now represent p as (t, p_0) – since p_0 is close to the origin, this avoids the problem of numerical precision issues being significantly higher for points far away. A similar method can be used for isometries.

If two tiles t, t' are adjacent, the isometry $X_{t,t'} = X_t X_{t'}^{-1}$ which maps t'-relative coordinates to t-relative coordinates equals $R^\alpha T^x R^\beta$, where the angles α and β correspond to the chosen orientations of t and t', and x is the distance between two adjacent tiles, which can be computed using hyperbolic trigonometry. The adjacency structure of tiles can be computed combinatorially [8,11].

In our tests, we use tessellations either to produce tests with known correct answers or (as described above) to enhance the numerical precision. See Fig. 2 for the tessellations we use.

5   Tests

We compare representations on the following tests (some parameterized by d).

LoopIso. In this test, we construct a path $t_0, \ldots, t_k$ in the tiling by always moving to a random adjacent tile until we get to a tile d afar; then, we return to the start (also randomly, may stray further from the path). We compose all the relative tile isometries $X_{t_k, t_{k-1}} \cdots (X_{t_2, t_1} X_{t_1, t_0})$ into f, which theoretically should equal identity. We see if $f(C_0) = C_0$. The test result is the first distance d for which $f(C_0)$ is not found to equal C_0 (distance > 0.1).

LoopPoint. Same as LoopIso, but we apply the consecutive isometries to point right away. We compute $h = X_{t_k, t_{k-1}} \cdots (X_{t_2, t_1} (X_{t_1, t_0} C_0))$. We see if $h = C_0$.

AngleDist. We construct a random path $t_0, \ldots, t_d$. This time, we do not loop. We compute $h = X_{t_d, t_{d-1}} \cdots (X_{t_2, t_1} (X_{t_1, t_0} C_0))$. We check whether the angle and distance of h from C_0 have been computed correctly. In the variant *AngleDist2*, we multiply the matrices in the opposite order. The correct angle and distance are computed using high-precision floating point numbers. The test result is the first distance d for which the error exceeds 0.1.

Distance. We compute the distance between two points in distance d from the starting point. Such computations are of importance in social network analysis applications. The angle between them is very small (similarity), or close to 180° (dissimilarity), close to 1° (other). The test result is the first distance d for which the error exceeds 0.1.

Walk. This test is based on an effect of numerical precision issues that is most visible in HyperRogue [15]. After walking in a small line, it can often be clearly observed that we have "deviated" from the original straight line. This test checks how long we can walk until this happens.

We construct an isometry A representing a random direction. In each step, we compose this isometry with a translation ($A := A T^{1/16}$). Whenever the point $A C_0$ is closer to the center of another tile, we rebase to that new tile. For a test, we do this in parallel with two isometries A and B, where

$B = AT^{1/32}$. We count the number of steps until the paths diverge. In the *WalkGood* variant, we instead compare to the result obtained using high-precision floating point numbers.

Close. Here, we see whether minor errors accumulate when moving close to the center. Like in loop tests, we move randomly until we reach distance $d + 1$, after which we return to the start (always reducing the distance). After each return to the start, we check if the representation is still fine; if yes, we repeat the loop, letting the errors accumulate over all loops. We stop when the error is high or after 10000 steps. The variants *Close* and *CloseInverse* differ in the order of multiplying X matrices.

6   Experimental Results

For our implementation and experimental results, see [9]. Our implementation is written in C++. All computations use the IEEE754 double precision format. The GMP library is used for the high-precision comparisons.

Figure 3 presents the heatmap of averages of running each representation on each test 20000 times. We observe significant differences between the performance of various representations. Some of these experimental results can be quite easily explained. It is clear from Fig. 1 that the Beltrami-Klein disk model is not good at representing points far away from C_0. In the Klein disk model, a point in distance d from C_0 is mapped to a point in distance $\tanh(d)$ from the center of the disk, which is $1 - \Theta(\exp(2d))$. Floating point numbers cannot express such a slight difference from 1. Conversely, in the Poincaré disk model, this is $1 - \Theta(\exp(d))$. Therefore, we can expect the effective distance represented accurately in flattened reduced to be double that of flattened linear. This issue carries over to most computations in non-flattened representations, although not all of them. This effect has been studied in [12].

In the invariant linear representation, the point $R^\alpha T^x C_0$ is represented as $(\cos(\alpha)\sinh(x), \sin(\alpha)\sinh(x), \cosh(x))$, and since floating-point numbers are good at representing large numbers, we can recover α and x even if x is very large (as long as $\cosh(x)$ fits in the range of our floating-point type). This makes invariant linear significantly better than the Poincaré disk model (flattened reduced) in some experiments, such as AngleDist. This relies on multiplying the matrices in the correct order and on the fact that we care only about distances – numerically computing AB^{-1} for two isometries moving C_0 to two points a and b which are closed to each other but far from C_0 is not likely to yield meaningful results. While polar representations can represent even larger distances x, this does not carry over to computations in our implementation, which have to compute $\exp(x)$, $\cosh(x)$ or $\sinh(x)$ anyway.

In three dimensions, flattened reduced fails due to some isometries of the $\{4, 3, 5\}$ honeycomb not being representable (due to not having the unit component). The gyro variant fixes this issue. In some tests, the lack of normalization causes the coordinates to quickly blow up exponentially, which is avoided in normalized variants. This happens, e.g., in the *Walk* test for polar invariant (avoided in polar forced) and in the *LoopPoint* test for reduced careless.

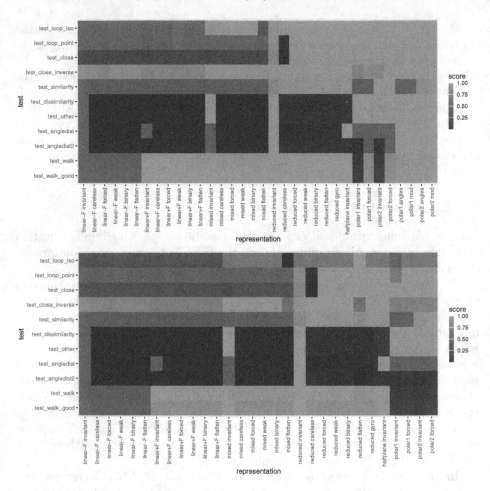

Fig. 3. Aggregate results of our tests, in 2 dimensions (above) and 3 dimensions (below). For each test, the score of the best representation is normalized to 1.

The invariant reduced representation shares both advantages of invariant and reduced representations. In the half-plane and half-space models, we no longer have the problem of faraway points getting close to 1 (typically, they are complex numbers with the imaginary coordinate close to 0 instead). So, these representations are quite good at representing large distances. We can expect these representations to be highly accurate; our experiments support that. Polar representations are highly accurate, too.

Figures 4 and 5 depict the accurate rankings. A representation A wins against the representation B if the probability that a randomly chosen simulation result obtained by A is greater than a randomly chosen simulation result obtained by B exceeds 0.5. If that probability is equal to 0.5, we have a tie between A and B; otherwise, A loses against B. We use the Condorcet voting rule (breaking ties

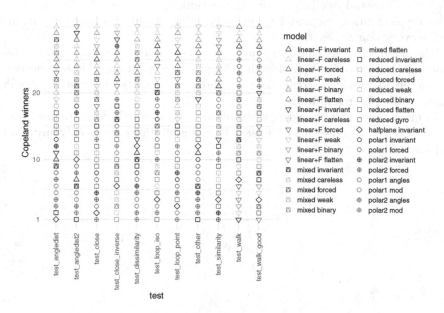

Fig. 4. Condorcet rankings, in 2 dimensions. Better representations are closer to the bottom.

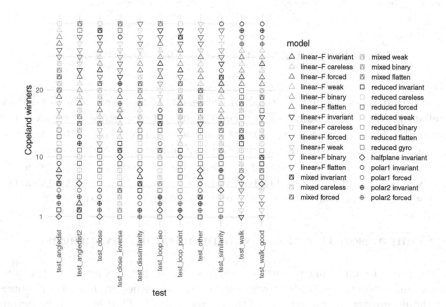

Fig. 5. Condorcet rankings, in 3 dimensions.

using the Copeland rule [19]) to obtain the ranking. To compute the score for a given representation, we add 1 for every winning scenario, 0 for every tie, and −1 for every losing scenario.

The simulations are conducted under specific conditions, including 2000 iterations and the consideration of both 2D and 3D tessellations. These conditions are chosen to provide a comprehensive and representative view of the performance of each representation.

In many tests, polar2 mod representation wins, although the halfplane invariant is also very successful and linear+F representations (especially invariant) perform well in walk tests.

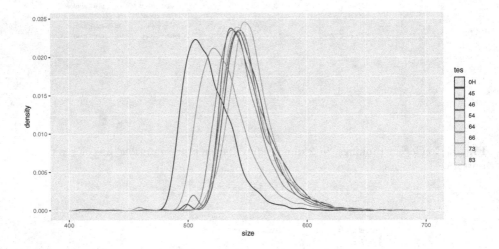

Fig. 6. The result of the *Walk* test, divided by tessellation.

Figure 6 depicts the density graph of the results of the *Walk* test, divided by tessellation. We use the linear+F invariant representation since it achieves the best results. For comparison, we also present the result of a similar test without using any tessellation; this result uses the half-plane representation (linear+F invariant is half as good in this case). From the tessellations we have considered, {7, 3} yields the best results; other tessellations get slightly worse results except {6, 6} which is significantly worse, and no tessellation is even worse.

7 Comparison Based on Non-numerical Advantages

We focused on comparing various representations of hyperbolic geometry concerning numerical precision. However, computational issues are not the only ones that should impact the final choice of the representation. That is why this section discusses the non-numerical advantages of particular representations.

The Poincaré model is often used to teach hyperbolic geometry, so many people use it for this reason – it is the one they know the best. However, the linear representation is more natural for those who understand Minkowski geometry. For example, the formulas for T^x and $\mathrm{mid}(x, y)$ are straightforward in Minkowski hyperboloid, but the respective Poincaré model formulas are not. Polar coordinates, based on the idea of angle and distance from the origin, are also intuitive for many people [23]. While the half-space models have great numerical properties, their lack of symmetry makes them less intuitive.

One advantage of matrix representations is that they can represent both orientation-preserving and orientation-reversing isometries. To represent orientation-reversing isometries in Clifford algebras, half-plane/half-space, or polar representations, we need to add an extra bit of information – whether the isometry should be composed with a mirror image. This introduces some extra complexity.

In some applications (e.g., to work with non-compact honeycombs in $\mathbb{H}^3$ [21]) we need to represent not only the *material* points discussed so far, but also *ideal* and *ultra-ideal* points. Intuitively, ideal points are the points on the boundary of the Beltrami-Klein model, and ultra-ideal points are outside of the boundary. In the Poincaré disk and half-plane model, we can represent ideal points (they are on the boundary), but we cannot represent ultra-ideal points (the points outside of the boundary are better interpreted as alternative representations of material points). Thus, only non-normalized (careless) linear/mixed can represent ultra-ideal points, while only non-normalized/flattened linear/mixed/reduced/half-plane/half-space can represent ideal points.

In some applications, it is useful to have a single implementation of all isotropic geometries (hyperbolic, Euclidean, and spherical, with varying curvatures). In [1,4], the Poincaré disk model (and its spherical analog, stereographic projection) is chosen for this reason. However, polar coordinates also work in all isotropic geometries, and a similar general model can be obtained for the hyperboloid ($\mathbb{H}^d$), sphere ($\mathbb{S}^d$), and plane model; furthermore, linear representations can also be used for other Thurston geometries [17]. From the representations we studied, it seems that only half-plane and half-space do not easily generalize beyond hyperbolic geometry.

There are applications of hyperbolic geometry where we focus on the neighborhood of a straight line L rather than the origin point C_0. In such applications, in $\mathbb{H}^2$, it is useful to use coordinates (x, y) where x is the distance along L, and y corresponds to the distance from L. Here, y can be simply the distance from L (*Lobachevsky coordinates*, where $T^x R^{\pi/2} T^y C_0$ gets coordinates (x, y), analogous to longitude and latitude in spherical geometry), or in some other way, like Bulatov's conformal band model [6,16]. Such a situation occurs in [16], where the conformal band model is used. There are also potential applications in video games (racing along a straight line) and data analysis (bi-polar data). This is a very specific potential application, with major advantages in specific circumstances and no advantages otherwise, so we do not compare such representations.

8   Conclusions

This paper aimed to determine which representation of hyperbolic geometry is best concerning numerical issues. To this end, we compared five representations (linear, mixed, reduced, halfplane/halfspace, and generalized polar), controlling for six variants of dealing with numerical errors (invariant, careless, flattened, forced, weakly forced, binary). We conducted six tests capturing different scenarios leading to accumulating numerical imprecisions. Our results suggest that polar representation is the best in many cases, although the halfplane invariant is also very successful. It is known that numerical errors can be successfully combated by combining representation with tessellations, so our research took that into account. From the tessellations we have considered, $\{7, 3\}$ yields the best results. Fixed linear representations (especially invariant) perform well in game-design-related scenarios (walk tests: how long we can walk until we observe that we have "deviated" from the original straight line).

Acknowledgments. This work has been supported by the National Science Centre, Poland, grant UMO-2019/35/B/ST6/04456.

Disclosure of Interests. Authors have no conflict of interest to declare.

References

1. Bachmann, G., Bécigneul, G., Ganea, O.E.: Constant curvature graph convolutional networks. In: Proceedings of the 37th International Conference on Machine Learning, ICML 2020. JMLR.org (2020)
2. Bläsius, T., Friedrich, T., Katzmann, M., Krohmer, A.: Hyperbolic embeddings for near-optimal greedy routing. In: Algorithm Engineering and Experiments (ALENEX), pp. 199–208 (2018)
3. Bläsius, T., Friedrich, T., Krohmer, A., Laue, S.: Efficient embedding of scale-free graphs in the hyperbolic plane. In: European Symposium on Algorithms (ESA), pp. 16:1–16:18 (2016)
4. Block, A., Skopek, O., Bachmann, G., Ganea, O., Bécigneul, G.: A universal model for hyperbolic, euclidean and spherical geometries (2019). https://andbloch.github.io/K-Stereographic-Model/
5. Boguñá, M., Papadopoulos, F., Krioukov, D.: Sustaining the internet with hyperbolic mapping. Nat. Commun. **1**(6), 1–8 (2010). https://doi.org/10.1038/ncomms1063
6. Bulatov, V.: Conformal models of hyperbolic geometry (2010). http://bulatov.org/math/1003/band-stretch1.html#(1)
7. Celińska-Kopczyńska, D., Kopczyński, E.: Discrete hyperbolic random graph model (2021)
8. Celińska-Kopczyńska, D., Kopczyński, E.: Generating tree structures for hyperbolic tessellations (2021)

9. Celińska-Kopczyńska, D., Kopczyński, E.: Numerical aspects of hyperbolic geometry (2024). https://figshare.com/articles/software/Numerical_Aspects_of_Hyperbolic_Geometry/25325338. Accessed 13 Apr 2024
10. CodeParade: Hyperbolica (2022)
11. Epstein, D.B.A., Paterson, M.S., Cannon, J.W., Holt, D.F., Levy, S.V., Thurston, W.P.: Word Processing in Groups. A. K. Peters Ltd., USA (1992)
12. Floyd, W.J., Weber, B., Weeks, J.R.: The achilles' heel of o(3, 1)? Exp. Math. **11**(1), 91–97 (2002). https://doi.org/10.1080/10586458.2002.10504472
13. Friedrich, T., Katzmann, M., Schiller, L.: Computing voronoi diagrams in the polar-coordinate model of the hyperbolic plane (2023)
14. Hart, V., Hawksley, A., Matsumoto, E.A., Segerman, H.: Non-euclidean virtual reality I: explorations of $\mathbb{H}^3$. In: Proceedings of Bridges: Mathematics, Music, Art, Architecture, Culture, pp. 33–40. Tessellations Publishing, Phoenix, Arizona (2017)
15. Kopczyński, E., Celińska, D., Čtrnáct, M.: HyperRogue: playing with hyperbolic geometry. In: Proceedings of Bridges: Mathematics, Art, Music, Architecture, Education, Culture, pp. 9–16. Tessellations Publishing, Phoenix, Arizona (2017)
16. Kopczyński, E., Celińska-Kopczyńska, D.: Conformal mappings of the hyperbolic plane to arbitrary shapes. In: Goldstine, S., McKenna, D., Fenyvesi, K. (eds.) Proceedings of Bridges 2019: Mathematics, Art, Music, Architecture, Education, Culture, pp. 91–98. Tessellations Publishing, Phoenix, Arizona (2019). http://archive.bridgesmathart.org/2019/bridges2019-91.pdf
17. Kopczyński, E., Celińska-Kopczyńska, D.: Real-time visualization in anisotropic geometries. Exp. Math. 1–20 (2022). https://doi.org/10.1080/10586458.2022.2050324
18. Lamping, J., Rao, R., Pirolli, P.: A focus+context technique based on hyperbolic geometry for visualizing large hierarchies. In: Proceedings of the SIGCHI Conference on Human Factors in Computing Systems, CHI 1995, pp. 401–408. ACM Press/Addison-Wesley Publishing Co., New York (1995). https://doi.org/10.1145/223904.223956
19. Maskin, E., Dasgupta, P.: The fairest vote of all. Sci. Am. **290**(3), 64–69 (2004)
20. Munzner, T.: Exploring large graphs in 3D hyperbolic space. IEEE Comput. Graphics Appl. **18**(4), 18–23 (1998). https://doi.org/10.1109/38.689657
21. Nelson, R., Segerman, H.: Visualizing hyperbolic honeycombs. J. Math. Arts **11**(1), 4–39 (2017). https://doi.org/10.1080/17513472.2016.1263789
22. Nickel, M., Kiela, D.: Poincaré embeddings for learning hierarchical representations. In: Guyon, I., Luxburg, U.V., Bengio, S., Wallach, H., Fergus, R., Vishwanathan, S., Garnett, R. (eds.) Advances in Neural Information Processing Systems 30, pp. 6341–6350. Curran Associates, Inc. (2017). http://papers.nips.cc/paper/7213-poincare-embeddings-for-learning-hierarchical-representations.pdf
23. Osudin, D., Child, C., He, Y.-H.: Rendering non-euclidean space in real-time using spherical and hyperbolic trigonometry. In: Rodrigues, J.M.F., et al. (eds.) ICCS 2019. LNCS, vol. 11540, pp. 543–550. Springer, Cham (2019). https://doi.org/10.1007/978-3-030-22750-0_49

24. Sala, F., De Sa, C., Gu, A., Re, C.: Representation tradeoffs for hyperbolic embeddings. In: Proceedings of ICML, pp. 4460–4469. PMLR, Stockholmsmässan, Stockholm Sweden (2018). http://proceedings.mlr.press/v80/sala18a.html

25. Ungar, A.: Gyrovector spaces and their differential geometry. Nonlinear Funct. Anal. Appl. **10**, 791–834 (2005)

26. Yu, T., De Sa, C.M.: Numerically accurate hyperbolic embeddings using tiling-based models. In: Wallach, H., Larochelle, H., Beygelzimer, A., d'Alché-Buc, F., Fox, E., Garnett, R. (eds.) Advances in Neural Information Processing Systems, vol. 32. Curran Associates, Inc. (2019). https://proceedings.neurips.cc/paper/2019/file/82c2559140b95ccda9c6ca4a8b981f1e-Paper.pdf

A Numerical Feed-Forward Scheme
for the Augmented Kalman Filter

Fabio Marcuzzi[(✉)] [iD]

Department of Mathematics "Tullio Levi Civita", University of Padova,
Via Trieste 63, 35121 Padova, Italy
marcuzzi@math.unipd.it

Abstract. In this paper we present a numerical feed-forward strategy
for the Augmented Kalman Filter and show its application to a diffusion-
dominated inverse problem: heat source reconstruction from boundary
measurements. The method is applicable in general to forcing term estimation in lumped and distributed parameters models and gives a significant contribution where, in industry and science, probing signals are
used through a diffusive material-body to estimate its localized internal
properties in a non-destructive test, like in ultrasound or thermographic
inspection.

Keywords: Kalman Filter · forcing term estimation · feed-forward
control · inverse heat transfer problems

1 Introduction

There is an increasing interest, in applications, at estimating forcing terms of
various nature (e.g. mechanical, thermal, etc.) from a relatively small set of data
obtained by measurements of physical quantities (e.g. displacements, temperatures, etc.) that are assumed to be sufficiently informative of the effect of these
forcing terms on the considered real system. This estimate represents a virtual
measurement of the (physically) unmeasurable forcing term, performed by an
algorithm which is usually called a *soft sensor* [4].

If the real system can be represented by a physico-mathematical model, then
the inverse problem of estimating the (input) forcing term from output measurements must be solved by taking into account the model, the ill-posedness
of the problem, and the uncertainty about both the data measured and the
model's parameters. These aspects call for an interdisciplinary approach and for
methods that are able to tackle with both the deterministic and the stochastic
aspects of the problem. For this reason, in this field of applications, Bayesian
methods have emerged as a main approach and a well known achievement in
this direction, widely used in applications, is the Kalman Filter (KF) [11,13].

To solve the problem of estimating the input forcing term, the reference
model for the KF is often configured by augmenting the state vector to include

The original version of the chapter has been revised. The reference 4 has been corrected.
A correction to this chapter can be found at
https://doi.org/10.1007/978-3-031-63778-0_30

the dynamics of the forcing term (input), and suppose the system be driven by white noise. In this way, the forcing term estimation problem becomes a state estimation problem, where the reference model represents the dynamics of two subsystems, i.e. the forcing term generator and the real system, and also of their interaction. This is usually uni-directional, i.e. the system's state does not affect the forcing term. This case is paradigmatic for the so-called Augmented Kalman Filter (AKF): the outputs are taken from the original state variables, i.e. the augmented state variables are unmeasurable, and the original variables does not affect the augmented variables.

In this paper we will consider this precise setting, where the reference model of the AKF represents explicitly the forward dependence between the forcing term and the system's state, while the AKF gain tries to estimate implicitly the backward dependence between them, in its effort to estimate the forcing term from the output prediction error. Our aim is to introduce an explicit representation of this backward dependence and feed it forward into the AKF state predictor. To do so, we compute a simulation of the inverse model and we will show and compare two possible strategies for using it to add a feed-forward term to the proportional control of the estimation error made by the standard AKF formulation. In this paper we will face diffusion dominated problems with distributed forcing terms, a difficult case for existing methods, as we will see.

In the literature, for source estimation in mechanical applications see e.g. Lourens [16] where the Augmented Kalman Filter (AKF) has been introduced for the purpose of mechanical load estimation, and the introduced covariance model that describes the forcing term modelling error is used as a regularizer in the force estimation problem, and precisely the diagonal elements of the covariance matrix are tuned as regularization parameters. Indeed, this covariance matrix can be seen also as the weighting matrix of a least-squares problem [13] and this is actually one way we formulate our feed-forward action in the KF, although not the best, as we will see. In [16] there is also an experimental comparison with other combined deterministic-stochastic techniques to force estimation problems. In Neats [17] there is an analysis of the stability of the AKF, which shows that there are common measurement configurations that exhibit a drift in the state estimates, due to unobservability issues and propose to add dummy-measurements; in [19] the sparse constraint is used to solve the drift problem in AKF with more generality. See [15] for a recent, general survey on mechanical load estimation techniques in both frequency and time domains.

For thermal applications, like the one here considered as a model problem, the application of the Kalman Filter is more problematic. The Augmented Kalman Filter is used e.g. in Qi et al. [18], where there is also a comparison and references with another approach: KF-RLSE, i.e. the combination of a Kalman Filter and a Recursive Least-Squares Estimator of the residuals obtained from the KF. In [18] the heat flux (forcing term) is simply a scalar term, not a distributed field like is set in this paper. Another approach is the Optimal two-stage Kalman filter (OTSKF) and is preferred to the AKF because of its poor ability for the simultaneous estimation of spatio-temporal heat flux and temperature field. In this paper we show that a feed-forward scheme improves dramatically this ability. Actually, some years ago the two-stage Kalman filter had been proposed in [14] as an efficient implementation of the Augmented Kalman Filter and, in principle, the feed-forward scheme could be applied also for this formulation, if preferred. Finally, we mention a fundamental work by Gillijns e De Moor [9], that is not applicable here, anyway, since it requires that the number of measurement points be greater or equal to the forcing term nodes/variables.

In this paper we assume that the input generator and the real system are modeled adequately, since our scope is on finding a correct and efficient formulation for the feed-forward action. First of all, we must legitimate this choice: note that the standard AKF operates a proportional control over the state estimation error and it is well known that in problems where the input-state dynamics are slow this results in poor performances. There are variants where a proportional-integral actions [2,3,8] is performed, at the cost of doubling the number of state variables, or a *smoothing* is done through post-processing [1]. Actually, a feed-forward [12] approach is a consolidated good practice in control theory in general and a combination of feed-forward with feedback to eliminate steady-state errors is mostly used: feed-forward is used for tracking capabilities, and feedback for steady-state accuracy.

The paper is organized as follows: in Sect. 2 we set model problem and AKF formulation; in Sect. 3 the feed-forward strategy is presented; in Sect. 4 we show some numerical results and a Discussion and Conclusions section ends the paper.

2   Problem Settings and Kalman Filter Estimation of the Forcing Term

As a model problem of a diffusive process, let us consider the heat equation

$$
\begin{cases}
\rho C\, \partial_t T_{(f_\vartheta)} = \kappa\, \Delta T_{(f_\vartheta)} + f_\vartheta, & in\ D_c^{(0)} \times [0, t_f] \\
\kappa\, \nabla T_{(f_\vartheta)} \cdot \mathbf{n}_S = q(t), & on\ S \times [0, t_f] \\
\kappa\, \nabla T_{(f_\vartheta)} \cdot \mathbf{n} = 0, & on\ \delta D_c^{(0)}/S \times [0, t_f] \\
T_{(f_\vartheta)}(0, \cdot) = T_0(\cdot), & in\ D_c^{(0)}.
\end{cases}
\tag{1}
$$

and suppose that the heat source term f_ϑ is an unknown function, in general, except that it is assumed different from zero only in a few disconnected regions of compact support. This is a common situation in many applications.

The aim of this paper is to estimate f_ϑ from a limited number of temperature measurements $\tilde{T}_{(f_\vartheta)}$, typically taken at the boundary. The restriction to a 2D problem is only for simplicity, the method we devise can be used in higher dimensions. The estimate of f_ϑ can be seen as an indirect measurement of f_ϑ from physical temperature measurements and we do this by exploiting the combination of the physico-mathematical model 1 and an abstract, data-driven model to describe f_ϑ. Therefore, the resulting method may be called a *physics-aware soft-sensor* [4].

Let us consider problem (1), discretized in space using the Finite Element Method (FEM) with Lagrangian elements P1, i.e. first-degree piecewise polynomials, and in time with the implicit Euler method, obtaining at iteration k:

$$M\frac{\tilde{T}_k - \tilde{T}_{k-1}}{d_t} = K\tilde{T}_k + f_k \quad \Rightarrow \quad (I - d_t\ M^{-1}K)\tilde{T}_k = \tilde{T}_{k-1} + d_t\ M^{-1}f_k$$

where $M \in R^{n \times n}$ and $K \in R^{n \times n}$ are the mass and stiffness matrices of the FEM discretization, d_t is the time step chosen in the time discretization and f_k is the heat source at time t_k. If we knew all the values of $T_{(f_\vartheta)}$ then the computation of the unknowns f_ϑ would be a simple algebraic reconstruction. Since we can measure only a few components of $T_{(f_\vartheta)}$, we can reformulate this model as a state-space dynamical system with the aim of estimating its state from output measurements. Let us consider the following state-space discrete model in physical coordinates:

$$\begin{aligned} x_m(k+1) &= A_m x_m(k) + B_m u_m(k) + v_m(k) \\ y_m(k) &= C_m x_m(k) + w(k) \end{aligned} \tag{2}$$

where $x_m(k) = \tilde{T}_k$, $u_m(k) = f_k$, $A_m = (I - d_t M^{-1}K)^{-1}$, $B_m = A_m d_t M^{-1} = (I - d_t M^{-1}K)^{-1} d_t M^{-1}$, C_m is a matrix built with the rows of the identity matrix corresponding to measured nodes, $v_m(k)$ is a stochastic term for model error and $w(k)$ for measurement error, both supposed Gaussian noise [11]. Now, we augment the state of (2) by adding a model for the forcing term f_k:

$$x(t) = \begin{bmatrix} x_f \\ x_m \end{bmatrix} \quad , \quad u(t) = \begin{bmatrix} 0 \\ 0 \end{bmatrix} \tag{3}$$

where $x(t)$ is the *augmented state vector*, $x_f = f_k$, $u(t)$ the new input vector, which can be now omitted, $y(t)$ the output vector and:

$$A = \begin{bmatrix} A_f & 0 \\ Z & A_m \end{bmatrix} \quad , \quad B = \begin{bmatrix} 0 \\ 0 \end{bmatrix} \quad , \quad C = \begin{bmatrix} 0 & C_m \end{bmatrix} \quad , \tag{4}$$

where $A_f = I$, $Z = B_m C_u$, $C_u = I$ and the augmented state-space model is:

$$\begin{aligned} x(k+1) &= Ax(k) + Bu(k) + v(k) \\ y(k) &= Cx(k) + w(k) \end{aligned} \tag{5}$$

Now, to estimate the state vector it is a common choice to adopt a Kalman Filter, that we recall here in its one-step version [13]:

$$P(k) = \left[\left(Q(k-1) + A(k-1)P(k-1)A(k-1)^T \right)^{-1} + C^T R^{-1} C \right]^{-1} \tag{6}$$

$$e_{outpred} = [C\left(A(k-1)\ \hat{x}(k-1) + B\ u(k-1)\right) - \bar{y}(k)] \tag{7}$$

$$\delta\hat{x}(k) = -P(k)\ C^T R^{-1} e_{outpred} \tag{8}$$

$$\hat{x}(k) = A(k-1)\ \hat{x}(k-1) + B\ u(k-1) + \delta\hat{x}(k) \tag{9}$$

where the state vector $x(k)$ is the concatenation of the temperatures $T_{(f_\vartheta)}$ computed at the mesh nodes and the source term f_ϑ at the same nodes

$$x(k) = \begin{bmatrix} f_\vartheta(k) \\ T_{(f_\vartheta)}(k) \end{bmatrix}, \tag{10}$$

while the vector $y(k)$ contains the measured temperatures. In the problem at hand, the reference model for the Kalman Filter is the union of a constant model for the heating source (this is the usual choice made in the literature for this kind of problem) and the FE discretization of (1), and we set the covariance matrix of the model error as

$$Q(k) = \begin{bmatrix} Q_f(k) & Q_{mf}^T(k) \\ Q_{mf}(k) & Q_m(k) \end{bmatrix} \tag{11}$$

where $Q_f(k) = \sigma_{Q_f}^2 I$, $Q_m(k) = \sigma_{Q_m}^2 I$ and $Q_{mf}(k) = \sigma_{Q_{mf}}^2 I$ in general, since we don't know the location of the forcing term. Then, set the initial covariance matrix of the state-estimation error as

$$P(0) = \begin{bmatrix} P_f(0) & P_{mf}(0)^T \\ P_{mf}(0) & P_m(0) \end{bmatrix} \tag{12}$$

where $P_f(0) = \sigma_{P_f}^2 I$ and $P_m(0) = \sigma_{P_m}^2 I$.

From Eq. (8) it is evident that the KF operates a proportional feedback action on the (scaled) output prediction error, with a gain

$$G_{KF}(k) = -P(k)\, C^T = \begin{bmatrix} P_{mf}(k)^T C_m^T \\ P_m(k) C_m^T \end{bmatrix} \tag{13}$$

and, since the product $P_{mf}(k)^T C_m^T$ here simply corresponds to a selection of columns of the matrix $P_{mf}(k)^T$, the proportional action made on the x_f variables can be understood better if we compute the recursive relation for $P_{mf}(k)$. At the limit $R = \infty$, from (6) we would have

$$\begin{aligned} P_{mf}^T(k) &= \left(\sigma_{P_f}^2 Z + A_m P_{mf}(k-1) \right)^T + Q_{mf}^T(k) \\ &= \left(\sigma_{P_f}^2 d_t A_m M^{-1} + A_m P_{mf}(k-1) \right)^T + Q_{mf}^T(k) \end{aligned} \tag{14}$$

Note that this would bring to $\delta\hat{x}(k) = 0$ since $C^T R^{-1} = 0$. Anyway, with realistic values of R, there is a rank-n_y modification of the inverse of $P(k)$ whose influence on $P_{mf}^T(k)$ is secondary in this analysis. Then, since A_m is the discretization of a diffusive operator, it will simply diffuse the output prediction error $e_{outpred}$ (7) to compute the state estimate update $\delta\hat{x}(k)$ (8), the matrix Q_{mf} cannot help, since we don't know the location of the sourcing term, and the poor result is depicted in Sect. 4.1, Fig. 2.

In general, it is well known that this proportional feedback may have strong limitations, e.g. for systems with a substantial inertia like thermal systems have, and for this reason a few algorithmic extensions have been developed in the literature, like e.g. a proportional-integral formulation [2], which requires a doubling of the state-variables, or like some additional post-processing (precisely, *smoothing* [1]) of the KF predictions. These are yet general improvements, that do not exploit the reference model as it could be done in the model problem here considered. For this reason we add instead a feed-forward action, described in the next section. At our knowledge, there is no contribution in the literature about introducing a feed-forward action into the Kalman Filter algorithm. In this paper, we do this for the Augmented Kalman Filter, with the assumption that the variable augmentation is done to describe the dynamics of an input forcing term. Therefore, there is a precise modelling assumption underlying this feed-forward scheme. In principle, the numerical scheme we propose may be used in general Kalman filtering, but a more general supporting assumption is not clear, at this moment, so we omit to discuss it.

3   A Feed-Forward Strategy for the Augmented Kalman Filter

The unknown forcing term can be interpreted as a deterministic load disturbance, whose behaviour can be at least partially known. This would call for a feed-forward mechanism added to the proportional feedback activity performed by the Kalman Filter but, since we cannot measure the forcing term, a standard feed-forward would be a trial-and-error with no value added, since it should assume to know the quantity that we are estimating. Indeed, our idea is to exploit the maximum principle: if the output prediction error has a mean value different from zero we attribute this fact to an internal heat source and use it to drive the feed-forward action, as it is described in the following subsections.

3.1   Modeling the Feed-Forward Action

Since we should obtain the feed-forward reference signal from the output prediction error (7), the most straightforward approach is to embed the feed-forward action into the Kalman gain (13). The idea is to include the relation between the output prediction error and the load disturbance (i.e. the inverse model) inside the covariance matrix $Q(k)$, as a deterministic least-squares weighting, following the interpretation of [13] of the KF. note that $Q_{mf}(k)$ in (14) translates the effect of a temperature error to an error in the forcing term, through the Kalman gain computation (13). Depending on how be build $Q_{mf}(k)$, this can be interpreted in different ways. Here we set each i-th row of $Q_{mf}(k)$ as the response of the model at a unit forcing term concentrated at the node i, after a number of time-steps that depends on the distance of the node from the measured boundary segment. Therefore, when the KF inverts the covariance matrix, this will translate a variation of the temperatures at the border into a variation of the forcing term additional to that operated by the standard Kalman Filter with $Q_{mf}(k) = 0$ (this would be a typical choice, since $Q_{mf}(k)$ classically expresses the mutual covariance between the model error of forcing term variables and temperature field variables, which can be assumed to be zero).

The action created by $Q_{mf}(k)$ is called *feed-forward* because it depends on the model between the load disturbance and the process output, and precisely on the inversion of that model; in this way, the apriori estimate of the forcing term is substituted by the output prediction error, under the assumption that a disturbance load is the main cause of it, as it is the case for the model problem here considered. Now the question becomes how to tune this feed-forward action. Note that $Q_f(k)$, $Q_m(k)$, $P_f(0)$, $P_m(0)$ and $P_{mf}(0)$ are tuned independently, e.g. following a best practice in Kalman filtering (see e.g. [11]). Let us consider Eqs. (8)-(9) and the augmented system's matrices (5). We have

$$\begin{bmatrix} \hat{x}(k) - \hat{x}(k-1) \\ \vdots \end{bmatrix} = \begin{bmatrix} -P_{mf}(k)^T C_m^T R^{-1} \left[C_m A_m \hat{x}_m(k-1) + C_m Z \hat{x}_f(k-1) - \bar{y}(k) \right] \\ \vdots \end{bmatrix}$$

(15)

Now, let us do n_f simulations where $\hat{x}_m(0) = x_m(0)$, $\hat{x}_f(0) = 0$ and $x_f(0)$ is zero everywhere except a single value which is equal to $L = 100$ (i.e. a pointwise forcing term), corresponding to a different mesh node at each experiment. In this way, from the first n_u equations of (15), we obtain a linear system with multiple right-hand side $M_{sl} = M_{PC} M_{ope}$ where each column of $M_{sl} \in R^{n_u \times n_f}$ has only one nonzero (the pointwise forcing term), $M_{PC} \in R^{n_u \times n_y}$ is supposed to be unknown and each column of $M_{ope} \in R^{n_y \times n_f}$ is the prediction error resulting from the pointwise forcing term represented at the corresponding column of M_{sl}. Noteworthy, for each experiment we take the relation (15) at only one time instant, which depends on the distance between the node and the measurement points; putting more equations in the system from the same experiment gives a final M_{ope} strongly ill-conditioned and we see no advantage, at the moment. Assume we transpose both sides and solve the resulting system, we will have a well-determined system if $n_f = n_y$, an over-determined system if $n_f > n_y$ and an underdetermined system if $n_f < n_y$:

$$M_{ope}^T M_{PC}^T = M_{sl}^T \tag{16}$$

Finally, the feed-forward formula is: recalling expression (14) at the limit $R = \infty$, we want $P_{mf}^T(k) = \bar{P}_{mf}^T(k)$, where the latter is computed from the solution of (16), i.e. $M_{PC}^T = \bar{P}_{mf}^T(k+1) \, C_m^T$. Therefore, we could set

$$Q_{mf}(k)^T C_m^T = \sigma_{Q_{mf}}^2 \, M_{PC}^T - \left(\sigma_{P_f}^2 \, Z + A_m P_{mf}(k) \right)^T C_m^T \tag{17}$$

where $Q_{mf}(k) C_m^T$ is simply a selection of columns of $Q_{mf}(k)$. The choice of the nodes where to apply the pointwise forcing term is crucial and depends on the application. In the model problem here considered, there is an interesting fact: pointwise forcing terms applied to nodes aligned on the same line orthogonal to the measurements boundary, give proportional responses, thus making M_{ope} singular and no unique solution of (16). To avoid it, let us consider to subdivide the nodes in unaligned grids of points, e.g. in Fig. 1: let us consider a two-dimensional domain with a regular discretization of points and group them as depicted, i.e. each i-th group of nodes has a different symbol, where only one node of the same group lies on each vertical line; assuming that the bottom horizontal line is the measurements boundary, we have that there aren't two nodes of the same group and aligned orthogonally to that boundary. Then, let us group accordingly the experiments and represent the i-th group of pointwise forcing terms as the columns of $M_{sl,i}$; in this way, we get a set $\{M_{ope,i}\}$ of much better conditioned matrices. Then, for each $M_{ope,i}$ build the linear system like (16) and solve it to get the solution $M_{PC,i}$. Then, formula (17) is implemented for each filter of a one-step filter bank, i.e. the feed-forward action $Q_{mf}(k)$ is computed as in (17) for each $M_{PC,i}$ and so the corresponding $\delta\hat{x}(k)$ from (8), that we call $\delta\hat{x}_i(k)$. At the end of this filter-bank computation, the applied $\delta\hat{x}(k)$ is the average of the $\delta\hat{x}_i(k)$ and so $P(k)$ is the average of the updated covariance matrices of each filter in the bank.

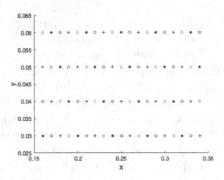

Fig. 1. Unaligned grids of points in the interval $(x, y) \in [0.16, 0.34] \times [0.03, 0.06]$.

Unfortunately, this strategy has some pitfalls and shows in practice a fundamental weakness: to be effective, the covariance matrix of (19), that should be inverted to obtain $P(k)$, becomes very ill-conditioned. However, it suggests an alternative strategy that, as we will, see, gives nice results. Therefore, keeping in mind the linear relation (16), in the next section we see how to extract a reference signal for the feed-forward action and then how to implement this action in addition to the state-update due to the proportional KF gain (13), as usual, indeed, in feedback control theory [12].

3.2 Feed-Forward Reference Extraction

We must obtain a feed-forward reference from the output prediction error, that actually contains also modelling and measurement errors, other than the effect of forcing term estimation error. For this reason, the scheme (15), that uses directly the output prediction error to drive the feed-forward action, is not accurate if entirely used to compute the feed-forward action, as actually done in (17) through the M_{PC} matrix. Here, instead, in order to extract a feed-forward reference signal from the output prediction error, and use it to drive the feed-forward action, we compute the coefficients c_{mq} of a linear combination of M_{ope} columns that approximates the output prediction error, i.e.:

$$M_{ope}c_{mq} = e_{outpred} \tag{18}$$

We must remember that M_{ope} columns should represent all the nodes where a forcing term should be estimated, and in this way it becomes singular. Note, however, that the columns of M_{ope} are output responses to point-wise loads; hence, each component of c_{mq} is the intensity of a point-wise load located in a mesh node. This suggests an easy regularization scheme on (18) that gives a solution c_{mq} corresponding to a forcing term with some spatial properties, like smoothness, sparsity, etc. In this way the regularization can be tuned accordingly with the specific application, e.g. by exploiting the physical insight. In Sect. 3.4 we will propose a general solution. The solution c_{mq}^{reg} of the regularized problem (21) is then multiplied by a gain G_{FF} and becomes the feed-forward contribution to the state estimate $\hat{x}(k-1)$ in (23), which is then used in the usual kalman Filter update (24)–(26). Therefore, the one-step Feed-Forward Augmented Kalman Filter (FF-AKF) algorithm, extended from (6)–(9), becomes:

$$P(k) = \left[\left(Q(k-1) + A(k-1)P(k-1)A(k-1)^T \right)^{-1} + C^T R^{-1} C \right]^{-1} \tag{19}$$

$$e_{outpred} = \left[C \left(A(k-1) \, \hat{x}(k-1) + B \, u(k-1) \right) - \bar{y}(k) \right] \tag{20}$$

$$M_{ope}^{reg} \, c_{mq}^{reg} = e_{outpred} \tag{21}$$

$$\delta \hat{x}_{FF}(k-1) = G_{FF} \, c_{mq}^{reg} \tag{22}$$

$$\hat{x}_{FF}(k) = A(k-1) \, \left(\hat{x}(k-1) + \delta \hat{x}_{FF}(k) \right) + B \, u(k-1) \tag{23}$$

$$e_{FFoutpred} = \left[C \hat{x}_{FF}(k) - \bar{y}(k) \right] \tag{24}$$

$$\delta \hat{x}(k) = -P(k) \, C^T R^{-1} e_{FFoutpred} \tag{25}$$

$$\hat{x}(k) = A(k-1) \, \hat{x}(k-1) + B \, u(k-1) + \delta \hat{x}(k) \tag{26}$$

Note that the feed-forward reference extraction made through the solution of system (18) is blind, i.e. purely algebraic. This means that components of the output prediction error which are due to modelling and/or measurement errors may be confused with pointwise forcing term responses and represented, at least partially, into the coefficients c_{mq}. This would lead to an overestimate of the overall forcing term and create an additional error diffused over the next output prediction errors. For this reason, a safe tuning of the gain G_{FF} is required and seems a practical way to prevent the feed-forward action to create additional noise. This will be confirmed also in the numerical experiments. The tuning of G_{FF} will be explained in the next Sect. 3.3 making use of the maximum principle for the heat equation.

3.3   Tuning the Feed-Forward Gain G_{FF}

It is easy to see that, according to the well known maximum principle for the heat equation, the effect of an underestimated internal heat source (which is a forcing term in the heat equation) of positive sign is a lower temperature predicted at the boundary with respect to the real, measured, one. Then, to tune conservatively the Feed-Forward gain G_{FF} it may be chosen in such a way that the integral of the output prediction error be positive and above a (safety) threshold. For negative heat sources it works the same way, symmetrically.

Moreover, an hysteresis mechanism is conveniently adopted to get a non-zero G_{FF} only if the output prediction error is significantly outside an interval of oscillation where, inside, it may be reasonably dictated by only modeling and measurement errors.

3.4   Regularization Issues

We already said that the matrix M_{ope} in (18) is singular and this is due to the fact that it has groups of columns strictly proportional to each other and, important, corresponding to nodes aligned orthogonally to the measured boundary. Therefore, any column in the same group can be picked up in the solution of (18), obtaining the same residual. This means that it will be difficult to estimate accurately the depth of the internal forcing term location, if the regularization is blind. Also, no *column selection* method (sparse recovery or NLLS, see [6], for a recent comparison, and references therein) can give us a good solution without regularization. Then, in this problem, regularization is necessary. Actually, with regularization we then gained a significant improvement from using NNLS, in particular when the heat source is very localized, the regularization parameter should be kept very small and column selection is useful also to solve the required QR factorizations [7], see Sect. 4.2.

For this reason, we adopt the subdivision of the nodes in unaligned sets, done in Sect. 3.1 (see Fig. 1) and consider to solve a system like (18) for each set. It means that each node in the same orthogonal line wrto the border is considered alone in the approximation of the output prediction error, i.e. each depth is considered independently from the others, and then the regularization optimize the shape of the forcing term according to a general criterion.

To do this, we do not solve the (18)-like systems independently, but build a block-diagonal matrix with the $M_{ope,i}$ matrices and regularize it. In this paper we aim at estimating smooth forcing terms, so the regularization is made with a discretization of the laplacian. Actually, we have tried also the TV regularization with quite worse results, as expected for the forcing terms adopted in these experiments, but not true in general. In Sect. 4.2 it will be shown the contribution of this block-diagonal formulation to the correct estimate of the depth of the internal heat source (forcing term). We show also that without regularization the solution concentrates in a greatly overestimated peak, not well centered with the true forcing term, and then wildly oscillates. Indeed, there is a variety of distributed forcing terms that can describe a generic output prediction error, in the problem here considered.

Finally, a consideration must be made about the λ coefficient which weights the regularization term: even knowing the true value of the forcing term, the optimization of λ is not trivial; in particular, optimizing λ at each iteration to minimize the estimation error of the forcing term brings to poor results. This will be future work.

4   Numerical Experiments

In this section, some numerical experiments are described to give a practical evidence of the algorithmic ideas previously presented. In all the following examples, experimental temperatures are simulated numerically, while intensive tests have been done in our previous work [5] to validate the model settings. A note about "inverse crimes": here we are solving the inverse problem using the same model that has generated the data, which is considered an inverse crime. Actually, we are interested in exact reconstructions which are non trivial, and the simplified setting we use is adequate to make significant comparisons, see the interesting discussion in the white-paper of Wirgin [20]. In a real, specific application one should then compare with data containing also model and measurement errors, to validate the practical accuracy of his method in the specific application.

Let us describe the model settings, where the following values of constants are used: $t_f = 1.51\ s$, $L = 0.1\ m$; $\rho C = 3.2 \cdot 10^6\ \frac{J}{m^3\ {}^\circ C}$, $k = 3.77 \cdot 10^3\ \frac{W}{m\ {}^\circ C}$. The initial condition is set to $T_0(\cdot) = 20^\circ C$. In this section an Implicit Euler method is adopted for the time discretization, using a temporal step $\Delta t = 0.0005$ in $(0, 0.1]$ and $\Delta t = 0.05$ in $(0.1, t_f]$. A P_1-FE method is used for space discretization, whose step length along y is $h_y = 0.01\ m$, $h_x = h_y$. The sensors are supposed to be in the middle of each mesh edge in the instrumented boundary segment. Numerical experiments have been carried out using Matlab. As a general forcing term we have used a gaussian forcing term f_ϑ with unknown variance and point of application. In the following subsections we see some relevant experiments.

4.1   Inadequacy of the KF Proportional Action with a Diffusive Gain

The ability of the proposed feed-forward scheme to adequately estimate the intensity and location, and in particular the depth, of localized internal forcing terms, is exemplified in Fig. 2. We see on the right that, without the contribution of the inverse model of the propagation between pointwise load and output prediction error, described from matrix M_{ope} in Sect. 3.1 and used in the FF-AKF algorithm (19)–(26), the estimate of the forcing term is concentrated close to the measured boundary, i.e. in completely wrong position, even if the output prediction error is converging (not shown). On the contrary, in the next section we will show the effectiveness of the feed-forward action, in the same experiment settings.

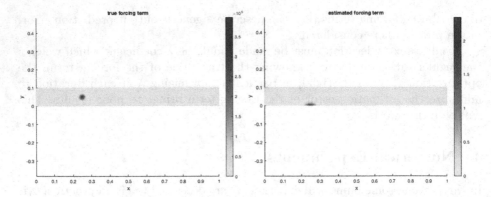

Fig. 2. The true forcing term f_ϑ (left) and its estimate by the standard Augmented Kalman Filter (right) on a rectangular domain $(x, y) \in [0, 1] \times [0, 0.1]$.

4.2   The Effect of Regularization and Block-Diagonal Reference Extraction

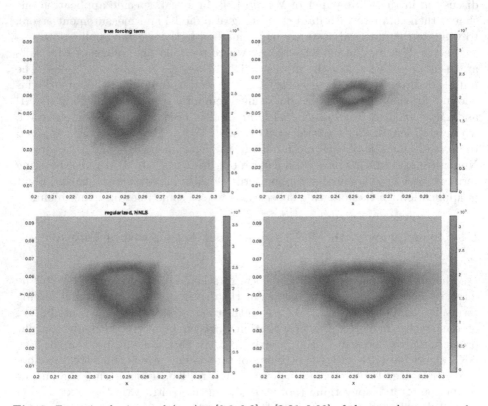

Fig. 3. Zoom in the interval $(x, y) \in [0.2, 0.3] \times [0.01, 0.09]$ of the true heat source f_ϑ (Top-left), its estimate with no regularization (Top-right), the estimate with a regularized NNLS solution (Bottom-left) and the estimate with a regularized OLS solution (Bottom-right).

In Fig. 3 we see a zoom of the small portion of the domain where is located the unknown heat source (the whole domain is shown in Fig. 2) and the corresponding estimates with the FF-AKF algorithm (19)–(26) using/not-using regularization and column selection algorithms. The best result is obtained with regularization and a NonNegative Least Squares (NNLS) solution (Bottom-left). Note that without regularization (Top-right) the magnitude and the support of the estimated forcing term are quite wrong, and with an Ordinary Least Squares solution (Bottom-right) the estimate leaks out from the support of the true forcing term; this is bad also for a correct reconstruction of the temperature field inside the body.

In Fig. 4 we see the effectiveness of the block-diagonal augmented scheme for feed-forward action determination. On the left, we see the estimated forcing term with a feed-forward action adopting the block-diagonal formulation described in Sect. 3.4, after 5 step (Top) and after 40 time-steps (Bottom), showing a good accuracy and stability of the estimate. On the right, we see the estimated forcing term without the block-diagonal formulation after 5 time-steps (Top): here

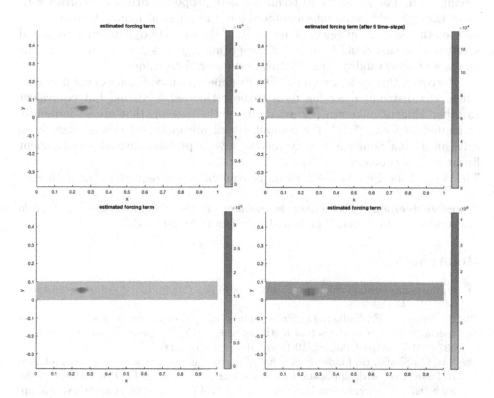

Fig. 4. Top-left: estimate of f_ϑ with a regularized OLS solution with the block-diagonal augmented scheme after a few time-steps (same as Fig. 3 Bottom-right); Bottom-left: the same estimate after 40 time-steps; Top/Bottom-right: same as Top/Bottom-left but without the block-diagonal augmented scheme.

it is evident that the nodes closer to the boundary are preferred, thus creating a strong bias on the values and location of the estimated forcing term. On the Bottom-right, we see the estimated forcing term without the block-diagonal formulation after 40 time-steps: here the effect of this bias is amplified by the fact that it has created a too big perturbation in the internal field temperature, that then induces an artifact in the forcing field estimation, in the process of minimizing the output prediction error.

5   Discussion and Conclusions

In this paper we have seen a class of models where many state-variables are not measured and many different combinations of them have quite similar effects on the measured ones. This is often the case with augmented state variables. These *observability* issues [11], cause a non-unique state estimate and the loss of physical interpretability for the computed one. For example, in Fig. 2 the standard AKF misses completely the right location and shape of the physical forcing term. The feed-forward technique here proposed, drives the correct estimate through additional information about the physical model. Observability, on the other side, is an algebraic property of the discrete dynamical system and a future direction could be to see if something rigorous can be said about the increase of observability given by this feed-forward technique.

Moreover, this feed-forward technique opens-up a broad range of applications, where the distributed forcing term to be estimated is virtual but equivalent to other kind of unknown perturbations of the system, that are so indirectly estimated, see e.g. [5,10]; the computational advantage of this analogy is to solve an original nonlinear (e.g. geometric) inverse problem through an equivalent linear inverse problem.

The code of the FF_AKF algorithm is available upon request to the author.

Acknowledgement. The author is member of the Gruppo Nazionale Calcolo Scientifico-Istituto Nazionale di Alta Matematica (GNCS-INdAM)

References

1. Optimal Smoothing, chap. 9, pp. 263–296. Wiley (2006). https://doi.org/10.1002/0470045345.ch9
2. Bakhshande, F., Saffker, D.: Proportional-integral-observer: a brief survey with special attention to the actual methods using ACC benchmark, vol. 28, pp. 532–537 (2015). https://doi.org/10.1016/j.ifacol.2015.05.049
3. Bas, O., Shafai, B., Linder, S.: Design of optimal gains for the proportional integral Kalman filter with application to single particle tracking. In: Proceedings of the 38th IEEE Conference on Decision and Control (Cat. No.99CH36304), vol. 5, pp. 4567–4571 (1999). https://doi.org/10.1109/CDC.1999.833262
4. Chinellato, E., Marcuzzi, F., Pierobon, S.: Physics-Aware soft sensors for embedded digital twins. In: Yang, X.S., Sherratt, S., Dey, N., Joshi, A. (eds.) Proceedings of Ninth International Congress on Information and Communication Technology. ICICT 2024, 2024. Lecture Notes in Networks and Systems, vol. 1013, pp. 417–427. Springer, Singapore (2024). https://doi.org/10.1007/978-981-97-3559-4_34

5. Dessole, M., Marcuzzi, F.: Accurate detection of hidden material changes as fictitious heat sources. Numer. Heat Transf. B Fundam. (2023). https://doi.org/10.1080/10407790.2023.2220905

6. Dessole, M., Dell'Orto, M., Marcuzzi, F.: The Lawson-Hanson algorithm with deviation maximization: finite convergence and sparse recovery. Numer. Linear Algebra Appl. **30**, e2490 (2023). https://doi.org/10.1002/nla.2490

7. Dessole, M., Marcuzzi, F.: Deviation maximization for rank-revealing QR factorizations. Numer. Algorithms (2022). https://doi.org/10.1007/s11075-022-01291-1

8. Farhat, A., Koenig, D., Hernandez-Alcantara, D., Morales-Menendez, R.: Tire force estimation using a proportional integral observer, vol. 783 (2017). https://doi.org/10.1088/1742-6596/783/1/012014

9. Gillijns, S., De Moor, B.: Unbiased minimum-variance input and state estimation for linear discrete-time systems. Automatica **43**(1), 111–116 (2007). https://doi.org/10.1016/j.automatica.2006.08.002

10. Giusteri, G.G., Marcuzzi, F., Rinaldi, L.: Replacing voids and localized parameter changes with fictitious forcing terms in boundary-value problems. Results Appl. Math. **20** (2023). https://doi.org/10.1016/j.rinam.2023.100402

11. Grewal, M.S., Andrews, A.P.: Kalman Filtering: Theory and Practice with MATLABÂ®, 4th edn., vol. 9781118851210. Wiley (2014). https://doi.org/10.1002/9781118984987

12. Guzman, J., Hagglund, T.: Tuning rules for feedforward control from measurable disturbances combined with PID control: a review. Int. J. Control (2021). https://doi.org/10.1080/00207179.2021.1978537

13. Humpherys, J., Redd, P., West, J.: A fresh look at the Kalman filter. SIAM Rev. **54**(4), 801–823 (2012). https://doi.org/10.1137/100799666

14. Keller, J., Darouach, M.: Optimal two-stage Kalman filter in the presence of random bias. Automatica **33**(9), 1745–1748 (1997). https://doi.org/10.1016/S0005-1098(97)00088-5

15. Liu, R., Dobriban, E., Hou, Z., Qian, K.: Dynamic load identification for mechanical systems: a review **29**(2), 831–863 (2022). https://doi.org/10.1007/s11831-021-09594-7

16. Lourens, E., Reynders, E., De Roeck, G., Degrande, G., Lombaert, G.: An augmented Kalman filter for force identification in structural dynamics. Mech. Syst. Signal Process. **27**(1), 446–460 (2012). https://doi.org/10.1016/j.ymssp.2011.09.025

17. Naets, F., Cuadrado, J., Desmet, W.: Stable force identification in structural dynamics using Kalman filtering and dummy-measurements. Mech. Syst. Signal Process. **50–51**, 235–248 (2015). https://doi.org/10.1016/j.ymssp.2014.05.042

18. Qi, H., Wen, S., Wang, Y.F., Ren, Y.T., Wei, L.Y., Ruan, L.M.: Real-time reconstruction of the time-dependent heat flux and temperature distribution in participating media by using the kalman filtering technique. Appl. Therm. Eng. **157**, 113667 (2019). https://doi.org/10.1016/j.applthermaleng.2019.04.077

19. Wei, D., Li, D., Huang, J.: Improved force identification with augmented Kalman filter based on the sparse constraint **167** (2022). https://doi.org/10.1016/j.ymssp.2021.108561

20. Wirgin, A.: The inverse crime (2004). https://doi.org/10.48550/ARXIV.MATH-PH/0401050

Calculation of the Sigmoid Activation Function in FPGA Using Rational Fractions

Pavlo Serhiienko[1] , Anatoliy Sergiyenko[1] , Sergii Telenyk[2] ,
and Grzegorz Nowakowski[2(✉)]

[1] Igor Sikorsky Kyiv Polytechnic Institute, Kyiv 03056, Ukraine
`aser@comsys.kpi.ua`
[2] Cracow University of Technology, 31-155 Cracow, Poland
`gnowakowski@pk.edu.pl`

Abstract. In this paper, we consider implementations of the sigmoid activation function for artificial neural network hardware systems. A rational fraction number system is proposed to calculate this function. This form of data representation offers several benefits, including increased precision compared to integers and more straightforward implementation in a field programmable gate array (FPGA) than the floating-point number system. In contemporary FPGA applications, rational fractions excel in regard to their compact hardware size, high throughput, and the ability to adjust the precision through the selection of the data width. The proposed module for calculation of the sigmoid activation function is shown to have high throughput and to occupy a relatively modest hardware volume compared to modules relying on piecewise polynomial approximation with fixed-point data.

Keywords: rational fraction · sigmoid function · FPGA · VHDL · artificial neural network

1 Introduction

The increasing number of publications devoted to artificial neural networks (ANNs) in recent years indicates a growing interest in implementing these in hardware. The main reason for this trend is the rapid development of the element base used in digital ANN implementations. One approach to both accelerating the operations of an ANN and minimising its power consumption is to exploit its parallelism through the use of a field programmable gate array (FPGA). The speed of the artificial neurons depends heavily on the speed with which the sigmoid activation function is calculated in the nodes of the ANN. However, implementing this function in an FPGA requires significant hardware resources. The choice of approximation method for this function and its hardware implementation are crucial factors that affect the accuracy and speed of the ANN algorithm. Previous studies have explored various approximation methods for the digital implementation of nonlinear activation functions, including tabular approaches, Taylor transformation, and piecewise polynomial approximation. The most commonly used method for implementing sigmoid-type activation functions is

L. Franco et al. (Eds.): ICCS 2024, LNCS 14837, pp. 146–157, 2024.
https://doi.org/10.1007/978-3-031-63778-0_11

a piecewise linear or quadratic approximation with integer data. This provides lower approximation accuracy, which may lead to higher performance, while reducing the approximation error increases the hardware resource usage and decreases the data processing speed [13–16]. Other methods, such as rational approximation and continued fractions, provide the highest precision results; however, these involve floating point data calculations, which require large amounts of hardware resources in FPGA.

This paper proposes a new algorithm for calculating the sigmoid activation function based on a rational fraction data representation.

2 Artificial Neural Nets and Data Representations

The most common data representation in an ANN is based on the floating point, and this is explained by the features of the ANN algorithm. The ith node of a typical ANN applies a signal multiplier by the weight $\theta_{i,j}$ at its jth input. During learning by the generalised delta rule (backpropagation algorithm), the weights $\theta_{i,j}$ are updated, taking into account the difference between the available and predicted results (recognition errors), and based on the results, the gradients δ_i^l are calculated. Each gradient indicates in which direction and at what speed the weights $\theta_{i,j}$ should be changed so that they eventually reach the optimal value.

A node calculates its activation a_i as the sum of the inputs x_i multiplied by the weights $\theta_{i,j}$, which is processed by a nonlinear activation function p_i. This function must be differentiable so that network parameters can be tuned using the error backpropagation algorithm. The most common is the sigmoid activation function:

$$p(x) = \frac{1}{1 + e^{-x}} \tag{1}$$

The weights $\theta_{i,j}$ of the nodes in the different layers are adjusted according to the chain rule. At the same time, the gradients δ_i^l, which are used to adjust the weights $\theta_{i,j}$ in the internal layers of the ANN, are calculated taking into account the gradients δ_i^{l+K}, which are propagated from all subsequent layers. The learning process involves several iterations in which the parameters $\theta_{i,j}$, δ_i^l are continuously updated until the network is optimised (for example, when the weights $\theta_{i,j}$ stop changing).

When the network is deep (more than three layers), the learning process may suffer from vanishing or exploding gradients δ_i^l depending on the choice of the activation function p_i. As a result, the weights of the initial layers cannot be properly adjusted [1].

To ensure both convergence of the weights and an ANN with the optimum throughput-cost ratio, the execution of an ANN usually has two stages. In the first, the ANN is trained using precise floating point calculations, meaning that the effect of vanishing or exploding gradients is minimised. In the second step, an effective data representation is selected, and the weights $\theta_{i,j}$ are truncated and rounded. The trained ANN is then used for a particular application. In this inference step, the ANN is effectively implemented in FPGA as well. Through this process, the unneeded relations in the ANN are removed, which significantly simplifies the structure configured in the FPGA (network pruning) [2]. However, a problem remains in terms of selecting both the data representation and the sigmoid function approximation that can provide the minimum degradation in the pattern recognition effectiveness [3, 4].

A wide range of data formats can be used for an ANN implementation. To compress the huge sets of weights $\theta_{i,j}$ to provide a sufficient data range, 16-bit floating point formats are used, such as fp16 and BFloat16 [2]. Moreover, the different ANN layers need different levels of precision, and specific floating point formats of different bit widths can be used with a FPGA [5].

The results of previous studies show that the data dynamic range for an ANN is more important than the mantissa range. This fact is taken into account in the BFloat16 format proposed by Google, in which the mantissa has only 7 bits while the exponent has 8 bits [6].

A logarithmic number system is sometimes used that can also provide a wide dynamic range [7] and multiplier-free structures for the node. However, a logarithmic arithmetic unit may be more expensive than a hardware multiplier due to the significant complexity of the logarithmic function approximation.

Other investigations have shown that the weights $\theta_{i,j}$ and outputs of the activation functions in a trained convolutional ANN such as AlexNet are distributed over a relatively small range [8]. To exploit this fact, the Posit format is used in some ANNs. This format is distinguished by the variable lengths of both the mantissa and exponent, which is coded by the additional regime field [9–11].

Although both the floating point and Posit formats provide the necessary computational precision, they yield lower throughputs, higher hardware volumes, and higher power consumption than integer formats for ANNs configured in FPGA [12]. FPGA is an excellent basis for the implementation of application-specific processors that calculate integer data. The data bit width can be tuned easily in an FPGA to provide the optimum ratios for the throughput, precision to hardware volume, and power consumption [4]. When performing calculations in the respective data ranges, integers are considered as fixed point data.

The position of the point in the fixed point format depends on the scale factor of the particular data. In most convolutional ANNs, the point in the data and weights $\theta_{i,j}$ is positioned just after the sign bit [8]; however, the input data for a sigmoid-type activation function are usually limited by a value of ± 8. The point therefore divides the word into an integer and its fractional parts. In this situation, the data distribution in the set of all integers of a particular bit width is not effective. Figure 1 illustrates the data distribution for a set of 2^8 unsigned integers. It is clear that in the case of the input data entering the sigmoid function, most of the uniformly represented data are concentrated in the range [4:8]. This example illustrates that the integer range is used ineffectively here.

Integers are used in convolution algorithms very well, but it is hard to find an effective algorithm to calculate a nonlinear function such as the sigmoid function using integers. As a result, most of the activation functions in ANNs are calculated using a piecewise linear or second-order polynomial approximation [13–17, 28]. Most of these have data representations based on 16 bits and limited precision, with a maximum error in the range [0.0005:0.07].

It is therefore preferable to use a data format that occupies a place between floating point and integer formats, and which provides high precision and dynamic range, high speed, and a low hardware volume for an ANN implementation as well as an effective sigmoid function calculation. The next section presents such a format.

3 Rational Fractions and Calculation of the Sigmoid Activation Function

3.1 Rational Fractions

A rational fraction is a numerical object that consists of an integer numerator and integer denominator, and represents a rational number. The rational fraction a/b has the characteristic that it can approximate a given transcendental number x. If $a/b < x$ and $c/d > x$, then the fraction $(a + c)/(b + d)$, called a medianta, is nearer to x than a/b or c/d. Hence, if a set of mediantes is built, then we can approximate a number x with any precision [18].

If a noninteger number x is represented by $2n$ digits in fixed point format with error ε_1, then it can be represented by the fraction a/b with error $\varepsilon_2 = \varepsilon_1$, and the numbers a and b have no more than n digits in their representation [19].

A representation based on a fractional number has a set of advantages. Firstly, any binary fraction or fixed point datum is dependent on the binary data representation, and does not exactly represent a real number. In binary representation, a floating point number is equal to a fraction where the denominator is a power of two, and is not equal to the respective decimal fraction. For example, the number $1/9 = 1/1001_2$ is an exact fraction in any numerical system, and can be represented with an error as the decimal fraction 0.111110 or binary fraction 0.111000111000011_2.

Secondly, the number distribution of rational fraction data is more effective for the implementation of an ANN compared to integers. This fact is illustrated in Fig. 1, where the charts represent a number n of different data samples in the range $[2^i, 2^{i+1}]$, $i = -6,\ldots,7$ for 16-bit fixed point data and rational fractions with an 8-bit numerator and denominator. From the graph, it is clear that the data are concentrated around a value of 1.0, as the data are for real ANNs.

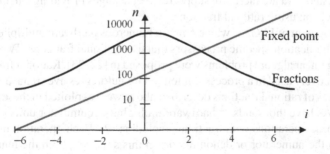

Fig. 1. Data distributions for sets of fixed point data and fractions.

Thirdly, rational fractions can help us to find an approximation for an irrational or transcendental number with a given level of precision. Many elementary functions are effectively calculated using suitable rational approximation formulas. When such a function is approximated using finite continued fractions or rational approximations, the rational fraction arithmetic fits very well. Many constants and constant tables are effectively stored as rational numbers.

Finally, rational fractions provide a comparatively simple set of arithmetic operations. The multiplication a/b to c/d and division of them are equal to $ac/(bd)$ and $bd/(ac)$, respectively. Note that the division of the numerator to the denominator is not calculated. The addition operation is equal to $(ad + bc)/(bd)$. For comparison of two numbers, it is sufficient to calculate $ad - bc$ [18]. Some specific calculations are simple, such as

$$1 + \frac{a}{b} = \frac{a+b}{b}; \text{ or } \frac{1}{1+\frac{a}{b}} = \frac{b}{a+b} \tag{2}$$

and these can help in calculating the continued fractions.

It should also be taken into account that the numerator and denominator bit width is less than half of the bit width of the fixed-point data that provides equal precision.

Hence, the hardware complexity of a fraction adder is similar to the complexity of an integer multiplier with the same precision, and the complexity of a fraction multiplier is two times less than that of an integer multiplier.

3.2 Rational Fractions in Processors

Some computers in the 1970s had already the rational fraction arithmetic except floating point one. The main disadvantage of rational fractions is that the number of bits increases dramatically when operations are implemented precisely [20]. To eliminate this disadvantage, the division of the numerator and denominator by their greatest common divisor was done, as in the rational fraction processor, which was proposed in [21]. However, when floating point coprocessors became widely used, fractional number processors dropped out of use.

Later, rational fractions were built into many mathematical CAD tools such as Maple, which are implemented in PC. Such fractions are widely used for calculations with unlimited precision, for solving modern cryptographic problems, and other tasks. Languages such as PERL and Java are therefore supported by packages providing calculations with unlimited precision using rational fractions.

The development of FPGAs, which contain numerous hardware multipliers, enabled the design of application-specific processors that used rational fractions. Two processors for computing linear algebra problems were proposed in [22, 23]. Rational fractions have also been effectively used in a processor intended for autoregressive signal analysis [24].

The features of rational fractions described above were exploited in these processors. Modern FPGAs have thousands of hardware multiplier-accumulator units (MPUs) that can perform base operations with fractions, including (2). Each operation may result in underflows in the numerator or denominator; in this situation, both the numerator and denominator are normalised by a left shift of an equal number of bits. This number of bits is equal to one after the addition operation and a maximum of $n/2$ after multiplication.

3.3 Calculation of the Sigmoid Function

In view of the features of the rational fraction operation, it is clear that the usual approximation methods such as piecewise approximation and power series are not effective, due to the large numbers of comparison and addition operations. In contrast, rational approximation and finite continued fractions are effectively calculated using rational fractions, as this approach requires much fewer elementary operations. We note that continued fractions frequently converge much more rapidly than power series expansions and in a much larger domain. A sigmoid activation function with a single precision floating point can be effectively calculated using only the rational Padè approximation [25].

The function in (1) can be calculated in two steps. In the first, the exponent function is calculated using the continued fraction approximation [26]

$$e^x = 1 + \cfrac{x}{1 - \cfrac{x}{2 + \cfrac{x}{3 - \ \cdots}}} \approx 1 + \frac{2x}{2 - x^2/6} \qquad (3)$$

while in the second, the formula in (1) is calculated.

To evaluate rational fraction calculations with different bit widths, a package of specific functions was designed in the VHDL language so that the functions marked as ' +', '−', '*', and '/' overload the respective operations but for the rational fraction data. Functions were also designed for calculating (2) as well as functions that transferred data from the floating point format to a rational fraction and back again.

The chart in Fig. 2 was created by evaluating the sigmoid function in (1) using the approximation in (3) on the basis of 16-bit rational fractions. This gives maximum errors of ±0.12 in the range [−6:6] and ± 0.002 in the range [−2:2]. It is clear from this example that a piecewise approximation would give good results. For instance, this function could be approximated by lines at the levels of ±1.0 when x is outside the range [−4:4]. However, the exponent function features provide a more effective means of improving the precision of the approximation.

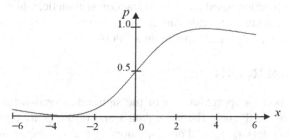

Fig. 2. Sigmoid function approximated using (3).

Consider the case where $y_0 = e^{x0}$ is the exact value of the function in (1) for an argument x_0. Then, for any argument x which is close to x_0, we have the approximation

$$e^x = y_0 e^z = y_0 \left(1 + \cfrac{z}{1 - \cfrac{z}{2 + \cfrac{z}{3 - \cdots}}}\right) \approx y_0(1 + z); \tag{4}$$

$$z = x - x_0$$

Here, x_0 is a rational fraction formed from the most significant bits of the input data x. The bits of x_0 serve as the address bits for the table, which stores the values y_0. Hence, the approximated exponent function is derived as the product of the table function $y_0(x_0)$ and the sum $1 + z$. A graph of the resulting sigmoid function when the numerator and denominator of x_0 have widths of 4 bits is shown in Fig. 3.

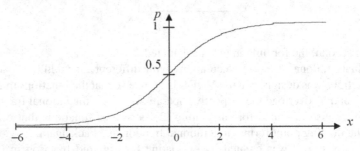

Fig. 3. Sigmoid function approximated using (4).

The resulting function gives a maximum error of –0.0088 in the range [−8:8]. This error can be reduced by increasing both the bit width of x_0 and the number of terms of the continued fraction in (4), and can be as small as necessary. When one additional term is used in (4), the maximum error is decreased to −0.0004.

In this section, we have proposed a new method for effective approximation of the sigmoid activation function based on rational fraction arithmetic, which gives moderate precision using a small number of calculations. In the next section, we present an example of a module used to calculate this function in a FPGA.

4 Experimental Results

The proposed method for approximation of the sigmoid activation function is implemented in FPGA as an IP core. The input data x and results p are represented by the 16-bit integer numerators x_n, y_n, and denominators x_d and y_d, respectively. A dataflow graph for the approximation algorithm is shown in Fig. 4. The input data are stored in the registers RXn and RXd, and are loaded and calculated in the pipeline mode. Their four most significant bits (including signs) form the value x_0, which serves as the address in the tables ROMEn and ROMEd, which store the values of the exponent coefficients $y_0 = y_{0n}/y_{0d}$.

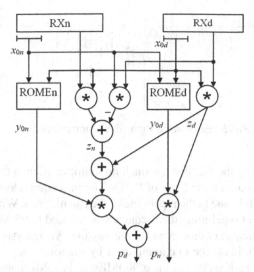

Fig. 4. Dataflow graph for calculation of the sigmoid function.

Five integer multipliers and three integer adders are used to perform the calculations shown in (1) and (4). The intermediate result z and result p are normalised by left-shifting the numerator and denominator by up to 2 bits. The module used for calculating this algorithm is described in VHDL language by the style for synthesis as the pipelined datapath. This datapath contains five pipeline register stages, and outputs the results in each clock cycle.

The parameters for the proposed module, as configured for different FPGA series, are shown in Table 1. The number of DSP blocks containing the multiplier is only two when the module is configured in a Xilinx Kintex7 FPGA.

Table 1. Parameters of the module for calculating the sigmoid function

Module	FPGA series	Hardware cost		Maximum clock frequency, MHz	Maximum error
		LUTs/ALMs	DSP blocks		
Proposed	Kintex7	381	2	238	0.0094
Proposed	Artix7	389	2	154	0.0094
Proposed	Cyclone V	151	4	117	0.0094
Tsmots [13]	Cyclone III	368	0	37	0.018
Campo [16]	Virtex6	232	6	53*	0.028
Gomar [27]	Virtex4	123	0	29*	0.087
Zhang [15]	Zed7	272	2	107	–
Li [14]	Virtex7	493	0	208	0.0078

* Data input frequency, as the algorithm is implemented in several sequential steps.

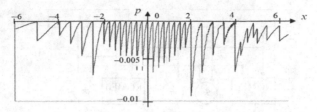

Fig. 5. Error in the sigmoid function calculation.

This is explained by the fact that the small bit multiplication in the data x0n and x0d is performed in hardware on the basis of LUTs. The maximum clock frequency in this FPGA reaches 238 MHz due to the pipelining of the calculations. When the calculations are carried out without pipelining, this frequency is reduced to 95 MHz.

Figure 5 shows the calculation error for the module. An analysis shows that there is the opportunity to minimise the maximum error by rounding the result. Moreover, the error depends on the address bit widths of ROMEn and ROMEd, and can be decreased dramatically when these bit widths are increased. Due to the hardware consumption in Table 1, these ROMs are performed in LUTs and there is the potential to increase their volume.

The different modules that have been used to calculate the sigmoid activation function using a piecewise polynomial approximation are presented in Table 1 for comparison. We note that they all use 16-bit fixed point arithmetic, and the maximum error is derived for the input range $[-8: 8]$. A comparison of these different modules shows that the proposed module has the highest speed, with a moderate value for the hardware volume and a comparatively small computational error.

Our module has the advantages of increasing the computational precision and minimising the hardware volume at the cost of bit width minimisation. The other modules of the ANN system can also use rational fraction arithmetic. However, in other situations, the proposed module has to be agreed upon both with the input and output floating point or integer data by attaching not complex wrapping hardware. The hardware attached to the output must have a division unit that calculates $p = p_n/p_d$.

5 Conclusion

A rational fraction number system has the advantage of providing higher precision than integers, and its FPGA implementation is simpler than that of a floating number system. The main advantages of using rational fractions in a modern FPGA implementation are small hardware volume, high throughput, and the possibility of regulating the precision by selecting the data width. It has been shown here that this data representation helps in designing effective modules for implementation of the sigmoid activation function. Our module for calculating the sigmoid activation function is shown to have high throughput and low hardware volume in comparison with modules based on a piecewise polynomial approximation using fixed point data. Future work on the use of rational fractions will focus on the implementation of an ANN system as a whole.

Acknowledgments. Funding: This research was funded by the Faculty of Electrical and Computer Engineering, Cracow University of Technology, and the Ministry of Science and Higher Education, Republic of Poland (grant no. E-1/2024).

References

1. Russell, S., Norvig, P.: Artificial Intelligence: A Modern Approach, 4th edn. Pearson, Boston (2022)
2. Young Kim, J-Y.: Chapter Five - FPGA based neural network accelerators. In: Kim, S., Deka, G.C. (eds.) Advances in Computers, vol. 122, pp. 135–165. Elsevier (2021). https://doi.org/10.1016/bs.adcom.2020.11.002
3. Bailey B., Machine Learning's Growing Divide. Semiconductor Engineering (2018). https://semiengineering.com/machine-learnings-growing-divide. Accessed 19 Apr 2024
4. Mahajan, R., Sakhare, D. Gadgil R.: Review of artificial intelligence applications and architectures. In: Thakare, A.D., Bhandari, S.U. (eds.) Artificial Intelligence Applications and Reconfigurable Architectures, pp. 25–34. Wiley Online Library (2023). https://doi.org/10.1002/9781119857891.ch2
5. Floating-Point Operator v7.1 PG060, Xilinx (2020). https://docs.amd.com/v/u/en-US/pg060-floating-point. Accessed 19 Apr 2024
6. Lai, L., Suda, N., Chandra, V.: Deep convolutional neural network inference with floating-point weights and fixed-point activations. In: Computer Science: Machine Learning, pp. 1–10 (2017). https://doi.org/10.48550/arXiv.1703.03073
7. Miyashita, D., Lee, E.H., Murmann, B.: Convolutional neural networks using logarithmic data representation. In: Computer Science: Neural and Evolutionary Computing (2016). https://doi.org/10.48550/arXiv.1603.01025
8. Zhang, H., Deivalakshmi Subbian, G., Lakshminarayanan, S.-B.K.: Application-specific and reconfigurable AI accelerator. In: Mishra, A., Cha, J., Park, H., Kim, S. (eds.) Artificial Intelligence and Hardware Accelerators, pp. 183–223. Springer, Cham (2023). https://doi.org/10.1007/978-3-031-22170-5_7
9. Johnson, J.: Rethinking floating point for deep learning. In: Computer Science: Numerical Analysis (2018). https://doi.org/10.48550/arXiv.1811.01721
10. Carmichael, Z., Langroudi, H.F., Khazanov, C., Lillie, J., Gustafson, J.L., Kudithipudi, D.: Deep positron: a deep neural network using the posit number system. In: Design, Automation & Test in Europe Conference & Exhibition (DATE), Florence, Italy, pp. 1421–1426. IEEE (2019). https://doi.org/10.23919/DATE.2019.8715262

11. Raposo, G., Tomás, P., Roma, N.: PositNN: training deep neural networks with mixed low-precision posit. In: ICASSP 2021 - 2021 IEEE International Conference on Acoustics, Speech and Signal Processing (ICASSP), Toronto, ON, Canada, pp. 7908–7912 (2021). https://doi.org/10.1109/ICASSP39728.2021.9413919
12. Nechi, A., Groth, L., Mulhem, S., Merchant, F., Buchty, R., Berekovic, M.: FPGA-based deep learning inference accelerators: where are we standing? ACM Trans. Reconfigurable Technol. Syst. **16**(4), 1–32 (2023). https://doi.org/10.1145/3613963
13. Tsmots, I., Skorokhoda, O., Rabyk V.: Hardware implementation of sigmoid activation functions using FPGA. In: IEEE 15th International Conference on the Experience of Designing and Application of CAD Systems (CADSM), Polyana, Ukraine, pp. 34–38 (2019). https://doi.org/10.1109/CADSM.2019.8779253
14. Li, Z., Zhang, Y., Sui, B., Xing, Z., Wang, Q.: FPGA implementation for the sigmoid with piecewise linear fitting method based on curvature analysis. Electronics **11**(9), 1365 (2022). https://doi.org/10.3390/electronics11091365
15. Zhang, L.: Implementation of fixed-point neuron models with threshold, ramp and sigmoid activation functions. In: 4th International Conference on Mechanics and Mechatronics Research, vol. 224. IOP Publishing (2017). https://doi.org/10.1088/1757-899X/224/1/012054
16. Campo, I., Finker, R., Echanobe, J., Basterretxea, K.: Controlled accuracy approximation of sigmoid function for efficient FPGA-based implementation of artificial neurons. Electron. Lett. **49**(25), 1598–1600 (2013). https://doi.org/10.1049/el.2013.3098
17. Laudani, A., Lozito, G.M., Fulginei, F.R., Salvini, A.: On training efficiency and computational costs of a feed forward neural network: a review. Comput. Intell. Neurosci. **2015**, 1–13 (2015). https://doi.org/10.1155/2015/818243
18. Kornerup, P., Matula, D.W.: Finite Precision Number Systems and Arithmetic. Cambridge University Press, Cambridge (2010). https://doi.org/10.1017/CBO9780511778568
19. Hintchin, A.Y.: Continued Fractions, 3rd edn. Nauka, Moscow (1978). (in Russian)
20. Horn, B. K. P. Rational arithmetic for minicomputers, Vol. 8, No. 2, pp. 171–176, Software Practice and Experience (1978)
21. Irvin M. J., Smith D. R.: A rational arithmetic processor. In: Proceedings of 5-th Symposium Computer Arithmetic (1981)
22. Maslennikow, O., Lepekha, V., Sergyienko, A.: FPGA implementation of the conjugate gradient method. In: Wyrzykowski, R., Dongarra, J., Meyer, N., Waśniewski, J. (eds.) Parallel Processing and Applied Mathematics, pp. 526–533. Springer, Heidelberg (2006). https://doi.org/10.1007/11752578_63
23. Maslennikow, O., Lepekha, V., Sergiyenko, A., Tomas, A., Wyrzykowski, R.: Parallel implementation of cholesky LL T -algorithm in FPGA-based processor. In: Wyrzykowski, R., Dongarra, J., Karczewski, K., Wasniewski, J. (eds.) Parallel Processing and Applied Mathematics, pp. 137–147. Springer, Heidelberg (2008). https://doi.org/10.1007/978-3-540-68111-3_15
24. Sergiyenko, A., Maslennikow, O., Ratuszniak, P., Maslennikowa, N., Tomas, A.: Application specific processors for the autoregressive signal analysis. In: Wyrzykowski, R., Dongarra, J., Karczewski, K., Wasniewski, J. (eds.) Parallel Processing and Applied Mathematics, pp. 80–86. Springer, Heidelberg (2010). https://doi.org/10.1007/978-3-642-14390-8_9
25. Hajduk, Z.: High accuracy FPGA activation function implementation for neural networks. Neurocomputing **247**, 59–61 (2017). https://doi.org/10.1016/j.neucom.2017.03.044
26. Roy, R., Olver, F.W.J.: Elementary functions. In: NIST Handbook of Mathematical Functions. Cambridge University Press, Cambridge (2010)

27. Gomar, S., Mirhassani, M., Ahmadi M.: Precise digital implementations of hyperbolic tanh and sigmoid function. In: 50th Asilomar Conference on Signals, Systems and Computers, Pacific Grove, USA, pp. 1586–1589 (2016). https://doi.org/10.1109/ACSSC.2016.7869646
28. Moroz, L., Samotyy, V., Gepner, P., Węgrzyn, M., Nowakowski, G.: Power function algorithms implemented in microcontrollers and FPGAs. Electronics **12**(16), 3399 (2023). https://doi.org/10.3390/electronics12163399

Parallel Vectorized Algorithms for Computing Trigonometric Sums Using AVX-512 Extensions

Przemysław Stpiczyński$^{(\boxtimes)}$ (iD)

Institute of Computer Science, Maria Curie–Skłodowska University, Akademicka 9/519, 20-033 Lublin, Poland
przemyslaw.stpiczynski@umcs.pl

Abstract. The aim of this paper is to show that Goertzel and Reinsch algorithms for computing trigonometric sums can be efficiently vectorized using Intel AVX-512 intrinsics in order to utilize SIMD extensions of modern processors. Numerical experiments show that the new vectorized implementations of the algorithms using only one core achieve very good speedup over their sequential versions. The new algorithms have been parallelized using OpenMP in order to utilize multiple cores. For sufficiently large problem sizes, the parallel implementations of the algorithms achieve reasonable speedup against the vectorized ones.

Keywords: Trigonometric sums · Goertzel and Reinsch algorithms · Vectorization · AVX-512 · Intrinsics · OpenMP

1 Introduction

For given real numbers $b_0, \ldots, b_n$, and x, let us consider the problem of computing trigonometric sums of the following forms

$$C(x) = \sum_{k=0}^{n} b_k \cos kx \quad \text{and} \quad S(x) = \sum_{k=1}^{n} b_k \sin kx, \qquad (1)$$

which appear in many numerical applications. For example, finding (1) is the central part of Talbot's algorithm [13,18,31] for computing the numerical inverse of the Laplace Transform. The sums are also used to compute individual terms of the Discrete Fourier Transform [32]. It is not recommended to use (1) directly due to large number of arithmetic operations and poor numerical properties [23]. Instead, one can calculate the sums $C(x)$ and $S(x)$ using the Goertzel algorithm [8,23], which is a special case of Clenshaw's algorithm [4] used for the summation of orthogonal polynomial series [2,3]. Actually, the main computational parts of both Goertzel's and Clenshaw's algorithms are the same. The Goertzel algorithm has also some other technical applications [29], especially in signal processing [10,12,15,20,30], thus its efficient implementations is highly desired [6,19]. Unfortunately, it can be numerically unstable for $|x| \ll 1$ [7]. Then it is

L. Franco et al. (Eds.): ICCS 2024, LNCS 14837, pp. 158–172, 2024.
https://doi.org/10.1007/978-3-031-63778-0_12

better to use the algorithm introduced by Reinsch [23], which is a little bit more complicated but has better numerical properties.

Although implementations of the Goertzel (Clenshaw) and Reinsch algorithms seem to be simple, their direct parallelization is not possible due to their recursive form. Papers [2,3] showed how to use Clenshaw's algorithm in order to develop a new method for the evaluation of the Chebyshev polynomials of the first kind and how to parallelize it using special properties of the polynomials. Numerical experiments performed on Cray T3D showed that the introduced approach allowed to achieve limited speedup and the efficiency up to 38%. Another approach [25,26] assumes that the Goertzel and Reinsch algorithms reduce to the problem of solving special narrow-banded systems of linear equations and introduces new *divide and conquer* parallel algorithms for solving such systems. The implementations of these algorithms achieve rather limited speedup on Intel processors: Pentium III, Pentium 4 and Itanium 2. Moreover, the new version of the Reinsch algorithm applied for finding the numerical inverse of the Laplace transform scales very well on Cray X1, but it cannot utilize vector extensions of its processors. Therefore, the approaches to parallelizing these algorithms described in the literature enable the use of parallel processors but do not allow to utilize vector extensions of modern multicore processors, which is crucial for achieving high performance [1,5,24,27,28,33].

In this paper we show how to modify and implement the *divide and conquer* parallel Goertzel and Reinsch algorithms [25,26] in order to develop almost fully vectorized versions of both algorithms. The new implementations utilize advantages of AVX-512 vector extensions [9] and can also be parallelized. The rest of the paper is organized as follows. Section 2 presents the ordinary Goertzel and Reinsch algorithms for computing (1). In Sect. 3 we briefly recall the *divide and conquer* versions of the algorithms introduced in [25]. Section 4 discusses the details of vectorized and parallel implementations of the algorithms based on AVX-512 intrinsics [9] and OpenMP [14] constructs. Section 5 shows the results of experiments performed on a machine with modern Intel multicore processors that confirm the efficiency of our new implementations. Finally we present some concluding remarks and plans for future studies.

2   Goertzel and Reinsch Algorithms

First let us observe that we can restrict our attention to the case where $x \neq k\pi$. If $x = k\pi$ then $S(x) = 0$ for all x and $\cos kx = \pm 1$, thus $C(x)$ can be computed using a simple summation algorithm. In case of the Goertzel algorithm [8,23] for finding (1), we need to compute two last entries (namely S_1 and S_2) of the solution of the following linear recurrence system with constant coefficients:

$$S_k = \begin{cases} 0, & k = n+1, n+2 \\ b_k + 2S_{k+1}\cos x - S_{k+2}, & k = n, \ldots, 1 \end{cases} \quad (2)$$

and then we have

$$C(x) = b_0 + S_1 \cos x - S_2$$
$$S(x) = S_1 \sin x. \quad (3)$$

To avoid the influence of rounding errors on the final computed solution, when x is closed to 0 [7], one can use the Reinsch algorithm [23]. In this case we set $S_{n+2} = D_{n+1} = 0$ and if $\cos x > 0$, then we solve the following linear recurrence system

$$\begin{cases} S_{k+1} = D_{k+1} + S_{k+2} \\ D_k = b_k + \beta S_{k+1} + D_{k+1} \end{cases} \tag{4}$$

for $k = n, n-1, \ldots, 0$, where $\beta = -4\sin^2 \frac{x}{2}$. If $\cos x \leq 0$, we solve

$$\begin{cases} S_{k+1} = D_{k+1} - S_{k+2} \\ D_k = b_k + \beta S_{k+1} - D_{k+1} \end{cases} \tag{5}$$

where $\beta = 4\cos^2 \frac{x}{2}$. Finally, we compute

$$S(x) = S_1 \sin x \text{ and } C(x) = D_0 - \frac{\beta}{2} S_1. \tag{6}$$

3   Divide-and-conquer Approach

Both algorithms presented in the previous section are very simple but it is clear that loops corresponding to (2) and (4–5) have obvious data dependencies, thus compilers cannot vectorize or parallelize them in order to utilize the underlying hardware of modern multicore processors. Now let us consider the *divide and conquer* approach [25] that forms the basis of new efficient implementations of the algorithms that will be discussed in Sect. 4. In this paper we will use a slightly different notation than that used in [25], but all given properties are equivalent to it.

3.1   Goertzel Algorithm

Let us consider the following approach for parallelizing the Goertzel algorithm. Equation (2) is equivalent to the following system of linear equations

$$\begin{bmatrix} 1 & -c & 1 & & & \\ & 1 & -c & 1 & & \\ & & \ddots & \ddots & 1 \\ & & & 1 & -c \\ & & & & 1 \end{bmatrix} \begin{bmatrix} x_1 \\ x_2 \\ \vdots \\ x_{n-1} \\ x_n \end{bmatrix} = \begin{bmatrix} b_1 \\ b_2 \\ \vdots \\ b_{n-1} \\ b_n \end{bmatrix}, \tag{7}$$

where $c = 2\cos x$ and $x_i = S_i$, $i = 1, \ldots, n$. For the sake of simplicity let us assume that $n = m \cdot r$, where integers $m, r \geq 2$. Then (7) can be rewritten as the following block system

$$\begin{bmatrix} U & L & & \\ & U & L & \\ & & \ddots & L \\ & & & U \end{bmatrix} \begin{bmatrix} \mathbf{x}_1 \\ \mathbf{x}_2 \\ \vdots \\ \mathbf{x}_r \end{bmatrix} = \begin{bmatrix} \mathbf{b}_1 \\ \mathbf{b}_2 \\ \vdots \\ \mathbf{b}_r \end{bmatrix}, \tag{8}$$

where $\mathbf{x}_j = [x_{(j-1)m+1}, \ldots, x_{jm}]^T \in \mathbb{R}^m$, $\mathbf{b}_j = [b_{(j-1)m+1}, \ldots, b_{jm}]^T \in \mathbb{R}^m$, $U \in \mathbb{R}^{m \times m}$ is of the same form as the matrix of the system (7), and

$$L = \begin{bmatrix} 0 \cdots \cdots 0 \\ \vdots \ddots \quad \vdots \\ 1 \quad 0 \ddots \vdots \\ -c \quad 1 \cdots 0 \end{bmatrix} \in \mathbb{R}^{m \times m}. \tag{9}$$

It is obvious that L can be rewritten as $L = \mathbf{e}_{m-1}\mathbf{e}_1^T - c\mathbf{e}_m\mathbf{e}_1^T + \mathbf{e}_m\mathbf{e}_2^T$, where $\mathbf{e}_k$ is k-th unit vector from $\mathbb{R}^m$. The solution to (7) can be expressed as

$$\begin{cases} \mathbf{x}_r = U^{-1}\mathbf{b}_r \\ \mathbf{x}_j = U^{-1}\mathbf{b}_j - U^{-1}L\mathbf{x}_{j+1}, \quad j = r-1, \ldots, 1, \end{cases} \tag{10}$$

Assuming that $U\mathbf{z}_j = \mathbf{b}_j$, $j = 1, \ldots, r$, and $\mathbf{y}_{m-1} = U^{-1}\mathbf{e}_{m-1}$, $\mathbf{y}_m = U^{-1}\mathbf{e}_m$, we get

$$\mathbf{x}_j = \mathbf{z}_j - x_{jm+1}\mathbf{y}_{m-1} + (cx_{jm+1} - x_{jm+2})\mathbf{y}_m. \tag{11}$$

It can be proved [25] that for $x \neq k\pi$, $k \in \mathbb{Z}$, entries of $\mathbf{y}_m = [y_1, y_2, \ldots, y_{m-1}, 1]^T$ satisfy

$$y_j = \frac{\sin(m+1-j)x}{\sin x}, \quad j = 1, \ldots, m-1. \tag{12}$$

Moreover, $\mathbf{y}_{m-1} = [y_2, y_3, \ldots, y_{m-1}, 1, 0]^T$, where all y_j also satisfy (12). Let $\mathbf{x}_j''$, $\mathbf{z}_j''$, $j = 1, \ldots, r$, denote first two entries of $\mathbf{x}_j$, $\mathbf{z}_j$, respectively, and

$$M = \begin{bmatrix} -y_1 & cy_1 - y_2 \\ -y_2 & cy_2 - y_3 \end{bmatrix} \in \mathbb{R}^{2 \times 2}. \tag{13}$$

Then we get

$$\begin{cases} \mathbf{x}_r'' = \mathbf{z}_r'' \\ \mathbf{x}_j'' = \mathbf{z}_j'' + M\mathbf{x}_{j+1}'', \quad j = r-1, \ldots, 1, \end{cases} \text{ and } \begin{bmatrix} S_1 \\ S_2 \end{bmatrix} = \mathbf{x}_1''. \tag{14}$$

The presented approach makes it possible to formulate a parallel algorithm. First we should find all vectors $\mathbf{z}_j$ in parallel, solving $U\mathbf{z}_j = \mathbf{f}_j$, and then we use (14) to find S_1, S_2.

3.2   Reinsch Algorithm

Now let us observe that Eqs. (4–5) are equivalent to the following system of linear equations

$$\begin{bmatrix} 1 & -\beta & \delta & & & & \\ & 1 & -1 & \delta & & & \\ & & 1 & -\beta & \delta & & \\ & & & \ddots & \ddots & \ddots & \\ & & & & 1 & -1 & \delta \\ & & & & & 1 & -\beta \\ & & & & & & 1 \end{bmatrix} \begin{bmatrix} x_1 \\ x_2 \\ \vdots \\ \vdots \\ \vdots \\ \vdots \\ x_{2n} \end{bmatrix} = \begin{bmatrix} f_1 \\ f_2 \\ \vdots \\ \vdots \\ \vdots \\ \vdots \\ f_{2n} \end{bmatrix}, \tag{15}$$

where
$$\delta = \begin{cases} -1, & \cos x > 0 \\ 1, & \cos x \le 0 \end{cases}, \quad x_k = \begin{cases} D_{\lfloor k/2 \rfloor}, & k = 1, 3, \dots, 2n-1 \\ S_{k/2}, & k = 2, 4, \dots, 2n \end{cases} \tag{16}$$

and

$$f_k = \begin{cases} b_{\lfloor k/2 \rfloor}, & k = 1, 3, \dots, 2n-3 \\ 0, & k = 2, 4, \dots, 2n-2 \\ b_{n-1} - \delta b_n, & k = 2n-1 \\ b_n, & k = 2n. \end{cases} \tag{17}$$

Similarly to the Goertzel algorithm, under the assumption that $n = m \cdot r$, the system (15) can be written in the following block form

$$\begin{bmatrix} U & L & & \\ & U & L & \\ & & \ddots & L \\ & & & U \end{bmatrix} \begin{bmatrix} \mathbf{x}_1 \\ \mathbf{x}_2 \\ \vdots \\ \mathbf{x}_r \end{bmatrix} = \begin{bmatrix} \mathbf{f}_1 \\ \mathbf{f}_2 \\ \vdots \\ \mathbf{f}_r \end{bmatrix}, \tag{18}$$

where all $\mathbf{x}_j, \mathbf{f}_j \in \mathbb{R}^{2m}$, and $U \in \mathbb{R}^{2m \times 2m}$ is of the same form as the matrix of the system (15), and

$$L = \begin{bmatrix} 0 & \cdots\cdots & 0 \\ \vdots & \ddots & \vdots \\ \delta & 0 & \ddots & \vdots \\ -1 & \delta & \cdots & 0 \end{bmatrix} \in \mathbb{R}^{2m \times 2m}. \tag{19}$$

This yields

$$\begin{cases} U\mathbf{x}_r = \mathbf{f}_r \\ U\mathbf{x}_j = \mathbf{f}_j + L\mathbf{x}_{j+1}, \quad j = r-1, \dots, 1, \end{cases} \tag{20}$$

so

$$\mathbf{x}_j = U^{-1}\mathbf{f}_j - U^{-1}L\mathbf{x}_{j+1} = \mathbf{z}_j - x_{jm+1}\mathbf{y}_1 - x_{jm+2}\mathbf{y}_2, \tag{21}$$

where $\mathbf{y}_1, \mathbf{y}_2$ satisfy $U\mathbf{y}_1 = [0, \dots, 0, \delta, -1]^T$ and $U\mathbf{y}_2 = [0, \dots, 0, \delta]^T$, respectively. Let us define

$$M = \begin{bmatrix} y_1^{(1)} & y_1^{(2)} \\ y_2^{(1)} & y_2^{(2)} \end{bmatrix} \in \mathbb{R}^{2 \times 2}, \tag{22}$$

where $y_1^{(1)}, y_2^{(1)}$, and $y_1^{(2)}, y_2^{(2)}$ denote first two entries of $\mathbf{y}_1, \mathbf{y}_2$, respectively. It can be proved [25] that the entries of M satisfy

$$\begin{aligned} y_1^{(1)} &= -\cos mx + (\beta/2)y_2^{(1)} \\ y_2^{(1)} &= -\sin mx / \sin x \\ y_1^{(2)} &= \delta \cos mx + \cos(m-1)x + (\beta/2)y_2^{(2)} \\ y_2^{(2)} &= (\delta \sin mx + \sin(m-1)x) / \sin x \end{aligned} \tag{23}$$

Finally, we get

$$\begin{cases} \mathbf{x}_r'' = \mathbf{z}_r'' \\ \mathbf{x}_j'' = \mathbf{z}_j'' - M\mathbf{x}_{j+1}'', \quad j = r-1,\ldots,1 \end{cases} \quad \text{and} \quad \begin{bmatrix} D_0 \\ S_1 \end{bmatrix} = \mathbf{x}_1''. \quad (24)$$

Similarly to the Goertzel algorithm, the parallel Reinsch algorithm consists of two parts: the parallel one, where we find all $\mathbf{z}_j$, and the sequential based on (24). It should be noted that increasing the value of r, i.e. potentially more processors operating in parallel, increases the number of operations in the sequential part. In Sect. 4 we will also show how to omit the assumption that $n = m \cdot r$.

It should be noted that good numerical properties of the parallel version of the Reinsch algorithm were confirmed empirically. The algorithm was used as the main part of the parallel version of Talbot's method for computing the numerical inverse of the Laplace transform [26]. No decrease in accuracy was observed compared to the sequential version of the algorithm.

4   Implementation of the Algorithms

Although the algorithms presented in Sect. 3 have potential parallelism, they do not explicitly utilize vector extensions of modern processors, like AVX-512 [9], what is crucial to achieve high performance [1,5,24,27,28,33]. Simply, the parallel (i.e. *divide*) parts of both algorithms are still based on recurrence computations.

In order to utilize advantages of AVX-512 vector extensions [9] and develop vectorizable implementations of the considered algorithms we will use intrinsics for SIMD instructions which allow to write constructs that look like C/C++ function calls corresponding to actual AVX-512 instructions. Such calls are automatically replaced with assembly code inlined directly into programs. Moreover, the use of intrinsics is a good choice to make sure that compilers do exactly what we want, especially in the case of complex nested loops that may prevent automatic vectorization.

Now let us consider the way how we can use these algorithms to obtain their efficient implementations that would operate on 512-bit vector registers as basic data structures. The __m512d datatype defined in AVX-512 can be used to store eight double precision floating point numbers. When we set $r = 8$ in (8), then the first *divide* part of the Goertzel algorithm is equivalent to the problem of solving the block system of linear equations $UZ = B$, where

$$B = \begin{bmatrix} b_1 & b_{m+1} & \cdots & b_{7m+1} \\ b_2 & b_{m+2} & \cdots & b_{7m+2} \\ \vdots & \vdots & \cdots & \vdots \\ b_m & b_{2m} & \cdots & b_{8m} \end{bmatrix} \in \mathbb{R}^{m \times 8}. \quad (25)$$

Then, the solution Z can be found using the following sequence of vector operations

$$Z_{k,*} \leftarrow B_{k,*} + cZ_{k+1,*} - Z_{k+2,*} \quad \text{for } k = m,\ldots,1, \quad (26)$$

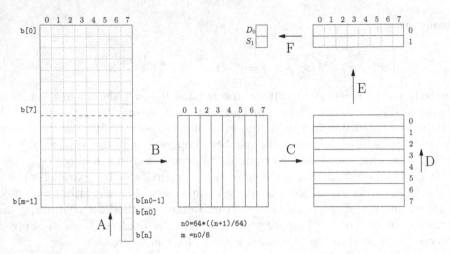

Fig. 1. Stages of the vectorized Reinsch algorithm using AVX-512

where $Z_{m+1,*}$ and $Z_{m+2,*}$ are zero vectors. Moreover, in order to use the *conquer* part of the algorithm (14), we only need to find the vectors $Z_{1,*}$ and $Z_{2,*}$, thus all necessary calculations based on (26) can be carried out using 512-bit vector registers filled with data loaded from memory.

In case of the Reinsch algorithm, we work similarly. We have to find $Z_{1,*}$ and $Z_{2,*}$, i.e. first two rows of the matrix Z defined by $UZ = F$, where

$$
F = \begin{bmatrix}
b_0 & b_m & \cdots & b_{7m} \\
0 & 0 & \cdots & 0 \\
b_1 & b_{m+1} & \cdots & b_{7m+1} \\
\vdots & \vdots & \cdots & \vdots \\
b_{m-1} & b_{2m-1} & \cdots & b_{n-1} - \delta b_n \\
0 & 0 & \cdots & b_n
\end{bmatrix} \in \mathbb{R}^{2m \times 8}, \tag{27}
$$

using the following sequence of vector operations

$$
\begin{cases}
Z_{k,*} \leftarrow Z_{k+1,*} - \delta Z_{k+2,*}, & k = 2m, 2m-2, \ldots, 2 \\
Z_{k,*} \leftarrow B_{k,*} + \beta Z_{k+1,*} - \delta Z_{k+2,*}, & k = 2m-1, 2m-3, \ldots, 1.
\end{cases} \tag{28}
$$

Note that the assumption $n = m \cdot r$ can be omitted. Then, for both algorithms, it is necessary to perform a number of sequential steps defined by (2) or (4–5) to initialize last entries of the vectors $Z_{m+1,*}$ and $Z_{m+2,*}$ (Goertzel), or $Z_{2m+1,*}$ and $Z_{2m+2,*}$ (Reinsch), respectively.

Stages of the Reinsch algorithm are shown in Fig. 1. The sequence of coefficients $b_0, \ldots, b_n$ stored in the array b of double precision numbers can be divided into two parts. The first one can be treated as a collection of $m/64$ square 8×8 blocks. Thus it contains the coefficients $b_0, \ldots, b_{n_0-1}$, $n_0 = 64 \cdot \lfloor (n+1)/64 \rfloor$, that should be processed using AVX-512 vector extensions. The second part, i.e. the numbers $b_{n_0}, \ldots, b_n$, should be processed first (Stage A) using (4–5). Then each column of the last block (i.e. eight consecutive numbers) is loaded into a vector register using the intrinsic _mm512_load_pd() (Stage B). Then (Stage C), such a block stored in eight registers is transposed so that each register contains one row of the block. During Stage D, we use the intrinsic _mm512_fmadd_pd() which performs *a fused multiply–add*, and depending on the sign of δ, the intrinsic _mm512_add_pd(), or _mm512_sub_pd() to process all rows of the block according to (28). Stages B–D are repeated for all blocks (Stage E). Finally (Stage F), we use (24) on the first two rows of the first block in order to get D_0 and S_1.

The vectorized Goertzel algorithm can be implemented similarly but with one difference. The intrinsic _mm512_load_pd() loads a 512-bits vector composed of eight packed double-precision floating-point numbers from memory but on condition that the memory address is aligned on a 64-byte boundary or a general-protection exception may be generated. In case of the Goertzel algorithm, the coefficient b_0 is only used in (3). Thus, the vectorized part will operate on blocks containing the coefficients $b_{64}, \ldots, b_{n_0-1}$, and the coefficients $b_1, \ldots, b_{63}$ should be processed sequentially using (2).

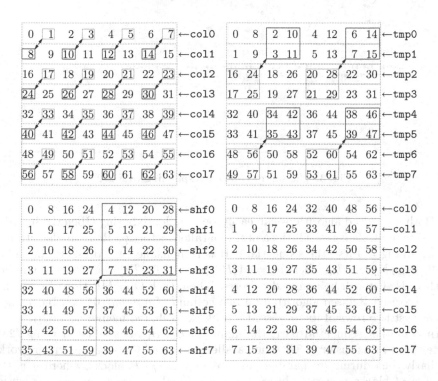

Fig. 2. Three steps and the final result of the transposition of 8×8 block stored in eight _m512d registers: red arrows indicate blue and green blocks that should be swapped

It should be noticed that Stage C, i.e. transpositions of 8×8 blocks stored columnwise in vector registers, can be performed in three steps using only AVX-512 intrinsics. First, we have to swap single entries between two consecutive vectors (Fig. 2, top-left). Then, 2×2 blocks are swapped between pairs of vectors (Fig. 2, top-right). Finally, we swap 4×4 blocks stored in first and second halves of quads of vectors (Fig. 2, bottom-left) in order to get the result (Fig. 2, bottom-right). The details can be found in Fig. 3.

```
_inline void vec8x8_transpose(__m512d *col0,__m512d *col1,__m512d *col2,__m512d *col3,
                              __m512d *col4,__m512d *col5,__m512d *col6,__m512d *col7)

{  __m512d tmp0, tmp1, tmp2, tmp3, tmp4, tmp5, tmp6, tmp7;
   __m512d shf0, shf1, shf2, shf3, shf4, shf5, shf6, shf7;

// Step 1: Unpack and interleave double-precision floating-point elements
//         from the low and high halves of each 128-bit lane of pairs of columns
   tmp0 = _mm512_unpacklo_pd(*col0,*col1);
   tmp1 = _mm512_unpackhi_pd(*col0,*col1);
   tmp2 = _mm512_unpacklo_pd(*col2,*col3);
   tmp3 = _mm512_unpackhi_pd(*col2,*col3);
   tmp4 = _mm512_unpacklo_pd(*col4,*col5);
   tmp5 = _mm512_unpackhi_pd(*col4,*col5);
   tmp6 = _mm512_unpacklo_pd(*col6,*col7);
   tmp7 = _mm512_unpackhi_pd(*col6,*col7);
// Step 2: Shuffle 128-bits (composed of 2 double-precision floating-point elements) selected from pairs,
//         store the results (elements are copied when the corresponding mask bit is not set)
   shf0 = _mm512_mask_shuffle_f64x2(tmp0,0b11001100,tmp2,tmp2,0b10100000);
   shf1 = _mm512_mask_shuffle_f64x2(tmp1,0b11001100,tmp3,tmp3,0b10100000);
   shf2 = _mm512_mask_shuffle_f64x2(tmp2,0b00110011,tmp0,tmp0,0b11110101);
   shf3 = _mm512_mask_shuffle_f64x2(tmp3,0b00110011,tmp1,tmp1,0b11110101);
   shf4 = _mm512_mask_shuffle_f64x2(tmp4,0b11001100,tmp6,tmp6,0b10100000);
   shf5 = _mm512_mask_shuffle_f64x2(tmp5,0b11001100,tmp7,tmp7,0b10100000);
   shf6 = _mm512_mask_shuffle_f64x2(tmp6,0b00110011,tmp4,tmp4,0b11110101);
   shf7 = _mm512_mask_shuffle_f64x2(tmp7,0b00110011,tmp5,tmp5,0b11110101);
// Step 3: Shuffle 128-bits (composed of 2 double-precision floating-point elements)
//         selected from pairs, and store the results
   *col0 = _mm512_shuffle_f64x2(shf0,shf4,0b01000100);
   *col1 = _mm512_shuffle_f64x2(shf1,shf5,0b01000100);
   *col2 = _mm512_shuffle_f64x2(shf2,shf6,0b01000100);
   *col3 = _mm512_shuffle_f64x2(shf3,shf7,0b01000100);
   *col4 = _mm512_shuffle_f64x2(shf0,shf4,0b11101110);
   *col5 = _mm512_shuffle_f64x2(shf1,shf5,0b11101110);
   *col6 = _mm512_shuffle_f64x2(shf2,shf6,0b11101110);
   *col7 = _mm512_shuffle_f64x2(shf3,shf7,0b11101110);
}
```

Fig. 3. The source code corresponding to Fig. 2

Both vectorized algorithms can easily be parallelized using OpenMP in order to utilize multiple cores. The idea of the parallel vectorized Reinsch algorithm is shown in Fig. 4. First, the sequential part is responsible for processing the tail of coefficients $b_{n_0}, \ldots, b_n$, where $n_0 = 64 \cdot P \cdot \lfloor (n+1)/(64 \cdot P) \rfloor$, and P is the number of threads that will be assigned to cores. Then, each OpenMP thread performs the vectorized Reinsch algorithm on one column of $q = n_0/(64 \cdot P)$ blocks and stores first two rows of its processed top block to shared memory. Finally, again during another sequential part, the numbers D_0, S_1 are evaluated using (24). Note that in this case $r = 8P$. The parallel vectorized Goertzel algorithm works similarly, but during its parallel part we process $q \cdot P$ blocks, where q is the number of blocks processed by each thread. Finally, during another sequential part, we use (14) and then process the coefficients $b_1, \ldots, b_{63}$ using (2) in order to evaluate S_1 and S_2.

It is clear that the use of the vectorized Reinsch algorithm is possible only for $n \geq 64$. The vectorized Goertzel algorithm requires $n \geq 127$, i.e. when

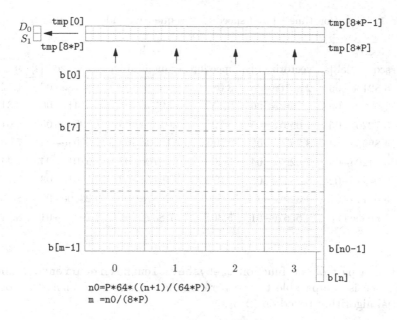

Fig. 4. Stages of the parallel vectorized Reinsch algorithm using OpenMP

vectorization can be applied. Moreover, the use of the parallel implementations is reasonable when a sufficient number of blocks can be processed in parallel.

All source codes of the discussed implementations have been made available in our GitHub repository https://github.com/pstpicz/trigsums. It contains sequential, vectorized, and parallel versions of both Goertzel and Reinsch algorithms, as well as test programs that can be used to evaluate performance and accuracy of the algorithms.

5   Results of Experiments

All considered implementations have been tested on a machine with two Intel *Xeon Platinum 8358* processors (totally 64 cores with hyperthreading, 2.6 GHz, 48 MB of cache memory), 256 GB RAM, running under Linux with Intel OneAPI version 2023 containing the C/C++ compiler.

Table 1 shows the execution time of both sequential and vectorized implementations of our versions Goertzel and Reinsch algorithms, as well as the speedup of the vectorized implementations compared to their sequential counterparts, obtained for various problem sizes. It can be observed that both vectorized algorithms' implementations are faster than the sequential ones, even for very small problem sizes, and the speedup increases as the problem size increases, up to 6.4 for Goertzel, and 8.57 for Reinsch. It is worth adding that the sequential implementation of the Goertzel algorithm has also been implemented in the *Boost*

Table 1. Execution time [s] and speedup of sequential and vectorized algorithms

n	Goertzel			Reinsch		
	sequential [s]	vectorized [s]	speedup	sequential [s]	vectorized [s]	speedup
$2 \cdot 10^2$	8.8215e−06	2.8610e−06	3.08	1.1921e−05	7.8678e−06	1.52
$2 \cdot 10^3$	1.5020e−05	4.7684e−06	3.15	1.5974e−05	6.9141e−06	2.31
$2 \cdot 10^4$	5.7220e−05	1.0967e−05	5.22	8.0109e−05	1.5974e−05	5.01
$2 \cdot 10^5$	4.8995e−04	8.2970e−05	5.91	7.1883e−04	1.0204e−04	7.04
$2 \cdot 10^6$	5.0120e−03	7.8297e−04	6.40	7.1170e−03	9.5701e−04	7.44
$2 \cdot 10^7$	4.9497e−02	8.2841e−03	5.97	7.1515e−02	9.1901e−03	7.78
$2 \cdot 10^8$	5.0171e−01	9.3292e−02	5.38	7.6957e−01	9.2379e−02	8.33
$2 \cdot 10^9$	5.035e+00	9.5067e−01	5.30	7.882e+00	9.1985e−01	8.57

library[1] as a part of the function `chebyshev_clenshaw_recurrence()` and its performance is comparable to the performance of our implementation of this sequential algorithm based on (2–3).

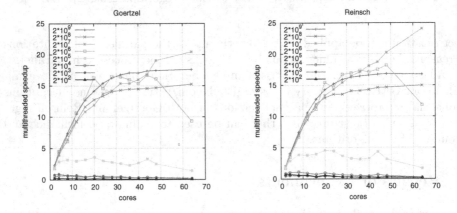

Fig. 5. Speedup of parallel Goertzel and Reinsch algorithms against vectorized ones

Figure 5 illustrates the speedup of the parallel vectorized algorithms over their vectorized counterparts calculated using $s_p(n) = t_v(n)/t_p(n)$, where $t_p(n)$ and $t_v(n)$ denote the execution time of the parallel vectorized and the vectorized algorithms, respectively, and p is the number of cores. Timing results have been collected using various numbers of sockets, cores, and threads per core, specified using the `KMP_HW_SUBSET` environment variable. We have also used `KMP_AFFINITY` to tell the OpenMP runtime how threads should be assigned to cores. The best

---

[1] https://www.boost.org/doc/libs/1_83_0/boost/math/special_functions/chebyshev. hpp.

results for $P = 2 \cdot X$ threads have been obtained for KMP_HW_SUBSET=2s,Xc,1t and KMP_AFFINITY=compact, what means that the best choice is to use one thread per core, distributing threads evenly between cores of two processors.

It can be observed that the qualitative behavior of both algorithms is almost the same. The use of multiple cores is not reasonable for smaller problem sizes. Then, most of cores remain idle during the parallel part of the execution. Speedup over the vectorized versions can be observed for $n > 10^5$. However in most cases, we can observe a peak for a number of cores for which the best performance is achieved. When we use more cores, the speedup does not increase or even becomes worse. The best speedup can be observed for $n = 2 \cdot 10^7$. It is approximately 20.4 for Goertzel and 24 for Reinsch, respectively. This means that the largest observed speedups of the parallel vectorized implementations achieved on 64 cores over the sequential implementations of the Goertzel and Reinsch algorithms are 122 and 187, respectively.

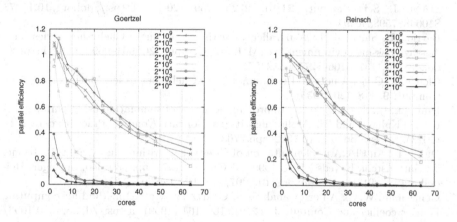

Fig. 6. Parallel efficiency of parallel Goertzel and Reinsch algorithms

Figure 6 shows the parallel efficiency of both parallel implementations calculated as $e_p(n) = s_p(n)/p$. For sufficiently large problem sizes, the parallel efficiency is very good but decreases as the number of cores increases. For the Goertzel algorithm, when a small number of cores is used (i.e. $p = 2, 4$) one can observe $e_p(n) > 1$, what is a fine example of the cache effect [21], when the use of multiple cores speeds up memory transfers over the sequential algorithm [17].

6   Conclusions and Future Works

We have demonstrated that both Goertzel and Reinsch algorithms, examples of recurrence computations, can be efficiently vectorized and parallelized when we apply vectorization and parallelization techniques to the algorithmic approach based on *divide and conquer* methods for solving narrow banded linear systems that arise for linear recurrences. It is clear that such approach can also be applied to develop fast implementations of other problems that reduce to narrow banded systems of equations. It should be noted that although the use of intrinsics significantly limits portability, the places where they are used are crucial for the performance and can easily be ported to other ISA extensions such as scalable vector extensions on ARM [22]. In the future we plan to develop portable implementations of the algorithms in SYCL [16] and study its performance portability on various CPU and GPU platforms [11].

References

1. Amiri, H., Shahbahrami, A.: SIMD programming using intel vector extensions. J. Parallel Distrib. Comput. **135**, 83–100 (2020). https://doi.org/10.1016/j.jpdc.2019.09.012
2. Barrio, R.: Parallel algorithms to evaluate orthogonal polynomial series. SIAM J. Sci. Comput. **21**(6), 2225–2239 (2000). https://doi.org/10.1137/S1064827598340494
3. Barrio, R., Sabadell, J.: A parallel algorithm to evaluate Chebyshev series on a message passing environment. SIAM J. Sci. Comput. **20**, 964–969 (1998). https://doi.org/10.1137/S1064827596312857
4. Clenshaw, C.W.: A note on the summation of Chebyshev series. Math. Tables Aids Comput. **9**, 118–120 (1955)
5. Dmitruk, B., Stpiczyński, P.: Improving accuracy of summation using parallel vectorized Kahan's and Gill-Møller algorithms. Concurr. Comput. Pract. Exper., 1–13 (2023). https://doi.org/10.1002/cpe.7763
6. Dulik, T.: An FPGA implementation of Goertzel algorithm. In: Lysaght, P., Irvine, J., Hartenstein, R. (eds.) FPL 1999. LNCS, vol. 1673, pp. 339–346. Springer, Heidelberg (1999). https://doi.org/10.1007/978-3-540-48302-1_35
7. Gentleman, W.M.: An error analysis of Goertzel's (Watt's) method for computing Fourier coefficients. Comput. J. **12**(2), 160–164 (1969). https://doi.org/10.1093/COMJNL/12.2.160
8. Goertzel, G.: An algorithm for the evaluation of finite trigonometric series. Am. Math. Mon. **65**, 34–35 (1958). https://doi.org/10.2307/2310304
9. Jeffers, J., Reinders, J., Sodani, A.: Intel Xeon Phi Processor High-Performance Programming. Knights Landing Edition. Morgan Kaufman, Cambridge (2016)
10. Kececioglu, O., Gani, A., Sekkeli, M.: A performance comparison of static VAr compensator based on Goertzel and FFT algorithm and experimental validation. Springerplus **5**, 391 (2016). https://doi.org/10.1186/s40064-016-2034-7
11. Marowka, A.: Reformulation of the performance portability metric. Softw. Pract. Exper. **52**(1), 154–171 (2022). https://doi.org/10.1002/spe.3002
12. Martinez-Roman, J., Puche-Panadero, R., Terron-Santiago, C., Sapena-Bano, A., Burriel-Valencia, J., Pineda-Sanchez, M.: Low-cost diagnosis of rotor asymmetries of induction machines at very low slip with the Goertzel algorithm applied to the rectified current. IEEE Trans. Instrum. Meas. **70**, 1–11 (2021). https://doi.org/10.1109/TIM.2021.3115216
13. Murli, A., Rizzardi, M.: Algorithm 682: Talbot's method for the Laplace inversion problem. ACM Trans. Math. Soft. **16**, 158–168 (1990)
14. van der Pas, R., Stotzer, E., Terboven, C.: Using OpenMP - The Next Step. Affinity, Accelerators, Tasking, and SIMD. MIT Press, Cambridge (2017)

15. Regnacq, L., Wu, Y., Neshatvar, N., Jiang, D., Demosthenous, A.: A Goertzel filter-based system for fast simultaneous multi-frequency EIS. IEEE Trans. Circuits Syst. II Express Briefs **68**, 3133–3137 (2021). https://doi.org/10.1109/TCSII.2021.3092069

16. Reinders, J., Ashbaugh, B., Brodman, J., Kinsner, M., Pennycook, J., Tian, X.: Data Parallel C++. Apress, Berkeley (2021). https://doi.org/10.1007/978-1-4842-5574-2

17. Ristov, S., Prodan, R., Gusev, M., Skala, K.: Superlinear speedup in HPC systems: why and when? In: Proceedings of the 2016 Federated Conference on Computer Science and Information Systems, FedCSIS 2016, Gdańsk, Poland, 11–14 September 2016, vol. 8, pp. 889–898. IEEE (2016). https://doi.org/10.15439/2016F498. Annals of Computer Science and Information Systems

18. de Rosa, M.A., Giunta, G., Rizzardi, M.: Parallel Talbot's algorithm for distributed memory machines. Parallel Comput. **21**, 783–801 (1995)

19. Seshadri, R., Ramakrishnan, S., Kumar, J.: Knowledge-based single-tone digital filter implementation for DSP systems. Pers. Ubiquit. Comput. **26**, 319–328 (2022). https://doi.org/10.1007/s00779-019-01304-2

20. Singh, B., Reddy, C.C.: Fast Goertzel algorithm and RLS-adaptive filter based reference current extraction for grid-connected system. In: 2020 IEEE PES Innovative Smart Grid Technologies Europe (ISGT-Europe), pp. 156–160 (2020). https://doi.org/10.1109/ISGT-Europe47291.2020.9248955

21. Speckenmeyer, E., Monien, B., Vornberger, O.: Superlinear speedup for parallel backtracking. In: Houstis, E.N., Papatheodorou, T.S., Polychronopoulos, C.D. (eds.) ICS 1987. LNCS, vol. 297, pp. 985–993. Springer, Heidelberg (1988). https://doi.org/10.1007/3-540-18991-2_58

22. Stephens, N., et al.: The ARM scalable vector extension. IEEE Micro **37**, 26–39 (2017). https://doi.org/10.1109/MM.2017.35

23. Stoer, J., Bulirsh, R.: Introduction to Numerical Analysis, 2nd edn. Springer, New York (1993)

24. Stojanov, A., Toskov, I., Rompf, T., Pueschel, M.: SIMD intrinsics on managed language runtimes, pp. 2–15 (2018). https://doi.org/10.1145/3168810

25. Stpiczyński, P.: Fast parallel algorithms for computing trigonometric sums. In: 2002 International Conference on Parallel Computing in Electrical Engineering (PARELEC 2002), Warsaw, Poland, 22–25 September 2002, pp. 299–304. IEEE Computer Society (2002). https://doi.org/10.1109/PCEE.2002.1115276

26. Stpiczyński, P.: A note on the numerical inversion of the Laplace transform. In: Wyrzykowski, R., Dongarra, J., Meyer, N., Waśniewski, J. (eds.) PPAM 2005. LNCS, vol. 3911, pp. 551–558. Springer, Heidelberg (2006). https://doi.org/10.1007/11752578_66

27. Stpiczyński, P.: Language-based vectorization and parallelization using intrinsics, OpenMP, TBB and Cilk Plus. J. Supercomput. **74**(4), 1461–1472 (2018). https://doi.org/10.1007/s11227-017-2231-3

28. Stpiczyński, P.: Algorithmic and language-based optimization of Marsa-LFIB4 pseudorandom number generator using OpenMP, OpenACC and CUDA. J. Parallel Distrib. Comput. **137**, 238–245 (2020). https://doi.org/10.1016/j.jpdc.2019.12.004

29. Syscl, P., Rajmic, P.: Design of high-performance parallelized gene predictors in MATLAB. BMC. Res. Notes **5**, 183 (2012). https://doi.org/10.1186/1756-0500-5-183

30. Sysel, P., Rajmic, P.: Goertzel algorithm generalized to non-integer multiples of fundamental frequency. EEURASIP J. Adv. Signal Process. **56** (2012). https://doi.org/10.1186/1687-6180-2012-56
31. Talbot, A.: The accurate numerical inversion of Laplace transforms. J. Inst. Maths. Applics. **23**, 97–120 (1979)
32. Vitali, A.: The Goertzel algorithm to compute individual terms of the discrete Fourier transform (DFT). Technical report. DT0089 Rev1, STMicroelectronics (2017)
33. Wang, H., Wu, P., Tanase, I.G., Serrano, M.J., Moreira, J.E.: Simple, portable and fast SIMD intrinsic programming: generic SIMD library, pp. 9–16 (2014). https://doi.org/10.1145/2568058.2568059

File I/O Cache Performance of Supercomputer Fugaku Using an Out-of-Core Direct Numerical Simulation Code of Turbulence

Yuto Hatanaka[1]($\boxtimes$), Yuki Yamane[2], Kenta Yamaguchi[2], Takashi Soga[3], Akihiro Musa[4,6], Takashi Ishihara[5], Atsuya Uno[7], Kazuhiko Komatsu[6], Hiroaki Kobayashi[8], and Mitsuo Yokokawa[1]

[1] Graduate School of System Informatics, Kobe University, Kobe, Japan
hatanaka-m-yuto@stu.kobe-u.ac.jp, yokokawa@port.kobe-u.ac.jp
[2] NEC Solution Innovators, Ltd., Koto-ku, Tokyo, Japan
{yamane.yuki,yamaguchi-zx}@nec.com
[3] Cybermedia Center, Osaka University, Osaka, Japan
soga.takashi.cmc@osaka-u.ac.jp
[4] NEC, Corp., Minato-ku, Tokyo, Japan
[5] Faculty of Environmental, Life, Natural Science and Technology, Okayama University, Okayama, Japan
takashi_ishihara@okayama-u.ac.jp
[6] Cyberscience Center, Tohoku University, Sendai, Japan
{musa,komatsu}@tohoku.ac.jp
[7] National Research Institute for Earth Science and Disaster Resilience, Tsukuba, Japan
a.uno@bosai.go.jp
[8] Graduate School of Information Sciences, Tohoku University, Sendai, Japan
koba@tohoku.ac.jp

Abstract. Turbulent flows play important roles in many flow-related phenomena that appear in various fields. However, despite numerous studies on turbulence, the nature of turbulence has not yet been fully clarified. Direct numerical simulation (DNS) of incompressible homogeneous turbulence in a periodic box is currently a powerful method for studying turbulent flows. However, even modern world-class supercomputers do not have sufficient computational resources to carry out DNS at very high Reynolds number (Re). Memory capacity constraints are particularly severe. Therefore, we have developed an out-of-core DNS (ooc-DNS) code that uses storage to overcome memory limitations. The ooc-DNS code can reduce memory usage by up to a quarter and allows DNS at a higher Re, which would be impossible under normal usage due to memory limitations. When implementing the ooc-DNS code, however, it is crucial to accelerate file input/output (I/O) because the I/O time for storage accounts for a large percentage of the execution time. In this paper, we evaluate the I/O performance of the ooc-DNS code when using a file system called the Lightweight Layered I/O Accelerator of the supercomputer Fugaku. We also evaluate the impact of the I/O cache

© The Author(s), under exclusive license to Springer Nature Switzerland AG 2024
L. Franco et al. (Eds.): ICCS 2024, LNCS 14837, pp. 173–187, 2024.
https://doi.org/10.1007/978-3-031-63778-0_13

and its size on I/O performance and show that the I/O processing can be accelerated by using the cache and optimizing its size. Finally, by taking on I/O cache size when executing the ooc-DNS code with $8,192^3$ grid points, the I/O speed and overall execution speed are increased by 2.4 times and 1.9 times compared to that without the I/O cache.

Keywords: Direct numerical simulation · Turbulent flows · Out-of-core implementation · Fugaku · I/O cache

1   Introduction

Turbulent flows are ubiquitous and play important roles in flow-associated phenomena that appear in various fields of science and technology. Despite numerous studies of turbulence, however, the nature of turbulence has not yet been fully clarified.

In turbulent flows, eddies of various spatial and temporal scales coexist. They non-linearly interact with each other to produce complex motions, so that a small difference in initial conditions can result in unpredictable different motions of individual eddies in turbulent flows. However, it is conceivable that statistically universal laws may exist in these seemingly completely unpredictable complex flows, independent of differences in boundary and initial conditions.

Direct numerical simulation (DNS) of turbulence in a periodic box is a suitable method by which to study the homogeneous isotropic equilibrium state of turbulence on small scales at sufficiently high Reynolds numbers. Direct numerical simulation of box turbulence is highly accurate under the simplest possible conditions and allows DNS of turbulence flows at higher Reynolds numbers. In fact, many DNSs have been conducted and have contributed to the development of turbulence theory [7,8], beginning with Orzag's DNS in 1969 [12].

However, even modern world-class supercomputers do not have sufficient computational resources to carry out DNSs of box turbulence at high Re. Figure 1 shows the trend in supercomputer computational performance and memory capacity and that the computational performance has increased approximately 11,574 times in eighteen years, while the memory capacity has increased only approximately 509 times. This fact suggests that memory capacity constraints will eventually become more severe. For example, a double precision DNS with $32,768^3$ grid points using the developed code requires at least 5.6 PiB of memory. However, there is no supercomputer in Japan that has this amount of memory capacity. Therefore, we have developed an out-of-core DNS (ooc-DNS) code such that a DNS with a large number of grid points can be executed [9,15]. The ooc-DNS code saves arrays containing variables to files on external storage devices. In fact, the code reduces memory usage by up to a quarter and allows DNS at higher Re, which would be impossible under normal usage due to memory limitations. When executing the ooc-DNS code on the supercomputer Fugaku, however, the file input/output (I/O) time accounts for a large percentage of the total execution time because the I/O speed is much slower than the computation speed.

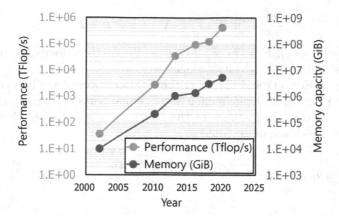

Fig. 1. Performance and memory capacity trends for computers listed in TOP500 [3]. The solid orange line represents the first-place performance in the TOP500. The blue dashed line represents the memory capacity of the computers that are ranked first in the TOP500 [2,6,10,11,13,14].

The purpose of this paper is to accelerate I/O processing and reduce the I/O time when ooc-DNS is executed on Fugaku, and we focus on the I/O cache in compute node (CN)-cache provided by a file system called Lightweight Layered I/O Accelerator (LLIO). Specifically, we evaluate the impact of the I/O cache size on I/O performance. We also evaluate the I/O performance of the ooc-DNS code when executed on Fugaku.

In the remainder of this paper, Sect. 2 describes the ooc-DNS code. Section 3 describes the architecture of Fugaku. In Sect. 4.1, we evaluate the performance of Fugaku's CN-cache using the IOR benchmark program [1]. In Sect. 4.2, we evaluate the performance of the CN-cache by using a benchmark program that simulates the behavior of ooc-DNS. In Sect. 4.3, we evaluate the performance of the CN-cache using the ooc-DNS code with $2,048^3$ and $4,096^3$ grid points. In Sect. 4.4, we implement the ooc-DNS code with $8,192^3$ grid points and confirm that optimizing the CN-cache size increases the I/O speed and overall execution speed. Conclusions are given in Sect. 5.

2   Out-of-Core Direct Numerical Simulation Code

2.1   Direct Numerical Simulation Code Implementation

We consider a cube with side length 2π and periodic boundary conditions as a computational domain. Within this computational domain, we consider homogeneous isotropic turbulence according to the Navier-Stokes equations under the incompressible condition with unit density, as follows

$$\nabla \cdot \boldsymbol{u} = 0, \tag{1}$$

$$\frac{\partial \boldsymbol{u}}{\partial t} + (\boldsymbol{u} \cdot \nabla)\boldsymbol{u} = -\nabla p + \nu \nabla^2 \boldsymbol{u} + \boldsymbol{F}, \tag{2}$$

where $\boldsymbol{u} = (u_1, u_2, u_3)$, $\boldsymbol{F} = (F_1, F_2, F_3)$, p, and ν are the velocity, external force, pressure, and kinematic viscosity, respectively.

Discretizing Eqs. (1) and (2) using the Fourier spectral method on the discretized grid points that divide a cube into N equal parts in each direction leads to ordinary differential equations for the Fourier coefficient $\hat{\boldsymbol{u}}_l$ of the velocities in real space, which are represented as

$$\frac{\mathrm{d}\hat{\boldsymbol{u}}_l}{\mathrm{d}t} + \nu\|\boldsymbol{l}\|_2^2 \hat{\boldsymbol{u}}_l = -((\widehat{\boldsymbol{u} \cdot \nabla)\boldsymbol{u}})_l + \boldsymbol{l} \cdot \frac{\boldsymbol{l} \cdot ((\widehat{\boldsymbol{u} \cdot \nabla)\boldsymbol{u}})_l}{\|\boldsymbol{l}\|_2^2} \qquad (1 \le l_1, l_2, l_3 \le N). \tag{3}$$

Here,

$$((\widehat{\boldsymbol{u} \cdot \nabla)\boldsymbol{u}})_l = il_1 \sum_{l=k+k'} \hat{u}_k \hat{u}_{k'} + il_2 \sum_{l=k+k'} \hat{v}_k \hat{u}_{k'} + il_3 \sum_{l=k+k'} \hat{w}_k \hat{u}_{k'}. \tag{4}$$

where $\boldsymbol{l} = (l_1, l_2, l_3)$, $\boldsymbol{k} = (k_1, k_2, k_3)$, and $\boldsymbol{k}' = (k_1', k_2', k_3')$ are the wave numbers in Fourier space, and the hat symbol denotes the Fourier coefficient. The ordinary differential equations are time evolved using the four-stage fourth-order Runge-Kutta-Gill (RKG) method.

Equations (4) are computed using a transform method based on the three-dimensional fast Fourier transform (3D-FFT). Aliasing errors introduced by the transform method are completely eliminated by the phase shift method and cutting modes larger than $\frac{\sqrt{2}N}{3}$ [5].

The original parallel code was developed for two-dimensional domain decomposition with Message Passing Interface (MPI) for data distribution, and the 3D-FFT parallelized by pencil decomposition is used in the implementation. Figure 2 illustrates how the computational domain is divided into pencils.

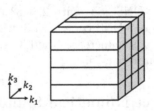

k_3
k_2
k_1

Fig. 2. Pencil domain decomposition in spectral space

2.2 Out-of-Core Implementation Concept

When carrying out DNS of box turbulence using our code by the spectral and RKG methods, a total of 18 variables are required, including the velocity field in each direction in spectral space, intermediate variables in the RKG method, the velocity field in real space, and two variables for the 3D-FFT.

The ooc-DNS code divides each array containing the variables into several subarrays and stores them in separate files. In addition, the code holds only as much memory as the size of one file for each subarray. Figure 3 illustrates how arrays to be stored in storage are divided within a process. When assigning or referencing variables stored in the files, they are processed one file at a time using arrays in memory that are prepared for the size of one file. Figure 4 describes how to change the original code to the ooc-DNS code using a Fortran-like description. The rank of the array variable is 3, and the last rank is partitioned into the number of files at declaration. Then, a loop is added to repeatedly read each portion of the variables, perform calculations, and write updated values to the file.

Pencil domain decomposition in spectral space   **Variables divided in the process that are stored in storage**

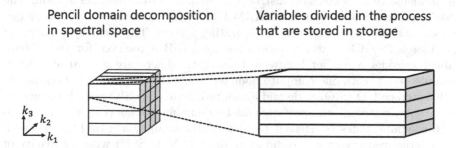

Fig. 3. Illustration of array division within the process

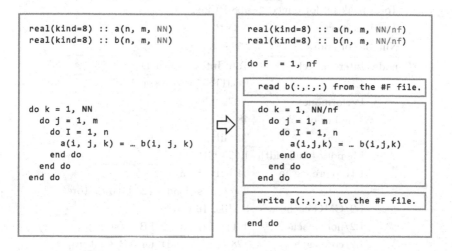

```
real(kind=8) :: a(n, m, NN)
real(kind=8) :: b(n, m, NN)




do k = 1, NN
  do j = 1, m
    do I = 1, n
      a(i, j, k) = … b(i, j, k)
    end do
  end do
end do
```

⇨

```
real(kind=8) :: a(n, m, NN/nf)
real(kind=8) :: b(n, m, NN/nf)

do F = 1, nf

  read b(:,:,:) from the #F file.

  do k = 1, NN/nf
    do j = 1, m
      do I = 1, n
        a(i,j,k) = … b(i,j,k)
      end do
    end do
  end do

  write a(:,:,:) to the #F file.

end do
```

Fig. 4. Pseudo kernel of the out-of-core direct numerical simulation (ooc-DNS) code

This method processes data that is too large to fit into the physical memory of a computer and allows for easy configuration of which of the 16 variables will be stored in storage. The two variables for the 3D-FFT are all placed in memory as in the original code because the 3D-FFT is performed more frequently.

Although the additional file I/O time increases the computation time, this implementation reduces the memory required to execute the code by up to a quarter and allows for the execution of DNSs with a size that was not possible by the original code.

3   Fugaku Architecture

The Fugaku is installed at RIKEN Center for Computational Science (R-CCS) in Kobe, Japan [11]. The system is built on the A64FX ARM v8.2-A, which uses scalable vector extension instructions with a 512-bit implementation. The A64FX processor is a many-core ARM CPU with 48 compute cores and two or four assistant cores used by the operating system. The memory capacity of one node is 32 GiB, of which approximately 5 GiB is reserved for the system. Table 1 provides a further hardware breakdown. There are a total of 158,976 compute nodes, with one compute node for every 16 compute nodes serving as both a storage I/O (SIO) node and a compute node. The SIO node is connected to a first-layer storage and performs file I/O for this storage (Fig. 5). The group of 16 compute nodes is referred to as the SIO group. Each SIO group has a non-volatile memory express solid state drive (NVMe SSD) with a capacity of approximately 1.6 TiB.

Table 1. Specifications of supercomputer Fugaku

Total peak performance		488 PFlops
Total memory		4.85 PiB
Number of nodes		158,976
node	Interconnect	Tofu Interconnect D
	CPU	FUJITSU Processor A64FX
	Performance	3.072 TFlops
	Number of cores	48
	Memory	32 GiB
	Memory bandwidth	1,024 GB/s
	L1D/core cache	64 KiB, 4 way
		256 GB/s (load), 128 GB/s (store)
	L2/CMG cache	8 MiB, 16 way
	L2/node cache	4 TB/s (load), 2 TB/s (store)
	L2/core cache	128 GB/s (load), 64 GB/s (store)
	I/O	PCIe Gen3×16

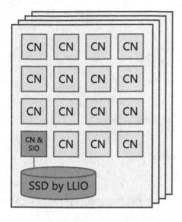

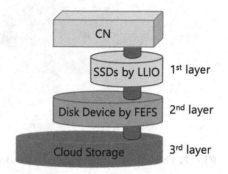

Fig. 5. Lightweight Layered I/O Accelerator (LLIO) configuration diagram. Here, CN indicates a compute node and CN&SIO indicates a compute node with a storage I/O function.

Fig. 6. Layered storage in supercomputer Fugaku

Figure 6 shows that the storage system consists of three primary layers. The first-layer storage consists of NVMe SSDs controlled by LLIO. The second-layer storage consists of multiple hard disk drives controlled by a Lustre-based global file system called the Fujitsu Exabyte File System (FEFS). The total capacities of the first- and second-layer storages are approximately 15.9 PiB and 150 PiB, respectively. The third-layer storage consists of commercial cloud storage services.

The LLIO provides three storage areas in the first-layer storage: a file cache area for the second-layer storage, a shared temporary area for compute nodes assigned a job, and a local temporary area for each compute node. The local temporary area has the largest bandwidth of these three areas [4].

The LLIO also provides an I/O cache function in compute nodes in order to accelerate the I/O from/to the first-layer storage. This function uses a portion of the compute node memory as an I/O cache, called the CN-cache. The CN-cache area is allocated in the size specified by the user in the memory area, and the size on one compute node that can be specified ranges from 4 MiB to 32 GiB. When this size is not specified explicitly, it is the default size of 128 MiB. Therefore, if there is remaining memory capacity out of the capacity that is used by a user program, then the CN-cache size can be increased accordingly. For example, if a program that uses 25 GiB of memory per compute node, considering that the system uses approximately 5 GiB of memory as previously stated, the remaining 2 GiB can be allocated as the CN-cache. Figure 7 illustrates the memory usage for one compute node in this example.

The CN-cache can be enabled and used without any modifications to the code or compiling the code again against specific libraries that overload read and write

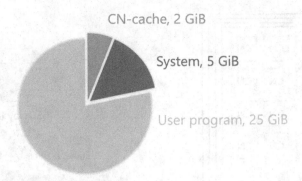

Fig. 7. Example of memory usage for one compute node on supercomputer Fugaku

posix calls. The CN-cache size can just be set as an environment variable when executing the program.

The CN-cache function allows data to be cached during read and write operations. Hereinafter, caching during read operations is referred to as "read cache" and caching during write operations is referred to as "write cache".

The purpose of the read cache is to store data that has been read from the first-layer storage into the CN-cache. This results in the acceleration of the input processing. The decision to use the read cache can be made by the user through a specified parameter. Similarly, the write cache is used to store data that is to be written to the first-layer storage in the CN-cache, resulting in the acceleration of the output processing. A threshold value can be specified for the write cache to determine whether it should be used. The write cache is used when the data size to be written is less than or equal to the specified threshold. It is important to note that the read cache is always enabled in this paper, and the write cache threshold is consistently set to the same size as the CN-cache size, unless otherwise stated.

In this paper, we use the first-layer storage for storing temporary files, measure the I/O time when the size of the CN-cache is changed, and evaluate its performance.

4   Evaluation of Compute Node (CN)-Cache Performance

In this section, we evaluate the CN-cache performance from three perspectives. First, we evaluate the performance of the CN-cache using IOR [1], a benchmark program for measuring parallel file system performance. Next, we evaluate the performance using a simple program similar to the I/O kernel of the ooc-DNS code. Finally, we evaluate the performance using the ooc-DNS code.

The CN-cache performance is evaluated by varying the CN-cache size. When the CN-cache performance of 0 MiB is measured, the read cache is set to disabled and the write cache threshold is to 0 MiB.

4.1 Performance Evaluation of the I/O Cache with IOR

There are more than 30 options that can be specified when executing an IOR program. Considering the I/O kernel behavior of the ooc-DNS code, we specified those options as shown in Table 2. Just as the ooc-DNS code divides a 512 MiB variable into 32 segments, the IOR code divides a 512 MiB block-sized file into 32 segments. However, the parameter "fsyncPerWrite" is set to 1 when the CN-cache size is 0 MiB. In contrast, "fsyncPerWrite" is set to 0 when the CN-cache size is not 0 MiB.

Table 2. Description of options and contents specified in IOR

Parameter	Value	Description
repetition	10	number of repetitions of test
fsync	1	performs fsync upon POSIX file close
SegmentCount	32	number of segments
intraTestBarriers	1	uses barriers between open, write/read, and close
blockSize	512 MiB	-
transferSize	16 MiB	-
fsyncPerWrite	0 or 1	performs fsync after each POSIX write

The IOR program was executed using 4,608 MPI processes on 1,152 compute nodes, with four processes per node. The sizes of the CN-cache were set to 0, 4, 16, 32, 64, 128, 192, 256, and 1,024 MiB. Figure 8 shows the bandwidth of first-layer storage in one SIO group, calculated based on the results of the IOR execution. The read and write bandwidths are depicted in orange and blue, respectively. The read and write bandwidths for the CN-cache size of 0 are depicted by the symbol "■" in the same figure.

It is found that the actual read and write bandwidths for the first-layer storage in one SIO group are approximately 4,500 MiB/s and 1,900 MiB/s, respectively. The values are nearly identical to those measured by Akimoto et al. [4].

The highest write bandwidth with the CN-cache size of 1,024 MiB is obtained because the data is transferred together in larger sizes. Since the CN-cache of 4 MiB is not used for transferring data of 16 MiB, the bandwidths with CN-cache sizes of 0 MiB and 4 MiB are approximately the same.

The read and write bandwidths are lowest when the CN-cache size is between 16 MiB and 32 MiB because the CN-cache size is too small to transfer the data continuously. However, the bandwidths gradually increase as the CN-cache size increases. Processing takes longer due to the lack of free CN-cache area, and increasing the CN-cache size improves the bandwidths.

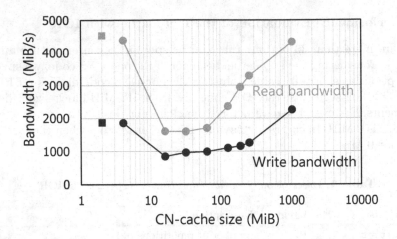

Fig. 8. Measurements of bandwidth per the storage I/O (SIO) group, the first-layer storage, using IOR

4.2 Performance Evaluation of the I/O Cache with a Simple Program Similar to the Ooc-DNS Code

In the ooc-DNS code, if variables in files are referenced and assigned, then the sequence of reading values from the file, calculating them, and writing them to the file is repeated until all subarrays are calculated. The benchmark program shown in Fig. 9 was generated to perform similar operations as the ooc-DNS code with three arrays: a, b, and c. Each array is divided into 32 subarrays, each of which is stored into separate files. The size of arrays a, b, and c in each process is 512 MiB, so that the size of the subarrays is 16 MiB. The input and output times were measured with barrier synchronization after each operation. The variable F in Fig. 9 is the number of subarrays. In the same way, the benchmark program was also generated with six arrays: a, b, c, d, e, and f.

These programs were executed using 192 MPI processes on 48 compute nodes, with four processes per node. The sizes of the CN-cache size were set to 0, 16, 32, 64, 96, 128, 192, 256, and 512 MiB. Figures 10 and 11 show the execution results when using three arrays and six arrays, respectively. For each configuration, we repeat the code execution 11 times and the figures show the average time of the last 10 iterations. The read and write times are depicted in orange and blue, respectively. The times for the CN-cache size of 0 MiB are depicted in the same figure using the symbol "■."

Figure 10 shows that when the CN-cache size is 0 MiB, the I/O time is longest and there is not much difference in the read time and write time. When the CN-cache size is 16 MiB or larger, the write time is approximately twice as long as the read time. The write time is shortest when the CN-cache size is 128 MiB. On the other hand, the read time is shortest when the CN-cache size is 96 MiB.

Figure 11 shows that the I/O time is longest when the CN-cache size is 0 MiB and that there is no difference in the read time and write time when the

```
do F = 1, 32
   call read_a_file(F)
   call read_b_file(F)
   call read_c_file(F)

   call calculation(a,b,c)

   call write_a_file(F)
   call write_b_file(F)
   call write_c_file(F)
end do
```

```
subroutine calculation(a,b,c)
   a = a*04 + a*0.3 + a*0.2
   b = b*04 + b*0.3 + b*0.2
   c = c*04 + c*0.3 + c*0.2
end subroutine calculation
```

Fig. 9. Simple program created to evaluate CN-cache performance

CN-cache size is 0 MiB. The write time is shortest when the CN-cache size is 128 MiB, and increases monotonically when the CN-cache size is 128 MiB or larger. On the other hand, the read times for the CN-cache size of 128 MiB and 192 MiB are approximately 2 and 1.8 times longer than for that of 96 MiB, respectively. However, no such significant difference in read time was observed when only reading and calculating were repeated without writing. Therefore, these spikes for 128 MiB and 192 MiB are caused by the sequence of reading and writing instructions. More specifically, when the CN-cache size is between 128 MiB and 192 MiB and the transfer data is 16 MiB, consecutive reads and writes will degrade the speed of reading data.

It is found that I/O time was reduced by using the CN-cache for both cases of three and six arrays. However, it is also found that there are spikes in read time, as shown in Fig. 11, and that increasing the CN-cache size results in longer I/O time in some cases. Therefore, optimizing the CN-cache size is important to further reduce I/O time.

4.3   Performance Evaluation of I/O Cache with the Ooc-DNS Code

We carried out the ooc-DNS code with two problem sizes, $N^3 = 2,048^3$ and $4,096^3$.

The number of nodes used in the execution with $N^3 = 2,048^3$ and $4,096^3$ are 32 and 256, respectively, with four MPI processes allocated per node. The DNSs with $N^3 = 2,048^3$ and $4,096^3$, however, require approximately 1,281 GiB and 10,256 GiB memory capacities, respectively, and it is necessary to store variables to files in storage considering that one node has only a 32-GiB memory capacity. We changed the number of variables to be stored in files on each problem size so that as few variables as possible are stored in files. In fact, 11 or 12 out of 16 variables are divided into 32 subarrays for the cases of $N^3 = 2,048^3$ and $4,096^3$, respectively. Then, the size of one file storing a subarray is 16 MiB, and the memory usage is approximately 0.52 times for N = 2,048 and approximately

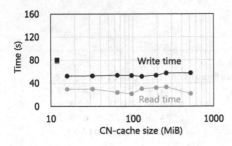
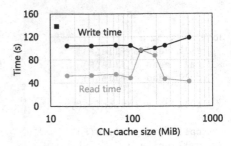

Fig. 10. Measured performance of CN-cache in simple program similar to ooc-DNS code with three arrays

Fig. 11. Measured performance of CN-cache in simple program similar to ooc-DNS code with six arrays

0.45 times for N = 4,096. Barrier synchronization is performed at the end of the interval for which time measurements were taken, and the I/O time required to proceed a time step is measured. The numbers of read and write operations in the ooc-DNS code are 4,069 and 3,008, respectively, for $N^3 = 2,048^3$, for one time step. In contrast, the numbers of read and write operations in the ooc-DNS code are 4,704 and 3,648, respectively, for $N^3 = 4,096^3$.

The times for read and write operations are measured by varying the CN-cache size as 0, 4, 16, 32, 64, 96, 128, 256, and 512 MiB. The times for the read and write operations and their sums for $N^3 = 2,048^3$ and $N^3 = 4,096^3$ are plotted in Figs. 12 and 13, respectively. The times for read, write, and their sum are depicted in orange, blue, and gray, respectively. The times for the CN-cache size of 0 MiB are plotted by the "■" symbol.

The difference in I/O time between the DNS sizes of $2,048^3$ and $4,096^3$ is due to the different number of variables stored in first-layer storage, and, regardless of CN-cache size, the I/O total time is shorter for $N^3 = 2,048^3$.

Figure 12 shows that the write performance when the CN-cache size is 96 MiB or larger is worse than when the CN-cache size is 0 MiB. Figure 13 shows that the write performance when the CN-cache size is 192 MiB or larger is similarly worse. These results indicate that the write performance is better with a small CN-cache size, despite the large number of write operations. The reasons for these are thought to be that the CN-cache improves the write speed, and that there is too much data in the CN-cache for the LLIO to manage.

Figures 12 and 13 also show that spikes are found for read performance when the CN-cache size is 128 MiB for the two cases. In Fig. 12, the read time at 128 MiB is approximately 1.5 times longer than that at 96 MiB. On the other hand, in Fig. 13, the read time at 128 MiB is approximately 1.8 times longer than that at 96 MiB. These behaviors are also confirmed in Fig. 11 in Sect. 4.2.

We found that the I/O total time for the CN-cache size of 0 MiB is the longest and that the CN-cache improved the I/O total speed and reduce I/O total time for practical applications.

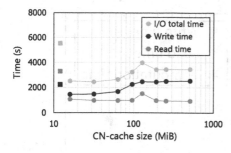

Fig. 12. Performance for $2,048^3$ grid points

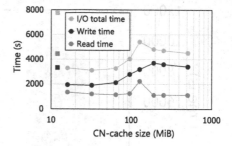

Fig. 13. Performance for $4,096^3$ grid points

4.4   Execution of Ooc-DNS Code with $8,192^3$ Grid Points

The DNS with $N^3 = 8,192^3$ was executed using 8,192 processes on 2,048 nodes with four processes per node. The DNS with $N^3 = 8,192^3$ requires a memory capacity of approximately 80-TiB, and it is necessary to store variables to files in storage. Twelve out of 16 variables are divided into 32 subarrays and sent to storage. Then, the size of one file storing a subarray is 16 MiB, and the memory usage is approximately 0.45 times as large as that without the out-of-core implementation. Barrier synchronization is performed at the end of the interval for which time measurements were taken, and the I/O time and the other time required to proceed a time step are measured. The numbers of read and write operations in the ooc-DNS code are 4,704 and 3,648, respectively, for one time step.

The times for reading, writing, and computation are measured at two CN-cache sizes of 0 and 32 MiB. These times are plotted in Fig. 14. The times for read, write, and other computation are depicted in orange, blue, black, respectively. It is found that there is little difference in computation time between the CN-caches of 0 MiB and 32 MiB. In addition, when the CN-cache size is set to 32 MiB, a two-fold speed up in overall execution time is achieved compared to the CN-cache size of 0 MiB.

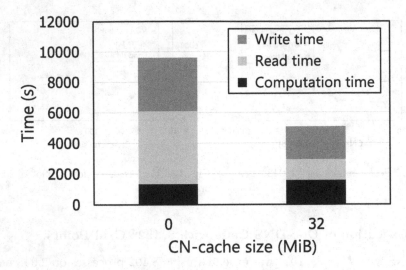

Fig. 14. Execution time for ooc-DNS with $N^3 = 8,192^3$

5   Conclusions

Modern world-class supercomputers do not have sufficient memory capacity to carry out DNS at very high Reynolds number. Therefore, we implemented the ooc-DNS code. When we ran the ooc-DNS code on Fugaku, the I/O time accounted for the majority of the execution time. In order to accelerate the I/O speed when executing the ooc-DNS code, we have focused on the CN-cache function for optimizing the I/O cache size on Fugaku and evaluated the CN-cache performance from three perspectives. First, the basic evaluation of the performance of the CN-cache was conducted using the IOR benchmark program. Next, we evaluated the performance with a simple program that is similar to the ooc-DNS code. The performance was then evaluated with the ooc-DNS code. Finally, we optimized the CN-cache size in order to accelerate the ooc-DNS code. As a result, it was found that specifying a CN-cache size larger than 512 MiB does not reduce the I/O time in the execution of the ooc-DNS code due to performance degradation of the write cache. In addition, the read performance degrades drastically without explicit specification of the CN-cache size. By specifying an appropriate CN-cache size when executing the ooc-DNS code with $N^3 = 8,192^3$, the I/O speed and the overall execution speed was increased by 2.4 times and 1.9 times, respectively, as compared to the CN-cache size of 0 MiB.

Acknowledgment. This study used the computational resources of the Supercomputer Fugaku provided by the RIKEN Center for Computational Science through the HPCI System Research project (Project ID No. hp230143). This work was partially conducted and funded at Joint-Research Division of High-Performance Computing (NEC) of Cyberscience Center at Tohoku University.

References

1. Github: IOR. https://github.com/hpc/ior
2. System Overview (ES). https://www.jamstec.go.jp/es/jp/es1/system/system. html. Accessed 12 Dec 2023
3. Performance development (2023). https://www.top500.org/statistics/perfdevel/. Accessed 12 Dec 2023
4. Akimoto, H., Okamoto, T., Kagami, T., Seki, K., Sakai, K., Imade, H., et al.: File system and power management enhanced for supercomputer Fugaku. Fujitsu Tech. Rev. **3**, 2020–2023 (2020)
5. Canuto, C., Hussaini, M.Y., Quarteroni, A., Zang, T.A.: Spectral Methods in Fluid Dynamics. Springer, Heidelberg (1988)
6. Fu, H., Liao, J., Yang, J., Wang, L., Song, Z., Huang, X., et al.: The Sunway TaihuLight supercomputer: system and applications. Sci. China Inf. Sci. **59**(7), June 2016. https://doi.org/10.1007/s11432-016-5588-7
7. Ishihara, T., Kaneda, Y., Morishita, K., Yokokawa, M., Uno, A.: Second-order velocity structure functions in direct numerical simulations of turbulence with R_λ up to 2250. Phys. Rev. Fluids **5**, 104608 (2020). https://doi.org/10.1103/ PhysRevFluids.5.104608
8. Kaneda, Y., Ishihara, T., Yokokawa, M., Itakura, K., Uno, A.: Energy dissipation rate and energy spectrum in high resolution direct numerical simulations of turbulence in a periodic box. Phys. Fluids **15**(2), L21–L24 (2003). https://doi.org/10. 1063/1.1539855
9. Komatsu, K., Momose, S., Isobe, Y., Watanabe, O., Musa, A., Yokokawa, M., et al.: Performance evaluation of a vector supercomputer sx-aurora tsubasa. In: SC18: International Conference for High Performance Computing, Networking, Storage and Analysis, pp. 685–696 (2018). https://doi.org/10.1109/SC.2018.00057
10. Liao, X., Xiao, L., Yang, C., Lu, Y.: MilkyWay-2 supercomputer: system and application. Front. Comp. Sci. **8**(3), 345–356 (2014). https://doi.org/10.1007/s11704-014-3501-3
11. Mitsuhisa, S., Yutaka, I., Hirofumi, T., Yuetsu, K., Tetsuya, O., Miwako, T., et al.: Co-design for A64FX manycore processor and "Fugaku". In: SC20: International Conference for High Performance Computing, Networking, Storage and Analysis p. 1, November 2020. https://doi.org/10.1109/sc41405.2020.00051, https://cir.nii. ac.jp/crid/1360013173149649280
12. Orszag, S.A.: Numerical methods for the simulation of turbulence. Phys. Fluids **12**(12), II–250–II–257 (1969).https://doi.org/10.1063/1.1692445
13. Vazhkudai, S.S., De Supinski, B.R., Bland, A.S., Geist, A., Sexton, J., Kahle, J., et al.: The design, deployment, and evaluation of the CORAL pre-exascale systems. In: SC18: International Conference for High Performance Computing, Networking, Storage and Analysis, pp. 661–672. IEEE (2018)
14. Yang, X.J., Liao, X.K., Lu, K., Hu, Q.F., Song, J.Q., Su, J.S.: The TianHe-1A supercomputer: its hardware and software. J. Comput. Sci. Technol. **26**(3), 344–351 (2011). https://doi.org/10.1007/s02011-011-1137-8
15. Yokokawa, M., Yamane, Y., Yamaguchi, K., Soga, T., Matsumoto, T., et al.: I/O Performance evaluation of a memory-saving DNS code on SX-aurora TSUBASA. In: 2023 IEEE International Parallel and Distributed Processing Symposium Workshops (IPDPSW), pp. 692–696 (2023). https://doi.org/10.1109/IPDPSW59300. 2023.00117

A Novel Computational Approach for Wind-Driven Flows over Deformable Topography

Alia Al-Ghosoun[1(✉)] and Mohammed Seaid[2]

[1] Mechatronics engineering department, Philadelphia University, Amman, Jordan
aalghsoun@philadelphia.edu.jo
[2] Department of Engineering, Durham university, South Road,
Durham DH13LE, UK
m.seaid@durham.ac.uk

Abstract. Single-layer shallow water models have been widely used for simulating shallow water waves over both fixed and movable beds. However, these models can not capture some hydraulic features such as small eddy currents and flow recirculations. This study presents a novel numerical approach for coupling multi-layer shallow water models with elastic deformations to accurately capture complex recirculation patterns in wind-driven flows. This class of multi-layer equations avoids the computationally demanding methods needed to solve the three-dimensional Navier-Stokes equations for free-surface flows while it provides stratified flow velocities since the pressure distribution is still assumed to be hydrostatic. In the current study, the free-surface flow problem is approximated as a layered system made of multiple shallow water equations of different water heights but coupled through mass-exchange terms between the embedded layers. Deformations in the topography are accounted for using linear elastostatic systems for which an internal force is applied. Transfer conditions at the interface between the water surface and the topography are also developed using frictional forces and hydrostatic pressures. For the computational solver, we implement a fast and accurate hybrid finite element/finite volume method solving the linear deformations on unstructured meshes and the nonlinear flows using well-balanced discretizations. Numerical results are presented for various problems and the computed solutions demonstrate the ability of the proposed model in accurately resolving wind-driven flows over deformable topography.

Keywords: Multi-layer shallow water · Elasticity · Finite volume method · Finite element method · Topography deformation · Wind-driven flows

1 Introduction

Free-surface models in hydraulic applications have gained an increasing interest during the last decades, see for example [13]. Ranging from flood forecasting [20] to monitoring hydraulic infrastructures such as dams and rivers [22]. Water

L. Franco et al. (Eds.): ICCS 2024, LNCS 14837, pp. 188–202, 2024.
https://doi.org/10.1007/978-3-031-63778-0_14

free-surface flows under the influence of gravity can be modelled using the well-established shallow water equations [3]. However, the main drawback of these equations lies in the lack of capturing some crucial physical dynamics in the vertical motion of water flow [1]. Moreover, under the impact of topography and wind forces, the hydrodynamics can be very complex and numerical modelling of such problems would require the use of the full three-dimensional Navier-Stokes equations [6]. Recently, multi-layer shallow water models have attracted enormous attention and become an important tool to capture many hydraulic problems such as the flow recirculations [4]. The multi-layer shallow water equations have also been subject of various research studies and have been used for modelling a wide variety of free-surface flows where water flows interact with the bed topography [21] and wind stresses [5]. For example, researchers in [12] have implemented the multi-layer model to study nonlinear internal wave propagation in shallow water flows, whereas in [10], authors have experimentally investigated the multi-layer flow field mapping in a small scale shallow water reservoir by coastal acoustic tomography. On the other hand, the incompressible smoothed particle hydrodynamics approach has been implemented in [14] to model dam-break flows over movable beds.

Different numerical methods were implemented in the literature to model the multi-layer shallow water flows. In [22], the finite difference method is used to solve a multi-layer model with non-flat bottom topography on both fixed and adaptive moving meshes [8]. A well-balanced Runge-Kutta discontinuous Galerkin method has also been proposed in [11] for the numerical solution of multi-layer shallow water equations with mass exchange and non-flat bottom topography. In recent years, a great amount of research effort has been devoted to developing consistent mathematical models and efficient numerical solvers for the interaction between topography deformation and water waves. In practice, modelling of wave flows by static deformation is based on two components including the description of topography deformation and the governing equations of the water flow. In [1,18], we have used the conventional single-layer shallow water equations for modelling flows over deformable beds. These models performed very well for simple flows such as dam-break and stream-flow problems as well as water waves generated by deformations on the topography. However, these models would fail to adequately represent flow circulations and do not account for the influences of wind effect and water layer densities in their formulations. Therefore, the present study is an improvement to our previous research in [1,18] using the multi-layer shallow water equations. In this work, the governing equations consist of the one-dimensional nonlinear multi-layer shallow water equations for the water flow and a two-dimensional linear elastostatic model for the deformation of topography. In addition to the internal stress applied to the bed, deformations in the topography can also be caused as a result of the hydrostatic pressure distribution and the frictional force obtained from the shallow water flow. These equations are fully coupled and solved simultaneously in time using transfer conditions at the interface between the water flow and the topography. This allows for hydrostatic pressure and friction forces to be

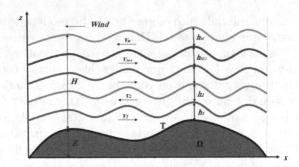

Fig. 1. Illustration of a coupled system for wind-driven flows over deformable topography.

implemented for the elastostatic equations whereas the deformed topography is accounted for in the multi-layer shallow water equations through the bathymetric forces. To solve the considered system we implement a well-balanced finite volume method for the multi-layer shallow water system and a stabilized finite element method for linear elasticity equations. This hybrid finite element/finite volume solver uses unstructured/structured meshes, respectively. The interfacial forces are sampled from the hydrostatic pressure and applied on the topography to be used in the stress analysis. To demonstrate the performance of the hybrid finite volume/finite element method, computational results obtained for wind-driven flows over deformable topography are presented. The effects of wind velocity, number of layers on the flow field and stress distributions are also investigated in this study. The rest of this paper is organized as follows: Formulation of mathematical models for the coupled system is presented in Sect. 2. Section 3 is devoted to the implementation of the numerical methods used for the solution procedure. Numerical results and examples for wind-driven flows over deformable topography are presented in Sect. 4. Concluding remarks are summarized in Sect. 5.

2   Governing Equations for Wind-Driven Flows over Deformable Topography

Considering the system illustrated in Fig. 1, the proposed coupled system consists of the two-dimensional constitutive relations of an isotropic elastic bed and the multi-layer shallow water equations. In elasticity theory [9], the bed deformation equations can be written as a relationship between the stress and strain in terms of the Lamé parameters as

$$\nabla \cdot \boldsymbol{\sigma} = \boldsymbol{f}, \qquad \boldsymbol{\sigma} = \lambda tr(\epsilon)\boldsymbol{I} + 2\mu\epsilon, \tag{1}$$

where $\boldsymbol{\sigma}$ is the stress tensor, $\boldsymbol{f}$ the body force per unit area, λ the first Lamé parameter (related to the bulk modulus), $tr(\epsilon)$ the trace of the strain tensor, $\boldsymbol{I}$

the identity matrix, μ the second Lamé parameter (related to the shear modulus). The displacement vector is denoted by $\mathbf{u} = (u_x, u_z)^{\mathsf{T}}$ and ϵ the infinitesimal strain tensor defined by

$$\epsilon = \frac{1}{2}\left(\nabla\mathbf{u} + (\nabla\mathbf{u})^{\mathsf{T}}\right). \tag{2}$$

Combining Eqs. (1) and (2), one obtains

$$\boldsymbol{\sigma} = \lambda(\nabla \cdot \mathbf{u})\boldsymbol{I} + \mu\left(\nabla\mathbf{u} + (\nabla\mathbf{u})^{\mathsf{T}}\right). \tag{3}$$

The system is equipped with the following boundary conditions

$$\boldsymbol{\sigma} = \boldsymbol{\sigma}_c, \qquad \mathbf{u} = \mathbf{0}, \qquad \text{on } \partial\Omega, \tag{4}$$

where $\partial\Omega$ is the fixed boundary of the topography domain Ω and $\boldsymbol{\sigma}_c$ is a pre-scribed stress. In the current study, the constitutive relation is defined as

$$\boldsymbol{\sigma} = \mathbf{D}\,\epsilon, \tag{5}$$

where the stress vector $\boldsymbol{\sigma}$ and the constitutive matrix $\mathbf{D}$ for a plane-strain case are given as

$$\boldsymbol{\sigma} = \begin{pmatrix} \sigma_x \\ \sigma_z \\ \tau_{xz} \end{pmatrix}, \qquad \mathbf{D} = \frac{E}{(1+\nu)(1-2\nu)}\begin{pmatrix} 1-\nu & \nu & 0 \\ \nu & 1-\nu & 0 \\ 0 & 0 & \dfrac{1-2\nu}{2} \end{pmatrix},$$

with E is the Young's modulus characterising the bed material and ν the Pois-son's ratio, the two Lamé parameters are given as

$$\lambda = \frac{E.\nu}{(1+\nu)(1-2\nu)}, \qquad \mu = \frac{E}{2(1+\nu)}.$$

For the wind-driven flows, we assume M layers of water bodies bounded at the bottom by the deformable topography and a free-surface subjected to wind stresses as illustrated in Fig. 1. For simplicity in the presentation, the governing equations for the multi-layer shallow water model considered in the present work read as

$$\frac{\partial\mathbf{W}}{\partial t} + \frac{\partial\mathbf{F}(\mathbf{W})}{\partial x} = \mathbf{Q}(\mathbf{W}) + \mathbf{R}(\mathbf{W}), \tag{6}$$

where $\mathbf{W}$ is the vector of conserved variables, $\mathbf{F}$ the vector of flux functions, $\mathbf{Q}$ and $\mathbf{R}$ are the vectors of source terms defined by

$$\mathbf{W} = \begin{pmatrix} H \\ Hv_1 \\ Hv_2 \\ \vdots \\ Hv_M \end{pmatrix}, \qquad \mathbf{F}(\mathbf{W}) = \begin{pmatrix} \displaystyle\sum_{\alpha=1}^{M} l_\alpha H v_\alpha \\ Hv_1^2 + \dfrac{1}{2}gH^2 \\ Hv_2^2 + \dfrac{1}{2}gH^2 \\ \vdots \\ Hv_M^2 + \dfrac{1}{2}gH^2 \end{pmatrix}, \qquad \mathbf{Q}(\mathbf{W}) = \begin{pmatrix} 0 \\ -gH\dfrac{\partial Z}{\partial x} \\ -gH\dfrac{\partial Z}{\partial x} \\ \vdots \\ -gH\dfrac{\partial Z}{\partial x} \end{pmatrix},$$

$$\mathbf{R(W)} = \begin{pmatrix} 0 \\ \frac{1}{l_1}\left(\mathcal{S}_1 - gn_b^2\frac{v_1|v_1|}{H^{1/3}} + 2\xi\frac{v_2-v_1}{(l_2+l_1)H}\right) \\ \frac{1}{l_2}\left(\mathcal{S}_2 + 2\xi\frac{v_3-v_2}{(l_3+l_2)H} - 2\xi\frac{v_2-v_1}{(l_2+l_1)H}\right) \\ \vdots \\ \frac{1}{l_{M-1}}\left(\mathcal{S}_{M-1} + 2\xi\frac{v_M-v_{M-1}}{(l_M+l_{M-1})H} - 2\xi\frac{v_{M-1}-v_{M-2}}{(l_{M-1}+l_{M-2})H}\right) \\ \frac{1}{l_M}\left(\mathcal{S}_M - \sigma^2\rho_a\frac{w|w|}{H} - 2\xi\frac{v_M-v_{M-1}}{(l_M+l_{M-1})H}\right) \end{pmatrix}. \quad (7)$$

where, $v_\alpha(t,x)$ is the local water velocity for the αth layer, $Z(x)$ the bed topography, g the gravitational acceleration, $H(t,x)$ the water height of the whole flow system, we refer the reader to [11] for the more details. In (7), the source term $\mathcal{S}_\alpha$ represents the momentum exchanges between the water layers defined as

$$\mathcal{S}_\alpha = u_{\alpha+\frac{1}{2}}\mathcal{M}_{\alpha+\frac{1}{2}} - u_{\alpha-\frac{1}{2}}\mathcal{M}_{\alpha-\frac{1}{2}}, \quad (8)$$

where the mass exchange terms $\mathcal{M}_{\alpha-\frac{1}{2}}$ and $\mathcal{M}_{\alpha+\frac{1}{2}}$ are evaluated using

$$\mathcal{M}_{\alpha-\frac{1}{2}} = \begin{cases} 0, & \alpha = 1, \\ \sum_{i=1}^{\alpha}\left(\frac{\partial(h_iu_i)}{\partial x} - l_i\sum_{j=1}^{N}\frac{\partial(h_ju_j)}{\partial x}\right), & \alpha = 2,3\ldots,N, \end{cases} \quad (9)$$

and

$$\mathcal{M}_{\alpha+\frac{1}{2}} = \begin{cases} \sum_{i=1}^{\alpha}\left(\frac{\partial(h_iu_i)}{\partial x} - l_i\sum_{j=1}^{N}\frac{\partial(h_ju_j)}{\partial x}\right), & \alpha = 1,2,\ldots,N-1, \\ 0, & \alpha = N, \end{cases} \quad (10)$$

respectively. Here, the interface velocities $u_{\alpha-\frac{1}{2}}$ and $u_{\alpha+\frac{1}{2}}$ are computed according to the sign of mass-exchange terms in (9) and (10) as

$$u_{\alpha-\frac{1}{2}} = \begin{cases} u_{\alpha-1}, & \mathcal{M}_{\alpha-\frac{1}{2}} \geq 0, \\ u_\alpha, & \mathcal{M}_{\alpha-\frac{1}{2}} < 0, \end{cases} \quad u_{\alpha+1/2} = \begin{cases} u_\alpha, & \mathcal{M}_{\alpha+\frac{1}{2}} \geq 0, \\ u_{\alpha+1}, & \mathcal{M}_{\alpha+\frac{1}{2}} < 0. \end{cases} \quad (11)$$

Note that a zeroth-order approximation of the $2(M-1)$ barotropic eigenvalues associated with $(M-1)$ interfaces gives

$$\lambda_{int}^{\pm,\alpha+\frac{1}{2}} = v \pm \sqrt{\frac{1}{2}g\sum_{\alpha=1}^{M}h_\alpha} + \mathcal{O}\left(|v_\beta - v|\right)_{\beta=1,\ldots,M}, \quad \alpha = 1,2,\ldots,M-1. \quad (12)$$

3   Hybrid Finite Element/Finite Volume Solver

To solve the above system we consider a finite element method for the two-dimensional elasticity equations. The starting point for the finite element method is the domain discretization. In the present study, we adopt the finite element method proposed in [1] using an unstructured mesh with quadratic triangular elements. Hence, the variational formulation of (1)–(2) consists of forming the inner product of Eq. (1) by a vector test function ϕ and integrate over the domain Ω

$$\int_\Omega (\nabla \cdot \boldsymbol{\sigma}) \cdot \phi \, d\boldsymbol{x} = \int_\Omega \boldsymbol{f} \cdot \phi \, d\boldsymbol{x}, \tag{13}$$

where $\mathbf{x} = (x, z)^\top$ and $\mathbf{n} = (n_x, n_z)^\top$ is the outward unit normal on $\partial\Omega$. Integrating the system by parts, since $\nabla \cdot \boldsymbol{\sigma}$ contains second-order derivatives of the primary unknown u

$$\int_\Omega (\nabla \cdot \boldsymbol{\sigma}) \cdot \phi \, d\boldsymbol{x} = \int_\Omega \boldsymbol{\sigma} : \nabla\phi \, d\boldsymbol{x} - \oint_{\partial\Omega} (\boldsymbol{\sigma} \cdot \mathbf{n} \cdot \phi) \, ds, \tag{14}$$

where the colon operator is the inner product between tensors (summed pairwise product of all elements). Here, $\boldsymbol{\sigma} \cdot \mathbf{n}$ is the traction or stress vector at the boundary, and is often prescribed as a boundary condition. Using that the traction stress vector $\boldsymbol{T} = \boldsymbol{\sigma} \cdot \mathbf{n}$, thus we obtain

$$\int_\Omega \boldsymbol{\sigma} \cdot \nabla\phi \, d\boldsymbol{x} = \int_\Omega \boldsymbol{f} \cdot \phi \, d\boldsymbol{x} + \oint_{\partial\Omega} \boldsymbol{T} \cdot \phi \, d\boldsymbol{x}, \tag{15}$$

which can be reformulated in a vector form as

$$\int_\Omega \widehat{\phi} \cdot \boldsymbol{\sigma} \, d\boldsymbol{x} = \oint_\Omega \phi^\top \cdot \boldsymbol{T} \, d\boldsymbol{x} + \int_\Omega \phi^\top \cdot \mathbf{f} \, d\boldsymbol{x}, \tag{16}$$

where $\phi = (\phi_x, \phi_z)^\top$, $\boldsymbol{T} = (\boldsymbol{T}_x, \boldsymbol{T}_z)^\top$ and $\widehat{\phi} = \left(\dfrac{\partial\phi_x}{\partial x}, \dfrac{\partial\phi_z}{\partial z}, \dfrac{\partial\phi_x}{\partial z} + \dfrac{\partial\phi_z}{\partial x} \right)^\top$.

To solve the weak form (16) with the finite element method, the domain Ω is discretized into a set of elements where the solution is approximated in terms of the nodal values U_j and the polynomial basis functions $\Psi_j(x, z)$ as

$$\mathbf{u}(x, z) = \sum_{j=1}^N U_j \Psi_j(x, z), \tag{17}$$

where N is the number of nodes. To solve the fully discretized problem, the elementary matrices are assembled into a global system of equations

$$\mathbf{Ku} = \mathbf{f}, \tag{18}$$

where $\mathbf{K}$ is the global stiffness matrix, $\mathbf{u}$ is the nodal displacement vector and $\mathbf{f}$ is the force vector.

Next, the multi-layer shallow water system with mass exchange was implemented for the water perturbations. For the spatial discretization of (6), we discretise the spatial domain into control volumes $[x_{i-1/2}, x_{i+1/2}]$ with uniform size $\Delta x = x_{i+1/2} - x_{i-1/2}$, $x_{i-1/2} = i\Delta x$ and $x_i = (i + 1/2)\Delta x$ is the center of the control volume. Integrating the Eq. (6) with respect to space over the control volume $[x_{i-1/2}, x_{i+1/2}]$, we obtain the following semi-discrete equations

$$\frac{d\mathbf{W}_i}{dt} + \frac{\mathcal{F}_{i+1/2} - \mathcal{F}_{i-1/2}}{\Delta x} = \mathbf{Q}(\mathbf{W}_i) + \mathbf{R}(\mathbf{W}_i), \qquad (19)$$

where $\mathbf{W}_i(t)$ is the space-averaged approximation of the solution $\mathbf{W}$ in the control volume $[x_{i-1/2}, x_{i+1/2}]$ at time t, $i.e.$,

$$\mathbf{W}_i(t) = \frac{1}{\Delta x} \int_{x_{i-1/2}}^{x_{i+1/2}} \mathbf{W}(t, x) \, dx,$$

and $\mathcal{F}_{i\pm1/2} = \mathbf{F}(\mathbf{W}_{i\pm1/2})$ are the numerical fluxes at $x = x_{i\pm1/2}$ and time t. Here, the time integration of (19) is performed using a second-order splitting method studied in [19]. Thus, to integrate the Eqs. (6) in time we divide the time interval into subintervals $[t_n, t_{n+1}]$ with length $\Delta t = t_{n+1} - t_n$ for $n = 0, 1, \ldots$. We also use the notation W^n to denote the value of a generic function W at time t_n. The considered operator splitting method consists of three stages as presented in [2]. The spatial discretization (19) is complete when a reconstruction of the numerical fluxes $\mathcal{F}_{i\pm1/2}$ and source terms $\mathbf{Q}(\mathbf{W}_i)$ and $\mathbf{R}(\mathbf{W}_i)$ are chosen. In the current work, the finite volume method of characteristics studied in [5] has been implemented and it can be rearranged in a compact form as

$$\frac{\partial U_\alpha}{\partial t} + \mathcal{U}_\alpha \frac{\partial U_\alpha}{\partial x} = S_\alpha(\mathbf{U}), \qquad \alpha = 0, 1, \ldots, M, \qquad (20)$$

where $q_\alpha = Hv_\alpha$ is the water discharge, $\mathbf{U} = (U_0, U_1, \ldots, U_M)^T$, $\mathbf{S}(\mathbf{U}) = (S_0, S_1, \ldots, S_M)^T$ with

$$\mathbf{U} = \begin{pmatrix} H \\ q_1 \\ q_2 \\ \vdots \\ q_M \end{pmatrix}, \qquad \mathbf{S}(\mathbf{U}) = \begin{pmatrix} -\sum_{\alpha=1}^{M} l_\alpha H \frac{\partial v_\alpha}{\partial x} \\ -Hv_1 \frac{\partial v_1}{\partial x} - gH \frac{\partial}{\partial x}(H + Z) \\ -Hv_2 \frac{\partial v_2}{\partial x} - gH \frac{\partial}{\partial x}(H + Z) \\ \vdots \\ -Hv_M \frac{\partial v_M}{\partial x} - gH \frac{\partial}{\partial x}(H + Z) \end{pmatrix},$$

and the advection velocity $\mathcal{U}_\alpha$ is defined as

$$\mathcal{U}_\alpha = \begin{cases} \sum_{\beta=1}^{M} l_\beta v_\beta, & \text{if} \quad \alpha = 0, \\ v_\alpha, & \text{if} \quad \alpha = 1, 2, \ldots, M. \end{cases} \qquad (21)$$

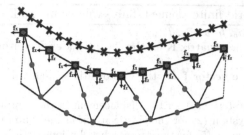

Fig. 2. Illustration of the finite element and finite volume nodes used for coupling conditions at the interface.

The characteristic curves associated with the Eq. (20) are solutions of the initial-value problems

$$\frac{dX_{\alpha,i+1/2}(\tau)}{d\tau} = \mathcal{U}_{\alpha,i+1/2}\Big(\tau, X_{\alpha,i+1/2}(\tau)\Big), \quad \tau \in [t_n, t_{n+1}],$$
$$X_{\alpha,i+1/2}(t_{n+1}) = x_{i+1/2}, \qquad \alpha = 0, 1, \ldots, M. \tag{22}$$

where $X_{\alpha,i+1/2}(\tau)$ are the departure points at time τ of a particle that will arrive at the gridpoint $x_{i+1/2}$ in time t_{n+1}. In our simulations we used the third-order Runge-Kutta method for the solution of the initial-value problems (22) (Fig. 2).

3.1   Coupling Conditions at the Interface

For the proposed model, coupling conditions are required to be transferred on the interface Γ at each time step between both the multi-layer and elasticity models. As illustrated in Fig. 1, the finite element and finite volume nodes on the interface do not coincide in general and therefore we employ a cubic spline interpolation to interchange the solutions between the two classes of nodes. At each time step, coupling conditions are required on the interface to transfer information between both models. In the present work, the deformed finite element nodes on the interface are used to reconstruct the bed Z for the shallow water Eq. (6). This bed profile is used in the finite volume solution of the flow system to obtain the water depth h^{n+1} and the water velocity v^{n+1}. On the interface, the horizontal x-direction and vertical z-direction forces are sampled from the water flow. Here, at each time step, the finite volume solutions of the multi-layer model are used to calculate the hydrostatic pressure and the friction distributions. These are then used to sample the horizontal and vertical forces at the interface to be used in the finite element solutions of the elasticity model. Thus, the horizontal force f_x in the x-direction is calculated using the friction term as

$$f_x = -gn_b^2 \frac{v^{n+1}\left|v^{n+1}\right|}{(h^{n+1})^{\frac{1}{3}}}. \tag{23}$$

Algorithm 1. Hybrid finite element/finite volume method.

Require: E, ν, ρ_w, n_b, h, u, C_r, T.

 1: Assemble the stiffness matrix $\mathbf{K}$ for the elastostatic system using the finite element method (16).
 2: Assemble the force vector $\mathbf{f}$ for the elastostatic system using the finite element method (1)-(2).
 3: Solve the linear system (16) for the displacement in the computational mesh.
 4: Solve the linear system (2) for the elastic strain in the computational mesh.
 5: Solve the linear system (5) for the stresses distributions in the computational mesh.
 6: Update the displacement of the finite element nodes on the interface.
 7: Solve the shallow water equations using:
 8: **for** each control volume $\left[x_i, x_{i+\frac{1}{2}} \right]$ **do**
 9:     Compute the numerical fluxes $\mathbf{F}^n_{i\pm\frac{1}{2}}$.
10:     Discretize the source term $\mathbf{Q}_i$ using the well-balanced discretization.
11:     Compute the solution in the first stage of the splitting $\mathbf{W}^{n+1}_i$.
12:     Compute the solution in the second and third stages of the splitting $\mathbf{W}^{n+1}_i$.
13: **end for**
14: Compute the horizontal force f_x using the bed friction according to (23).
15: Compute the vertical force f_z using the hydrostatic pressure according to (24).
16: Update the time step Δt according to the CFL condition (26).
17: Overwrite $t_n \longleftarrow t_n + \Delta t$ and go to step 2.

Similarly, the vertical force f_z in the z-direction is computed at each time step using the variation in the hydrostatic pressure as

$$p^{n+1} = -\rho g \frac{h^{n+1} - h^n}{\Delta t_n}. \tag{24}$$

Therefore, the vertical force f_z at each node on the interface Γ is reconstructed using the integral form as [15]

$$f_z^{(1)} = \int_{-1}^{1} -\frac{1}{2}\xi\,(1-\xi)\,p^{n+1}\frac{\hbar}{2}\,d\xi \approx \frac{\hbar}{6}p^{n+1},$$

$$f_z^{(2)} = \int_{-1}^{1} (1-\xi^2)\,p^{n+1}\frac{\hbar}{2}\,d\xi \approx \frac{2\hbar}{3}p^{n+1},$$

$$f_z^{(3)} = \int_{-1}^{1} \frac{1}{2}\xi\,(1+\xi)\,p^{n+1}\frac{\hbar}{2}\,d\xi \approx \frac{\hbar}{6}p^{n+1}, \tag{25}$$

where $\hbar$ is the length of the edge in the considered triangular element on the interface. Hence, once the element forces $f_z^{(1)}$, $f_z^{(2)}$ and $f_z^{(3)}$ are calculated according to (25), the global force f_z to be applied in the z-direction is calculated by accumulating the elemental forces on the overlapped nodes. Note that it is easy to verify that the element forces (25) satisfy the relation

$$f_z^{(1)} + f_z^{(2)} + f_z^{(3)} = \hbar p^{n+1}.$$

In present work, we refer to our approach as fully coupled, because at each time step the solution of the finite volume method depends on the solution of finite element method and vice versa. While the discretized systems are not assembled into a monolithic system, the coupling is achieved through mutual dependencies between these separate solvers, these steps are described in Algorithm 1.

4   Computational Results

The main goals of this section are to illustrate the numerical performance of the techniques prescribed in the previous sections and verify numerically their capability to capture the wind circulation and accurately calculate the stresses induced in the deformable beds. In all the computations reported herein, the Courant number is set to $Cr = 0.7$ and the timestep size Δt is adjusted at each time step according to the following CFL stability condition

$$\Delta t = Cr \frac{\Delta x}{\max\limits_{\alpha=1,\ldots,M}(|\lambda_\alpha^n|)}, \tag{26}$$

where M is the number of layers, and λ_α^n is the corresponding eigenvalues in each layer given in (12).

4.1   Accuracy Results

In this example, we investigate the accuracy of techniques proposed in the present work. Firstly, we examine the performance of the considered multi-layer finite volume method to the analytical solution of the wind flow problem over a flat bed presented in [17]. To this end, the velocity of $5, 10, 20$ and 50 layers free-surface over a flat bed (*i.e.* $Z = 0$) with 3400 m length and 10 m height are compared to the analytical solutions. The domain is divided into 100 control volumes and the following parameters were implemented: the air density, $\rho_a = 1.2$ kg/m^3, water density $\rho = 1025$ kg/m^3, friction coefficient $k = 0.1$ m/s, wind stress coefficient $\sigma^2 = 0.0015$, viscosity coefficient $\xi = 0.1$ m^2/s and acceleration of gravity $g = 9.81$ m/s^2. Figure 3a exhibits the velocity profile obtained at time $t = 1000$ s compared to the analytical solution. A good agreement between the numerical and analytical solutions are shown in this figure. The convergence to the analytical solution can be clearly seen as the number of layers increases.

For a numerical validation of the finite element method, the model presented in [16] has been compared to our finite element numerical solution. In this model, a squared plate foot is used in the test with a width of $R = L = 150$ mm and thickness of 20 mm. A vertical stress of 47 KN/m^2 is implemented at the center of the squared domain. In this example, a finite element linear elastic model with Young's modulus $E = 28000$ KN/m^2 and Poisson's ratio $\nu = 0.20$ is implemented. A comparison between the numerical and experimental vertical stresses at different points is shown in Fig. 3b. It can be seen that experimental results are in good agreement with numerical results. Under the considered elasticity condition, it has been found that the difference between the analytical and numerical results is only 6.5%.

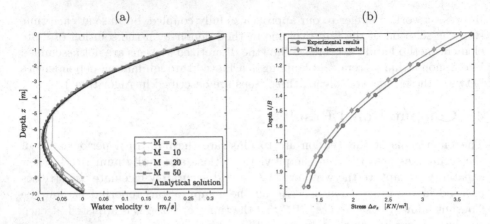

Fig. 3. Accuracy results for the finite volume method solving the multi-layer flow problem (a) and accuracy results for the finite element method solving the elastostatic bed problem (b).

4.2 Wind-Driven Circulation Flow by Pipe Failure in the Topography

The problem of pipe failures are among the common examples that received attention from civil and geotechnical engineers, see for example [7]. In these failures, the source of deformation comes from below the seabed surface. In this example, we consider shallow water waves generated by failure of two pipes in the bed topography. A rectangular domain 50 m long and 7 m high including two circular pipes with radius $R = 2.1$ m each and the initial water height is 5 m above the bed. A compressive force of 200 N is applied only at the upper half boundary of the pipes. Initially, the system is at rest and at time $t = 50$ s the constant force is applied on the upper surface of the pipes. Consequently, a deformation is expected on the pipes and therefore transmitted to the shallow water bed which generates water waves on the free-surface. The finite volume domain is discretized into 200 control volumes and the following parameters were implemented: the air density, $\rho_a = 1.2$ kg/m^3, water density $\rho = 1000$ kg/m^3, friction coefficient $k = 1 \times 10^{-1}$ m/s, wind stress coefficient $\sigma^2 = 0.0015$, viscosity coefficient $\xi = 0.01$ m^2/s, wind speed 5 m/s, acceleration of gravity $g = 9.81$ m/s^2, modulus of elasticity $E = 10000$ KN/m^2 and Poisson's ratio $\nu = 0.3$. A finite element mesh with 2471 elements and 5160 nodes is implemented in our simulations. In Fig. 4, the velocity fields obtained at time steps $t = 50$ s, 100 s and 120 s using 50 layers are presented. In this figure, the wind flows toward the right side of the domain leading to an initial water circulation with a vortex close to the right side of the domain. At a later time $t = 100$ s, the bed deformation occurred causing the water to perturb and so the water flow field is directed toward the maximum deformation in the bed. This model accurately responds to the sudden bed deformation and captures the water perturbation without any oscillations on the surface. The finite element method represents the bed deformation without

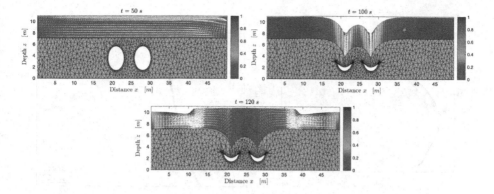

Fig. 4. Bed deformations and water velocity fields at three different times.

Table 1. Effects of wind speed and number of layers on the circulation centers.

$t = 1000$ s, $w = 5$ m/s		$t = 1000$ s, 50 layers	
Number of layers	Center of circulation	Wind speed	Center of circulation
$M = 50$	(38.75, 9.72)	5 m/s	(38.75, 9.72)
$M = 30$	(42.25, 9.722)	10 m/s	(43.25, 9.723)
$M = 20$	(43.25, 9.723)	15 m/s	(45.25, 9.724)
$M = 10$	(43.55, 9.723)	20 m/s	(47.25, 9.886)

any mesh distortion and without the need for a very refined mesh in the domain. An upward reflected wave is detected at time $t = 120$ s, the waves remain perturbed till they reach stability at time $t = 1000$ s. As can be seen in these results, a central circulation has been generated in the flow channel due to these deformations in the bed. It is worth mentioning that these circulations cannot be captured using the single-layer shallow water models, we refer the reader to [1,2] to compare the single-layer coupling results of this problem. It is clear that the center of this vortex is affected by the number of layers used in the computations and the wind speed. To further demonstrate this effect, centers of these recirculations using different numbers of layers and different wind speeds are summarised in Table 1. The center of the circulation is shifted to the right when the wind speed increases and when the number of layers decreases.

For further investigations, we presented in Fig. 4 the velocity profiles at the location $x = 40$ m obtained at time $t = 1000$ s. We also include in this figure cross-sections results for the vertical stress σ_z, the horizontal stress σ_x and the shear stress τ_{xz} at the location $x = 40$ m. Here, we considered 10, 20, 30 and 50 layers. For comparison, we also include a reference velocity field obtained for the 50 layers model using a very fine mesh of 1200 control volumes. It can be clearly shown from this figure, that an increase of the number of layers in the model illustrate a perfect convergence to the reference solution. The model

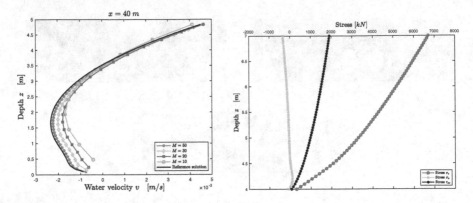

Fig. 5. Velocity profile at $x = 40$ m using different layers (left) and cross-sections of the stresses at $x = 40$ m (right).

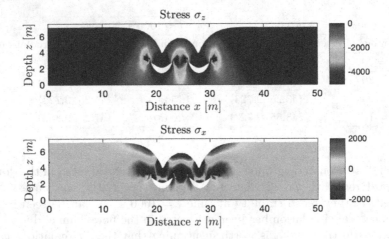

Fig. 6. Distribution for the stresses σ_z (top) and σ_x (bottom) obtained for shallow water waves generated by a pipe failure in the topography at time $t = 50$ s.

also accurately captures the velocity flow using a different number of layers and without any spurious oscillations or excessive numerical diffusion.

To investigate the distributed stresses resulting from the pipes failure underground. The two main dialary stresses σ_z and σ_x are shown in Fig. 6. It is clear from this figure that high stresses are distributed around the highest deformation in the bed. It is clear from these results that the stresses are distributed symmetrically around the vertical centerline of the mesh. The finite element method performs well and it reproduces stable solutions without nonphysical oscillations at stresses distributions. To examine the stress distributions with the bed depth, plots of vertical cross-sections in the stresses at the point $x = 21$ m for the vertical stresses σ_z, horizontal stresses σ_x and the shear stress τ_{xz} are presented in Fig. 5b. It can be noted from this figure that the vertical stresses have the highest values and the shear stresses are the lowest. It is also clear that the stress values are decreasing far from the location of the applied force and the maximum values are detected near the point of force effects.

5 Conclusions

A simple and accurate approach to couple free-surface multi-layer flows with bed deformations has been presented. The governing equations consist of coupling the nonlinear shallow water equations for water flow to the linear equations for elasticity. The coupling conditions between the two models is achieved through the interface between the two bodies and only the updated topography is required for the free-surface simulations. The hydrostatic pressure from water flow is also accounted for in the bed deformation and it is applied as external force on the elasticity model. As numerical solvers, we have considered a conservative finite volume method for the free-surface flow and a robust finite element method for the bed deformation. The new method has several advantages: First, it can solve steady flows over irregular beds without large numerical errors. Second, it can compute the numerical flux corresponding to the real state of water flow without relying on Riemann problem solvers. Third, reasonable accuracy can be obtained easily and no special treatment is needed to maintain a numerical balance, because it is performed automatically in the integrated numerical flux function. Furthermore, it has strong applicability to various problems in shallow water flows over deformed beds as shown in the numerical results. The proposed approach has been numerically examined for the test example of free-surface flow problems. The results make it promising to be applicable also to real situations where, beyond the many sources of complexity, there is a more severe demand for accuracy in predicting free-surface waves induced by sudden bed deformations, which must be performed for a long time. Future research will focus on the extension of these techniques to nonlinear plasticity in the bed deformation to allow strong interactions of water flows on largely deformed soils.

References

1. Al-Ghosoun, A., Osman, A., Seaid, M.: A hybrid finite volume/finite element method for shallow water waves by static deformation on seabeds. Eng. Comput. **38**, 2434–2459 (2021)
2. Al-Ghosoun, A., Osman, A.S., Seaid, M.: A computational model for simulation of shallow water waves by elastic deformations in the topography. Commun. Comput. Phys. **29**, 1095–1124 (2021)
3. Alongi, F., Puma, D., Nasello, C., Nizzo, S., Ciraolo, G., Noto, L.: An automatic ANN-based procedure for detecting optimal image sequences supporting LS-PIV applications for rivers monitoring. J. Hydrol. **626**, 130–167 (2023)
4. Audusse, A.: A multilayer Saint Venant system: derivation and numerical validation. Discrete Contin. Dynam. Systems **5**, 189–214 (2005)
5. Benkhaldoun, F., Sari, S., Seaid, M.: A simple multi-layer finite volume solver for density driven shallow water flows. Math. Comput. Simul. **99**, 170–189 (2014)

6. Duan, Q.: On the dynamics of Navier-Stokes equations for shallow water model. J. Differential Equations **250**, 2687–2714 (2011)
7. Gao, F., Jeng, D., Sekiguchi, H.: Numerical study on the interaction between non-linear wave, buried pipe line and non-homogenious porous seabed. Comput. Geotech. **30**, 535–547 (2003)
8. Higdon, R.: An automatically well-balanced formulation of pressure forcing for discontinuous Galerkin methods for the shallow water equations. J. Comput. Phys. **458**, 205–221 (2022)
9. Holtz, R., Kovacs, W., Sheahan, T.: Introduction to geotechnical engineering. J. Hydraul. Eng. **130**, 689–703 (2004)
10. Huang, J., Xie, X., Gao, Y., Xu, S., Zhu, M., Hu, Z., Xu, P.: Multi-layer flow field mapping in a small-scale shallow water reservoir by coastal acoustic tomography. J. Hydrol. **617**, 233–265 (2023)
11. Izem, N., Seaid, M.: A well-balanced Runge-Kutta discontinuous Galerkin method for multi-layer shallow water equations with non-flat bottom topography. Adv. Appl. Math. Mech. **14**, 725–758 (2022)
12. Liu, P., Wang, X.: A multi-layer model for nonlinear internal wave propagation in shallow water. J. Fluid Mech. **695**, 341–365 (2012)
13. Ortega, E., Onate, E., Idelsohn, S.: Method for shallow water equations. Int. J. Numer. Meth. Eng. **88**, 180–204 (2011)
14. Ran, Q., Tong, J., Shao, S., Fu, X., Xu, Y.: Incompressible SPH scour model for movable bed dam break flows. Adv. Water Resour. **82**, 39–50 (2015)
15. Sairajan, K., Deshpande, S., Patnaik, M., Poomani, D.: Base force and moment based finite element model correlation method. Adv. Space Res. **68**, 4056–4068 (2021)
16. Saleh, M., Laman, M., Baran, T.: Experimental determination and numerical analysis of vertical stresses under square footings resting on sand. Digest **19**, 4521–4538 (2008)
17. Shankar, N., Cheong, H., Sankaranarayanan, S.: Multilevel finite-difference model for three-dimensional hydrodynamic circulation. Ocean Eng. **24**, 785–816 (1997)
18. Stewart, A., Dellar, P.: An energy and potential enstrophy conserving numerical scheme for the multi-layer shallow water equations with completer Corilios force. J. Comput. Phys. **313**, 99–120 (2016)
19. Strang, G.: On the construction and the comparison of difference schemes. SIAM J. Numer. Anal. **5**, 506–517 (1968)
20. Teja, K., Manikanta, V., Das, J., Umamahesh, N.: Enhancing the predictability of flood forecast by combining numerical weather prediction ensembles with multiple hydrological models. J. Hydrol. **625**, 130–167 (2023)
21. Tubbas, K., Tsai, F.: Multilayer shallow water flow using lattice Boltzmann method with high performance computing. Adv. Water Resour. **32**, 1767–1776 (2009)
22. Zhang, Z., Tang, H., Duan, J.: High-order accurate well-balanced energy stable finite difference scheme for multi-layer shallow water equations on fixed and adaptive moving meshes. J. Hydrol. **617**, 811–823 (2023)

Unleashing the Potential of Mixed Precision in AI-Accelerated CFD Simulation on Intel CPU/GPU Architectures

Kamil Halbiniak[1]([✉]) [iD], Krzysztof Rojek[1] [iD], Sergio Iserte[2] [iD],
and Roman Wyrzykowski[1] [iD]

[1] Institute of Computer and Information Sciences, Częstochowa, Poland
{khalbiniak,krojek,roman}@icis.pcz.pl
[2] Barcelona Supercomputing Center, Barcelona, Spain
sergio.iserte@bsc.es

Abstract. CFD has emerged as an indispensable tool for comprehending and refining fluid flow phenomena within engineering domains. The recent integration of CFD with AI has unveiled novel avenues for expedited simulations and computing precision. This research paper delves into the accuracy of amalgamating CFD with AI and assesses its performance across modern server-class Intel CPU/GPU architectures such as the 4th generation of Intel Xeon Scalable CPUs (codename Sapphire Rapids) and Intel Data Center Max GPUs (or Ponte Vecchio).

Our investigation focuses on exploring the potential of mixed-precision techniques with diverse number formats, namely, *FP32*, *FP16*, and *BF16*, to accelerate CFD computations through AI-based methods. Particular emphasis is given to validating outcomes to ensure their applicability across a CFD motorBike simulation.

This research explores the performance/accuracy trade-off for both AI training and simulations, including OpenFOAM solver and interference with the trained model, across various data types available on Intel CPUs/GPUs. We aim to provide a thorough understanding of how different number formats impact the performance and accuracy of the DNN-based model in various application scenarios running on modern HPC architectures.

Keywords: HPC · CFD · AI/ML · DNN · mixed precision · CPU/GPU · Intel architectures

1 Introduction

Computational Fluid Dynamics (CFD) has emerged as a cornerstone in understanding and optimizing fluid flow phenomena across various engineering disciplines. In recent years, the intersection of CFD with artificial intelligence (AI) has sparked new possibilities, promising both accelerated simulations and enhanced

L. Franco et al. (Eds.): ICCS 2024, LNCS 14837, pp. 203–217, 2024.
https://doi.org/10.1007/978-3-031-63778-0_15

accuracy. This paper addresses a crucial amalgamation of topics at the forefront of this intersection-floating-point number formats, mixed precision in AI, and the utilization of Intel CPU and GPU architectures in CFD simulations.

In this paper, our focus centers on unraveling the intricacies of AI-accelerated CFD simulations by concentrating specifically on the Intel CPU and GPU architectures. Notably, we deliberately abstain from exploring the extensively studied NVIDIA GPU platform due to the wealth of existing literature in this domain. By directing our attention to the Intel ecosystem, we aim to contribute a distinctive perspective on mixed-precision computations, recognizing the diverse range of processors and graphics units integral to many computational environments.

Our study leverages the capabilities of OpenFOAM, a versatile and open-source CFD software, to conduct a comprehensive simulation of steady flow around a motorcycle and rider. OpenFOAM's flexibility and robust numerical algorithms make it an ideal tool for capturing the intricate fluid dynamics involved in complex scenarios. By utilizing OpenFOAM, we aim to not only investigate the impact of mixed-precision computations and Intel CPU/GPU architectures on the simulation accuracy and efficiency but also to showcase the practical application of these advancements in a real-world CFD scenario.

Our paper includes a list of contributions that we have made. These contributions are outlined below:

- *Utilizing different number formats for improved performance and validated accuracy.* Investigating the impact of floating-point data formats — float32 (*FP32*), float16 (*FP16*), and bfloat16 (*BF16*) —in CFD AI acceleration is a central focus of this work. We systematically explore the performance gains and trade-offs associated with each datatype. Additionally, we emphasize the crucial aspect of validating the accuracy of results obtained using these datatypes to ensure their applicability across diverse engineering scenarios.
- *AI acceleration of OpenFoam CFD simulation.* The adaptation of AI-accelerated techniques in CFD simulations, particularly through the integration of machine learning models into OpenFOAM, represents a significant stride towards more efficient and adaptive fluid flow predictions. Our paper demonstrates how AI can catalyze accelerating OpenFOAM simulations, leading to quicker turnaround times without compromising accuracy.
- *Verification on Intel CPU and GPU Platforms.* In pursuit of a holistic understanding, our paper extends beyond theoretical considerations. We delve into the practical implementation of our findings on Intel CPU and GPU architectures, offering a comparative analysis of performance and accuracy. By verifying the results on these widely used platforms, we bridge the gap between theoretical advancements and real-world applicability.

2　Related Work

In the realm of CFD simulations, the choice of data formats significantly influences both the accuracy and efficiency of computations. This work delves into the fundamental aspects of floating-point data formats, with a particular focus on *FP32*, *FP16*, and *BF16*. Additionally, we explore the application of mixed precision in deep learning techniques to accelerate CFD simulations, shedding light on the trade-offs between precision and computational efficiency in the context of AI-driven fluid dynamics simulations.

In recent years, there has been a notable surge in research exploring the integration of AI techniques within CFD simulations. These efforts aim to enhance the efficiency and precision of CFD simulations [16], thereby enabling the handling of more intricate and realistic problems [10,26].

A prevalent approach involves leveraging machine learning algorithms to model fluid behavior. For instance, neural networks have been successfully employed to simulate turbulence [2], forecast drag and lift forces on aircraft, and optimize the design of turbulent flow control devices [20]. Additionally, researchers have delved into utilizing AI techniques for optimizing CFD simulation parameters and settings [15]. Methods such as genetic algorithms have been utilized to determine the optimal mesh size and solver configurations for specific simulations, with the ability to dynamically adjust these parameters based on simulation outcomes [3].

Furthermore, investigations have explored employing AI techniques to analyze and interpret CFD simulation results [27]. Clustering algorithms have been effective in grouping similar flow patterns [21], while classification algorithms have been instrumental in identifying and categorizing various flow types.

To the best of our knowledge, there is a lack of literature exploring the usage of the newest Intel server-class GPUs (as well as CPUs) for CFD simulations incorporating AI techniques. In particular, in our previous works [16,17], we used NVIDIA V100 GPUs, while Graphcore IPU accelerator was employed in paper [18]. Despite the growing interest in leveraging AI for CFD acceleration and the increasing utilization of NVIDIA and AMD GPUs, there appears to be a gap in research addressing the optimization and performance assessment of Intel GPUs in this domain. The same conclusion is pertinent for the newest Intel data center CPUs, introducing AMX accelerators.

3 Floating-Point Data Formats and Mixed Precision in AI-Accelerated CFD Simulation

Precision in deep learning models is a critical factor influencing performance, memory utilization, and overall computational efficiency. This section provides an in-depth analysis of different floating-point data formats, namely *FP32*, *FP16*, and *BF16*, and their implications for mixed precision deep learning. By understanding the nuances of these formats, we aim to unravel the potential advantages and challenges associated with employing mixed precision techniques in the context of accelerating AI algorithms for CFD simulations.

The *FP32* format was the backbone of deep learning for a long time [14]. It offers a high range of representable values, making it suitable for a wide array of AI computations, particularly in problems where accuracy is crucial. On the other hand, *FP32* requires relatively a lot of memory and computing resources.

In contrast, lower precision formats such as *FP16* and *BF16* sacrifice precision in favor of reduced memory usage and computational demands [12], making them an attractive choice for large-scale AI computations (e.g., training of large models). Nonetheless, reduced precision may lead to numerical instability in certain computations. Therefore, the optimal choice of floating-point precision depends on the specific application and the trade-off between precision needs, memory constraints, and computational efficiency.

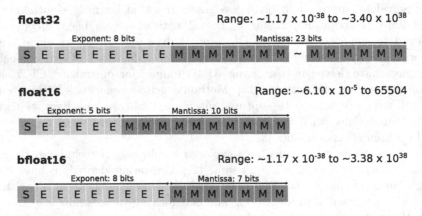

Fig. 1. Comparison of *FP32*, *FP16* and *BF16* data formats (ranges for normalized values) [12]

Figure 1 illustrates the comparison of *FP32*, *FP16* and *BF16* data formats. The *FP32* uses 32 bits to represent a floating-point number. In this format, a single bit is allocated for the sign, 8 bits for the exponent, and 23 bits for the mantissa. Half-precision *FP16* format utilizes 16 bits to represent floating-point values. Within these 16 bits, a single bit is reserved for sign, 5 bits for exponent, and 10 bits for mantissa [12]. This format offers a reduced precision compared to *FP32*, making it more memory-efficient and computationally faster. However, it comes with the cost of the smaller range of representable values [12]. The *BF16* also uses 16 bits while balancing precision and efficiency. It provides the approximate dynamic range of *FP32* format by retaining eight exponent bits but supports only a 7-bit mantissa rather than the 23-bit [4,12]. The *BF16* is a replacement for the *FP16* format. It allows for fast conversion to and from *FP32*. Unlike *FP16*, which usually requires special techniques, conversion from *FP32* to the *BF16* is performed by truncating the mantissa field [12]. This difference can be seen in the training process, where the use of *FP16* forces adding a loss scaling process to preserve small gradient values [14]. Another advantage of *BF16* is the hardware cost. By having three fewer mantissa bits, the *BF16* multiplier takes up about half of the hardware area (number of transistors) against the *FP16* unit [25]. The *BF16* data format is implemented in modern Intel CPUs and GPUs, Google's TPU, and NVIDIA GPUs [24].

Mixed-precision is a technique used to optimize the performance of numerical computations using a combination of lower and higher-precision data formats. It has become a powerful optimization to accelerate AI computations, especially deep learning training and inference processes [12]. Employing a combination of 16-bit and 32-bit floating-point data formats in a model during training aims to run it faster and reduce memory utilization. Additionally, accelerating the computations and reducing the execution time with lower-precision number formats allows decreasing energy consumption. By preserving certain parts of the model in the *FP32* format for numeric stability, the model will have a lower step time and train equally as well in terms of the evaluation metrics (e.g., accuracy) [24]. Modern deep learning frameworks and libraries, such as TensorFlow or PyTorch, provide tools to facilitate the implementation of mixed precision in both model training and inference. Although mixed precision can be run on most hardware, it will only speed up the computations on the devices that support mixing 16-bit and 32-bit data formats [24].

4   Intel CPU and GPU Architectures in Deep Learning Mixed-Precision Computation

As the landscape of deep learning accelerates, the choice of hardware architectures plays a pivotal role in achieving optimal performance [19]. Focusing on modern Intel's CPU and GPU architectures, this section investigates their role in facilitating mixed-precision computations for deep learning tasks. In particular, we investigate how the features and capabilities of the 4th Generation of Intel Xeon Scalable CPUs (codename Sapphire Rapids) [6] and Intel Data Center Max GPUs (codename Ponte Vecchio) [9] can be used in practice to accelerate AI computations.

The Intel Xeon Sapphire Rapids processors revolutionize a landscape of AI computations on general-purpose processors by introducing a built-in Intel Advanced Matrix Extension (Intel AMX) accelerator. Intel AMX is a dedicated hardware block of the processor core that helps optimize and accelerate deep learning training and inferencing workloads relying on matrix operations [6]. It allows running AI computations directly on the CPU instead of offloading them to a discrete accelerator (e.g., GPU). The Intel AMX architecture consists of two components [6]: (i) tiles consisting of two-dimensional registers (each 1KB in size) that store large chunks of data; (ii) Tile Matrix Multiplication (TMUL) which is an accelerator engine attached to the tile that performs AI matrix multiplication.

The AMX accelerator supports *INT8* (8-bit integer) and *BF16* data formats. While the first is a data type used for inferencing, the second can be used for both training and inference. Using the AMX, Intel Sapphire Rapids-based processors can quickly pivot between optimizing the AI workloads and general computing. In practice, the programmers can code AI computations to take advantage of the AMX instruction set and implement non-AI functionality based on the processor instruction set architecture [6].

Intel Data Center GPU Max is a series of general-purpose discrete GPUs, designed for breakthrough performance in data-intensive computing models used in AI and HPC [9]. The GPUs are available as PCIe cards and OpenCompute Accelerator Modules to offer remedies for servers with high GPU density. Intel Data Center GPUs are based on the X^e HPC architecture with a compute-focused, programable, and scalable element named the X^e-core [9]. Each core consists of 512-bit wide vector engines, 4096-bit wide matrix engines called Intel Xe Matrix eXtensions (Intel XMX), L1 data cache and shared local memory [9]. The vector engines of X^e-core support *FP64*, *FP32* and *FP16* vector computations. Simultaneously, the XMX is engineered to accelerate AI computations and provides support for *TF32* (19-bit tensor float format), *BF16*, *FP16* and *INT8* data formats.

The Intel Data Center GPUs are built upon the X^e HPC Stacks, which are made up of various tiles stacked on top of each other within a single package. The Xe HPC Stack includes of the X^e-core Tile (Compute Tile), L2 Cache Tile, Base Tile (PCIe paths, media engine, etc.), High Memory Bandwidth Tile, Xe Link Tile for scale-up and scale-out, and the Embedded Multi-Die Interconnect Bridge for communication between X^e HPC Stacks [9]. These components all reside within a Multi Tile Package. The top-of-the-line Intel Data Center Max 1550 GPU contains two X^e HPC Stacks with 64 X^e-cores, 204MB of L2 cache and 64GB HBM2e each. This Intel GPU offers the peak performance of 832 TFLOP for *BF16* and 52 TFLOPS for both *FP64* and *FP32* computations [9].

To leverage Intel CPUs/GPUs architectures in deep learning mixed-precision computation, the programmers may use the TensorFlow framework together with Intel Extension for TensorFlow (ITEX in short). ITEX is a heterogeneous, high-performance, deep-learning extension plugin based on the TensorFlow Pluggable Device interface [7]. It is designed to optimize the performance of TensorFlow-based applications on Intel computing architectures. The ITEX provides a feature called Advanced Auto Mixed Precision (AMP in short), which allows users to enable mixed-precision computations. The AMP is similar to stock Tensor-Flow Auto Mixed Precision but offers better usage and performance on Intel CPUs and GPUs [8]. Listing 1.1 shows the snippet of code that enables mixed-precision computing on Intel CPUs and GPUs with ITEX. The presented code sets the global policy leading to the mixed-precision computations based on *BF16* data format. As a result, it allows AI computations to be performed using AMX and XMX instruction sets on Intel Xeon Sapphire Rapids CPUs and Intel Data Centers Max GPUs, respectively. The programmer can easily change the mixed-precision policy to *FP16* data format by replacing `itex.BFLOAT16` with `itex.FLOAT16`.

Beyond specifying the data format used in mixed-precision computing, the AMP also allows controlling operators that can be converted to lower-precision data types. TensorFlow provides *Allow*, *Deny*, *Clear*, and *Infer* list of operators to classify operators based on the operation's numerical safety [5,7]. The numerical safety corresponds to how the accuracy of the model is affected by using lower precision. The exact lists could be found in `auto_mixed_precision_lists.h`

file in the TensorFlow GitHub repository [23]. Listing 1.2 presents the snippet
of code corresponding to the usage of the AMP to add and remove operators
from the TensorFlow lists. This presented code adds the `Conv2D` operator to
the *Deny* list which holds operations considered to be numerically dangerous
in lower precision [5]. Simultaneously, the `Conv2D` operator is removed from the
list of operators considered as numerically safe for execution (*Allow* list). The
modification of the remaining lists is analogous. It is important to note that it
is not allowed for the same operator to be in more than one list [24].

```
1 import intel_extension_for_tensorflow as itex
2
3 amp_options = itex.AutoMixedPrecisionOptions()
4 amp_options.data_type = itex.BFLOAT16
5
6 graph_options = itex.GraphOptions(
      auto_mixed_precision_options=amp_options)
7 graph_options.auto_mixed_precision = itex.ON
8
9 config = itex.ConfigProto(graph_options=graph_options)
10 itex.set_config(config)
```
Listing 1.1. Activation of Advanced Automatic Mixed Precision on Intel CPUs and
GPUs using Intel Extension for TensorFlow

```
1 import intel_extension_for_tensorflow as itex
2
3 amp_options = itex.AutoMixedPrecisionOptions()
4 amp_options.denylist_add= "Conv2D"
5 amp_options.allowlist_remove = "Conv2D"
```
Listing 1.2. Modification of TensorFlow lists of operators using advanced Automatic
Mixed Precision of Intel Extension for TensorFlow

5 AI-Accelerated CFD Simulation

5.1 CFD Simulation of Steady Flow Around a Motorcycle and Rider

Applying the basic approaches discussed in the preceding sections, we shift our
focus to a practical application - CFD simulation of steady flow around a motorcycle and rider. This job uses OpenFOAM [1] to calculate the steady flow. The
simpleFoam solver is used for this case, performing steady-state, incompressible
Reynolds-Averaged Navier-Stokes calculations (RANS) over the mesh. RANS is
a widely used approach in CFD for simulating turbulent flows. In fluid dynamics, the Navier-Stokes equations describe the motion of fluid, taking into account
viscosity, pressure, and velocity.

  In this study, we aim to harness the power of AI to streamline and enhance
the evaluation of crucial fluid dynamic parameters, specifically velocity U and

pressure p, within the framework of RANS. By integrating AI methodologies into the evaluation process, we seek to improve the accuracy and efficiency of estimating U and p fields.

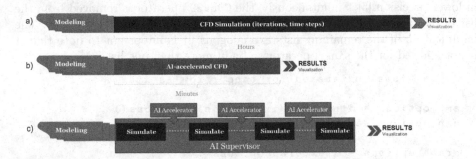

Fig. 2. The idea of integrating the AI module with CFD simulation, where a) presents standalone CFD simulation, b) idea of time reduction by combining CFD simulation with AI, c) the mechanism of management of the CFD simulation by AI supervisor.

5.2 Integration of AI Acceleration with CFD Solver

This study exploits the methodology proposed in our previous work [17] aimed at enhancing CFD simulations by integrating two main components: the AI supervisor and the AI accelerator. During the inference stage, the AI accelerator uses the previously trained AI model to predict the simulation results based on a relatively small number of iterations generated by the CFD solver. The AI supervisor dynamically switches between traditional CFD simulation and AI predictions, while the AI accelerator module expedites the process by extrapolating simulation results. Initially, the traditional CFD solver runs for a predetermined number of iterations to establish initial data points, subsequently utilized by a machine learning model to forecast fluid flow dynamics. Upon generating output, the AI supervisor directs the CFD solver to resume simulation based on the predicted data, iteratively alternating between CFD and AI components until convergence is reached. The convergence threshold hinges on factors such as the complexity of simulated flow dynamics and the quality of training data. The idea of our method is presented in Fig. 2.

The AI supervisor's role in this methodology is pivotal, discerning data patterns within the simulation to gauge if a steady state has been attained. By analyzing CFD simulation output, it determines whether to invoke the AI accelerator or conclude the simulation. The specifics regarding the AI model, including its architecture and training methodology, are presented in our prior publications [11,17]. The model primarily comprises convolutional layers, forming the backbone of its architecture. These convolutional layers play a central role in extracting features from the input data, enabling the AI model to learn and predict fluid flow dynamics effectively. All the operators within the model can be safely converted and used with lower precision.

5.3   Training Dataset

Machine learning algorithms, trained on vast datasets generated through RANS simulations, are employed to infer and optimize the estimation of U and p parameters. These datasets are carefully constructed, with each simulation representing various operational conditions, such as different motorbike speeds, alongside additional parameters like ambient temperature, and turbulence intensity. In our case, 50 simulations of a motorbike at different speeds were conducted, each comprising a window of 5 consecutive iterations. They are spaced at intervals of 20 timesteps, with steady-state conditions achieved after 500 iterations. To ensure robustness and diversity, $500/20 = 25$ samples are extracted from each simulation, resulting in the generation of $20 * 25 = 1250$ samples in total.

6   Experimental Results

6.1   Testing Platforms

In the tests, we use the following computing platforms:

1. a server with the two 4-th generation 56-core Intel Xeon Platinum 8480 CPUs and 512 GB DDR5 main memory;
2. a server with two 32-core Intel Xeon Platinum 8462Y CPUs, 1024 GB DDR5 main memory and four Intel Data Center Max 1550 GPUs.

Both servers are empowered with TensorFlow 2.13.0, Intel Extension for TensorFlow* 2.13.0.1 and Intel oneAPI Base Toolkit 2023.2.0. The computing platform with Intel Xeon 8480 CPUs has installed Python 3.9.16, while the server with Intel Max GPUs uses Python 3.11.5.

Table 1. Execution times (in seconds) and speedups achieved for AI training with different data formats running on two Intel Xeon 8480 CPU and a single Intel Data Center Max 1550 GPU

Data format	Intel Xeon 8480 CPUs		Intel Max 1550 GPU		S_{GPU}
	T_{CPU}	S_{FP32}	T_{GPU}	S_{FP32}	
FP32	2326	1	408	1	5.7x
BF16	1544	1.51x	238	1.71x	6.48x
FP16	—	—	430	0.95x	—

6.2 Accuracy and Performance Results

Table 1 shows the execution times T_{CPU} and T_{GPU} obtained for the AI training workloads running on two Intel Xeon 8480 CPUs and a single Intel Data Center Max 1550 GPU. In the benchmarks, we focus on pure *FP32* and the mixed-precision training with *BF16* and *FP16* formats, respectively. In the case of Intel GPU, the AI workloads are executed using a single Compute Tile from both available in the accelerator. This is because the TensorFlow environment, by default, treats every Intel Max 1550 GPU tile as a TensorFlow individual device. To maximize the performance of training, different batch sizes for Intel Xeon CPUs and Intel Max GPU were evaluated.

Intel Extension for TensorFlow employs OpenMP to parallelize the computations across cores/threads of Intel CPUs. To maximize the performance of the AI training, we also investigate different setups of OpenMP environment variables. Among the considered environment variables are [23]:

- OMP_NUM_THREADS which sets the number of OpenMP threads;
- KMP_AFFINITY which bind OpenMP threads to physical cores;
- KMP_BLOCKTIME which sets the time (in milliseconds) that a thread should wait, after completing the execution of a parallel region, before sleeping.

When benchmarking Intel CPUs, these variables were set as follows:

- OMP_NUM_THREADS=112;
- KMP_AFFINITY=fine,compact,1,0;
- KMP_BLOCKTIME=0.

The performance results presented in Table 1 are obtained for the training process with batch sizes equal to 64 and 8 for Intel Xeon 8480 CPUs and Intel Data Center Max 1550 GPU, respectively. Besides the execution times, the table presents also speedup S_{FP32} obtained against *FP32* data format. The last column of Table 1 illustrates the speedup $S_{GPU} = T_{CPU}/T_{GPU}$ obtained for Intel Max 1550 GPU against two Intel Xeon 8480 CPUs. The performance results in Table 1 indicate significant benefits of using mixed-precision training based on *BF16* number format. The speedup achieved against the pure *FP32* AI computations is about 1.5 and 1.7 times for Intel Xeon CPUs and Intel Max GPU, respectively. The Advanced Auto Mixed Precision feature does not support mixed-precision *FP16* AI computations on Intel processors [23]. Thus, we are not able to measure the performance of mixed-precision training for this data format. At the same time, the utilization of *FP16* during training on Intel Max GPU yields even longer execution time than *FP32* data format. The explanation for unexpected low performance may be the cost of casting tensors from *FP32* to *FP16*, or vice versa. In some cases (especially for operations on huge tensors), the casting cost surpasses the performance benefit of using low precision [5].

Based on Table 1, we conclude that Intel Data Center Max 1550 GPU outperforms two Intel Xeon 8480 CPUs. Already for *FP32* computations, a single tile of Intel GPU allows executing the training process 5.7 times faster. Even greater performance gain is notable for *FP16*, where the Intel GPU allows accelerating the AI computations about 6.5 times against two Intel CPUs.

Figure 3 displays the mean square error (MSE) for CFD simulations using various configurations of the CFD solver in conjunction with the AI accelerator. The MSE serves as a crucial metric indicating the deviation between predicted and actual values during the convergence process toward the steady state. This figure provides insights into the number of iterations required by each data precision to meet the designated MSE threshold, thereby reaching the desired accuracy and correlation levels.

Each configuration aims to achieve a specific MSE threshold predetermined by our AI supervisor. This threshold is selected to ensure that the result accuracy surpasses 97%, and the Pearson correlation coefficient exceeds 0.98 compared to the CFD standalone version.

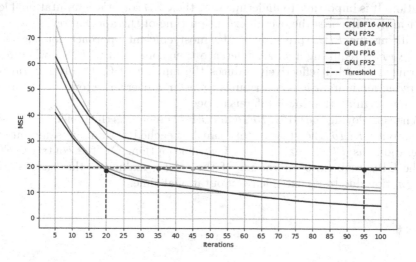

Fig. 3. MSE comparison and the number of iterations required to execute by the CFD solver to achieve the steady state with the AI accelerator.

Table 2 compares accuracy metrics for various configurations of the CFD solver integrated with the AI accelerator executed on both CPU and GPU platforms. It outlines the number of iterations required for convergence, MSE, Pearson correlation coefficient, and accuracy percentages for each configuration. Notably, it highlights the trade-offs between precision levels provided with various data formats (*FP16*, *BF16*, and *FP32*) for the studied platforms (CPU and GPU) in terms of computational model accuracy.

The last row of Table 2 represents the speedup determined as the ratio of iterations for the standalone CFD solver and AI-accelerated simulation. This way of calculating the speedup is justified by the fact that the execution time for performing CFD solver iterations is much longer than the execution time of AI predictions.

Table 2. Accuracy comparison and the number of iterations required to execute by the CFD solver to achieve the steady state with the AI accelerator

	CPU *BF16 AMX*	CPU *FP32*	GPU *BF16*	GPU *FP16*	GPU *FP32*
Iterations	45	35	20	95	20
MSE	19.32	19.34	19.61	19.24	18.54
Pearson	0.983	0.983	0.983	0.986	0.984
Accuracy[%]	96.9	96.9	96.9	96.9	97.0
Speedup	11.1	14.3	25.0	5.3	25.0

Figure 4 illustrates the number of iterations required by the CFD solver in conjunction with an AI accelerator to achieve a steady state during application execution. It is important to underline that the CFD solver's computational load significantly influences the overall execution time of the application.

Utilizing the *FP32* precision is the most efficient approach for the CPU. Despite *FP32* requiring a longer inference time compared to *BF16*, when using the trained model, it effectively reduces the number of CFD solver iterations. This reduction in iterations contributes to faster convergence towards the steady state, outweighing the slower inference speed.

When employing the GPU, *BF16* and *FP32* emerge as the preferred precision choice. With *BF16*, the steady state can be reached within the same number of iterations as *FP32*, making it a favorable option. On the contrary, *FP16* for the GPU requires 95 iterations of the solver, making it the least efficient configuration due to its extended computational time.

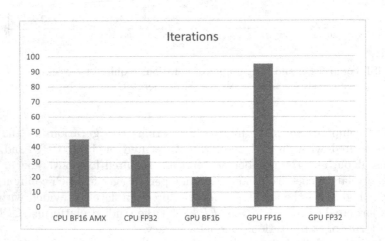

Fig. 4. The number of iterations required to execute by the CFD solver to achieve the steady state with the AI accelerator.

7   Conclusion

In this paper, we investigate the benefits of using modern server-class Intel CPU and GPU architectures for AI-accelerated simulations. This study focuses on unveiling the potential of using mixed-precision arithmetic based on *FP32*, *FP16*, and *BF16* data formats to accelerate CFD computations through AI techniques. We aim to not only assess the impact of the mixed-precision approach on the performance of computations, but also on the simulation accuracy based on the motorBike use-case scenario.

The performance results achieved in the paper show the performance benefits of using mixed-precision on both Intel Xeon 8480 CPUs and Intel Data Center Max 1550 GPUs for model training. The utilization of *BF16* format allows accelerating the training process about 1.5 and 1.7 times against pure *FP32* computations for Intel CPUs and GPU, respectively. At the same time, the *FP16* mixed-precision training workloads give even longer execution time than in the case of *FP32* data format. An additional conclusion resulting from the benchmarks is that a single tile of Intel Data Center Max 1550 GPU outperforms two Intel Xeon 8480 CPUs during the training of DNN models. In fact, Intel Max GPU allows executing the training process 5.7 and 6.5 times faster against two Intel Xeon CPUs, for *FP32* and *BF16* data formats, respectively.

The results achieved for the whole simulation underscore the nuanced balance between computational performance and model accuracy, as configurations diverge across precision levels (*FP16*, *BF16*, and *FP32*) and platforms (CPU and GPU). Specifically, for the CPU, the most efficient configuration emerges as *FP32* due to its ability to achieve convergence with fewer iterations of the CFD solver compared to other precision types. For the GPU, the optimal configurations are *BF16* and *BF32* that provide a comparable performance.

The presented analysis focuses on the convergence behavior of the CFD solver accelerated with AI. The MSE error serves as a crucial indicator of convergence, with each configuration aiming to meet a specific threshold determined by the AI supervisor. This threshold is carefully selected to ensure the result accuracy of about 97% and the Pearson correlation coefficient surpassing 0.98 compared to the standalone CFD version.

In our future work, we plan to investigate the differences observed in the experiments (see Fig. 2 and Fig. 4) when implementing the simulation on Intel CPU and GPU. These differences occur for both the *FP32* and *BF16* data formats, and are reflected in the increased number of iterations on CPU compared to GPU. We guess that the reason is the aggressive performance optimizations [22] performed by TensorFlow. This optimization is forced on the CPU by its oneDNN option, which is set by default for Intel CPUs (and can be turned off). This option is not used for Intel GPUs. The side effect of using TensorFlow with oneDNN optimizations are changes in the execution order of operations and greater sensitivity to floating-point round-off errors [22]. Consequently, this improves the performance at the cost of the reduced accuracy, which leads to an increased number of iterations during the inference stage.

Another direction of future work includes investigating the usage of mixed-precision on the newest NVIDIA H100 and H200 GPUs as providing innovative hardware features [13] compared to previous Volta and Ampere GPUs. We also plan to incorporate more application examples to show the benefits of using the mixed-precision approach for CFD simulations.

Acknowledgements. We extend our sincere gratitude to byteLAKE. Their support and collaboration have been instrumental in our research and development endeavors. The researcher from BSC is involved in the project The European PILOT which has received funding from the European High-Performance Computing Joint Undertaking under grant agreement No. 101034126 and No. PCI2021-122090-2A under the MCIN/AEI and the EU NextGenerationEU/PRTR.

References

1. OpenFOAM. https://www.openfoam.com. Accessed 23 Feb 2024
2. Berkooz, G., Holmes, P., Lumley, J.L.: The proper orthogonal decomposition in the analysis of turbulent flows. Annu. Rev. Fluid Mech. **25**(1), 539–575 (1993)
3. Bhatt, D., Zhang, B., Zuckerman, D.: Steady-state simulations using weighted ensemble path sampling. J. Chem. Phys. **133**(1) (2010)
4. Dörrich, M., Fan, M., Kist, A.M.: Impact of mixed precision techniques on training and inference efficiency of deep neural networks. IEEE Access **11**, 57627–57634 (2023)
5. He, X., Sun, J., Chen, H., Li, D.: Campo: Cost-Aware performance optimization for Mixed-Precision neural network training. In: 2022 USENIX Annual Technical Conference (USENIX ATC 22), pp. 505–518. USENIX Association, Carlsbad, CA (2022). https://www.usenix.org/conference/atc22/presentation/he
6. Intel: Accelerate Artificial Intelligence (AI) Workloads with Intel Advanced Matrix Extensions (Intel AMX) (2022). https://www.intel.com/content/dam/www/central-libraries/us/en/documents/2022-12/accelerate-ai-with-amx-sb.pdf
7. Intel: An Easy Introduction to Intel Extension for TensorFlow (2022). https://www.intel.com/content/www/us/en/developer/articles/technical/introduction-to-intel-extension-for-tensorflow.html
8. Intel: Intel Extension for TensorFlow: Advanced Auto Mixed Precision (2022). https://intel.github.io/intel-extension-for-tensorflow
9. Intel: Intel Data Center GPU Max Series Technical Overview. https://www.intel.com/content/www/us/en/developer/articles/technical/intel-data-center-gpu-max-series-overview.html (2023)
10. Iserte, S., et al.: Modeling of wastewater treatment processes with HydroSludge. Water Environ. Res. **93**(12), 3049–3063 (2021)
11. Iserte, S., Macías, A., Martínez-Cuenca, R., Chiva, S., Paredes, R., Quintana-Ortí, E.S.: Accelerating urban scale simulations leveraging local spatial 3D structure. J. Comput. Sci. **62**, 101741 (2022)
12. Kalamkar, D., et al.: A study of bfloat16 for deep learning training (2019). https://arxiv.org/abs/1905.12322
13. Luo, W., et al.: Benchmarking and Dissecting the Nvidia Hopper GPU Architecture (2024). https://arxiv.org/abs/2402.13499v1
14. Micikevicius, P., et al.: Mixed precision training (2018). https://arxiv.org/abs/1710.03740

15. Rojek, K., Wyrzykowski, R.: Performance and scalability analysis of AI-accelerated CFD simulations across various computing platforms. In: Singer, J., Elkhatib, Y., Heras, D., Diehl, P., Brown, N., Ilic, A. (eds.) Euro-Par 2022: Parallel Processing Workshops, pp. 223–234. Springer, Cham (2023). https://doi.org/10.1007/978-3-031-31209-0_17

16. Rojek, K., Wyrzykowski, R., Gepner, P.: AI-Accelerated CFD Simulation Based on OpenFOAM and CPU/GPU Computing. In: Paszynski, M., Kranzlmüller, D., Krzhizhanovskaya, V.V., Dongarra, J.J., Sloot, P.M.A. (eds.) Computational Science - ICCS 2021, pp. 373–385. Springer International Publishing, Cham (2021)

17. Rojek, K., Wyrzykowski, R., Gepner, P.: Chemical Mixing Simulations with Integrated AI Accelerator. In: Mikyška, J., de Mulatier, C., Paszynski, M., Krzhizhanovskaya, V.V., Dongarra, J.J., Sloot, P.M. (eds.) Computational Science - ICCS 2023, pp. 494–508. Springer Nature Switzerland, Cham (2023)

18. Rościszewski, P., Krzywaniak, A., Iserte, S., Rojek, K., Gepner, P.: Optimizing throughput of Seq2Seq model training on the IPU platform for AI-accelerated CFD simulations. Futur. Gener. Comput. Syst. **147**, 149–162 (2023)

19. Silvano, C., et al.: A Survey on Deep Learning Hardware Accelerators for Heterogeneous HPC Platforms (2023). https://arxiv.org/abs/2306.15552

20. Srivastava, S., Damodaran, M., Khoo, B.C.: Machine Learning Surrogates for Predicting Response of an Aero-structural-sloshing System (2019). https://arxiv.org/pdf/1911.10043

21. Sun, P., Gao, L., Han, S.: Identification of overlapping and non-overlapping community structure by fuzzy clustering in complex networks. Inf. Sci. **181**, 1060–1071 (2011)

22. TensorFlow: What's new in TensorFlow 2.9? (2022). https://blog.tensorflow.org/2022/05/whats-new-in-tensorflow-29.html

23. TensorFlow: TensforFlow Official GitHub Repository. https://github.com/tensorflow/tensorflow (2024)

24. TensorFlow: TensorFlow Guide: Mixed Precision (2024). https://www.tensorflow.org/guide/mixed_precision

25. Verheyde, A.: BFloat16 Deep Dive: ARM Brings BF16 Deep Learning Data Format to ARMv8-A (2019). https://www.tomshardware.com/news/bfloat16-deep-dive-arm-bf16-support-armv8-a,40305.html

26. Vinuesa, R., Brunton, S.L.: The Potential of Machine Learning to Enhance Computational Fluid Dynamics (2021). https://arxiv.org/pdf/2110.02085

27. Zhang, S., Wang, R., Zhang, X.: Identification of overlapping community structure in complex networks using fuzzy c-means clustering. Phys. A **374**, 483–490 (2007)

Quantum Computing

The Significance of the Quantum Volume for Other Algorithms: A Case Study for Quantum Amplitude Estimation

Jins de Jong[✉] and Carmen R. Hoek

TNO, Unit ISP, Zernikelaan 14, 9747AA Groningen, The Netherlands
jins.dejong@tno.nl

Abstract. The quantum volume is a comprehensive, single number metric to describe the computational power of a quantum computer. It has grown exponentially in the recent past. In this study we will assume this remains the case and translate this development into the performance development of another quantum algorithms, quantum amplitude estimation. This is done using a noise model that estimates the error probability of a single run of an algorithm. Its parameters are related to the quantum volume under the model's assumptions.

Applying the same noise model to quantum amplitude estimation, it is possible to relate the error rate to the generated Fisher information per second, which is the main performance metric of quantum amplitude estimation as a numerical integration technique. This provides a prediction of its integration capabilities and shows that quantum amplitude estimation as a numerical integration technique will not provide an advantage over classical alternatives in the near future without major breakthroughs.

Keywords: Quantum volume · Quantum amplitude estimation · Noisy quantum computing

1 Introduction

The promise of the quantum computer is that it can speed up some computation routines quadratically or can even solve classically intractable problems with an exponential speed-up. The capabilities of current quantum computers are limited by errors rather than by size or speed. The origin of these errors lies in the details of their physical realization. Given that there are multiple physical approaches to build a quantum computer, each with different components, strengths and weaknesses, it is hard to compare quantum computers' performance on an algorithm from the reported error characteristics of the physical components. To obtain more understanding of the execution quality, performance metrics focus

---

This work was supported by the Dutch National Growth Fund (NGF), as part of the Quantum Delta NL programme.

L. Franco et al. (Eds.): ICCS 2024, LNCS 14837, pp. 221–234, 2024.
https://doi.org/10.1007/978-3-031-63778-0_16

increasingly on application benchmarks [16,18,19]. These use the size, speed or solution quality of an application on a quantum machine as metric. Currently, the most reported metric of this kind is the quantum volume (QV) [7].

This paper studies the implications of the development of the QV for the quantum amplitude estimation (QAE) algorithm. Can the available QV be translated into a performance estimate for QAE?

1.1   Quantum Volume

Using application-based metrics rather than component properties to benchmark the performance of a quantum processing units (QPU) provides a view on the performance that is of direct interest for the user. The best-known metric of this kind today is the quantum volume. It gives in a single number the size of a specific computational problem, finding the heavy outputs of a randomized circuit, that can be run reliably.

The QV metric considers both the width of the machine and the depth of the circuit that is run. Circuits that are too wide, do not fit onto the available number of qubits. If the circuit becomes too deep, a run of the circuit becomes unreliable as a result of errors in the gate execution. The QV is the largest size of the problem that can be run on a quantum machine, without being dominated by noise. As such, it is a property of the hardware. In most current designs of quantum computers, the depth is the bottleneck for the development of the QV.

The QV will not remain the common performance metric in the future. Noiseless executions of the circuits are needed to determine the QV. In the near future these can only be performed using long-lasting simulations on classical hardware. These simulations will take so lang that for such QPU's alternative metrics, such as the Error per layered gate [30], will be needed.

1.2   QAE as a Numerical Integration Technique

On a quantum computer, numerical integration can be performed with quantum amplitude estimation (QAE) algorithms [11,15,25,27]. Compared to the classical algorithms, QAE enjoys faster convergence rates in ideal, noiseless settings. Furthermore, it may function on near-term NISQ devices [22,26]. This paper aims to study if these convergence rates are feasible on near-term quantum hardware.

For this we compare QAE with the Monte Carlo integration (MCI) technique for classical computers. In MCI a random sample from the domain is chosen and its average function value is multiplied by the volume of the domain as an estimate of the integral. Since the function is evaluated once for each sample point, the number of function evaluations N is the same as the number of samples. The error of MCI decreases as $N^{-1/2}$ [24]. The accuracy of this method is determined only by the number of samples that is processed.

Both MCI and QAE treat a finite number of samples. Thus, the sampling error is present too in QAE. In the approach [25] used here, a circuit of $n + 1$ qubits generates 2^n samples. QAE can only be beneficial, if $2^n > N$, where N

is the number of MCI samples. The classical example in Sect. 2.6 suggests 2^{26} samples are needed for this, so quantum computers with at least 27 qubits are needed [25]. This puts a lower bound on the QV required to perform competitive QAE.

Besides this sampling error, the quantum computer suffers from other errors inherent to the quantum process. The challenge is thus to device a method for which the quantum errors decay faster than $N^{-1/2}$, where N is interpreted as the number of function evaluations.

1.3   Related Work

The quantum volume [4,7] is shown to be a suitable candidate for a quantum metric by [3]. Limitations of randomized application-centric benchmarks have been studied in [23]. An exponential growth of the quantum volume has been observed, since its conception [9].

The QAE algorithm is based on the quantum phase estimation algorithm [5,8]. Replacing the quantum Fourier transform with maximum likelihood estimation [25] simplifies the circuit considerably. A method without post-processing is Iterative QAE [12] at the price of a varying execution length. Generalizations for real-valued integrands [17] or minimal circuit depth [13] have also been constructed. To avoid very deep circuits, maximum-depth approximations have been tested in the noiseless setting [21]. The performance of QAE in the presence of noise has been studied in [6,26,28].

1.4   Our Contributions

How can a user estimate the performance of a quantum algorithm from the available QV? We try to provide a first answer to this question. To this end a noise metric will be used that can be derived from the QV. It gives for a single circuit run an estimate of the error rate as a function of the size of the transpiled and decomposed circuit and the number of measurements.

A second contribution lies in the application of this method to QAE. The QAE algorithm of [25] looks very promising [20], but it is unclear when it will outperform classical integration methods. The observed exponential growth of the quantum volume can be translated into error parameters for QAE runs. Using these, predictions for the Fisher information per second can be made, which is the metric that determines the accuracy. These shows that major breakthroughs are needed to make QAE competitive with classical methods in the near future.

This article is structured as follows. In Sect. 2 an introduction of the QV, our noise model and QAE is given. The theoretical relation between the noise model and the quantum volume, as well as an applicability test of our noise model are given in Sect. 3. These are combined with the size characteristics of the QAE and QV circuits in Sect. 4 to provide an estimate of the integration power of QAE in the near future. The paper is concluded with Sect. 5.

2   Preliminaries

This section briefly treats the concept of QV, introduces our noise model with its assumptions and summarizes the framework of QAE and Fisher information for noiseless and noisy quantum processors.

The quantum volume of a QPU with N available qubits is determined from a set of experiments on squared circuits, meaning that the number of qubits q used is equal to the number of randomized layers, where $q = 1, 2, \ldots, N$. In each layer, the qubits are permuted randomly, divided into pairs that are rotated with a random element of $SU(4)$. Such a circuit on q qubits will return on measurement each q-bit answer with a certain probability. By definition, half of the answers will have a probability larger than the median. These are called the heavy outputs. The idea of a QV experiment is to reproduce as many of these heavy outputs using the noisy quantum computer. An ideal machine would produce 84.66% of the heavy outputs [3] and guessing would yield half of the heavy outputs. A quantum volume experiment is defined [7] to be successful if 2/3 of the heavy outputs is found with a one-sided 2σ certainty. The quantum volume is the largest layer depth q for which the experiments is successful.

To estimate when certain algorithms become feasible on a quantum computer one could look to the evolution of quantum volume. IBM reached a quantum volume of 512 with its Prague machine [10], which corresponds approximately to a doubling of the quantum volume every 9 months since 2018 [14]. Based on [1] one would deduce that the quantum volume doubles every 4 months for Quantinuum systems since June 2020. There are estimates [9] that the quantum volume has doubled approximately every 6 months since its conception, which is what we will use as a forecast

$$V_Q(T) = 2^{2+T/6} , \tag{1}$$

where T is the time in months since November 2018. In Sect. 1.2 it was estimated that 27 qubits may suffice. Based on (1) a reliable machine to execute QV circuits with 27 qubits may be available in May 2031. According to IBM's roadmap [2], circuits of 10^9 gates should be feasible then. However, the error sensitivity and implementation size of QAE circuits is different.

2.1   The Noise Model

Quantum processors are not deterministic, so statements regarding the quality of an algorithm run must be statistical. In addition, current QPU's are noisy, meaning that the execution of gates succeeds with a probability strictly smaller than 1. It follows that results on the probability of an error occurrence must be derived from statistics of the observations, the expected noiseless outcomes and the expected outcomes with noise. However, describing the outcome in the case of a faulty gate execution is very cumbersome, if possible at all. To avoid this, we make the simplifying assumption that a uniformly random answer is measured upon an error. In other words, if an error occurs, we assume that all structure is lost and a random output will be measured.

Half of the outcomes of a QV circuit are heavy outputs and in the QAE circuits of [25] only one qubit is measured. So, in both algorithms errors reduce the assumed success probability of a measurement to 0.5. Under this assumption, if i should be measured with probability p_i in a noiseless case and the probability of an error occurring is $1 - a$, upon measurements i will be found with probability

$$p'_i = ap_i + 0.5 \cdot (1 - a) .$$

Most sources of errors lead to similar behavior of QV experiments [4]. As a simplification, only two sources of error will be considered, the execution of the gates and the measurements. The probability of a successful measurement of a qubit will be given by b and the probability that 100 gates are executed correctly by c. Based on two parameters, the number m of measurements and the size s of the circuit, decomposed in the basis gates and transpiled into the topology of the hardware, this yields a total probability of an error-free run of

$$a = b^m \cdot c^{s/100} . \tag{2}$$

In practice, b will be close to 1.0 and the number of measured qubits q small, so that the formula is dominated by the second factor. In a different context, such assumptions have been applied to QAE in [6, 26, 28].

2.2   QAE Without Noise

To apply the noise model to the QAE algorithm, we recall some results on the Fisher information for the noiseless and noisy case. The implementation used in this study is based on [25], because its implementation is shallow and has a size that is known upfront.

The numerical integral we study is given by

$$I[f] = \int_0^1 \mathrm{d}x\, f(x)p(x)$$

$$\approx \mathbb{E}_p[f] = \frac{1}{2^q} \sum_{x=0}^{2^q-1} f(\frac{x}{2^q}) p(\frac{x}{2^q}) \tag{3}$$

$$=: \sin^2 \vartheta . \tag{4}$$

It is assumed that the function f can be shifted and scaled, so that $0 \leq I[f] \leq 1$ and an angle $\vartheta \in [0, \pi/2]$ exists satisfying (4). It is this angle that the QAE algorithm computes.

The probability distribution on the q qubits is implemented by the unitary

$$\mathcal{P} : |0\rangle_q \mapsto \sum_{x=0}^{2^q-1} \sqrt{p(x)} |x\rangle_q$$

and the function f is implemented by a unitary operator $\mathcal{R}$ on $q + 1$ qubits

$$\mathcal{R} : |x\rangle_q |0\rangle \mapsto |x\rangle_q (\sqrt{f(x)}|1\rangle + \sqrt{1 - f(x)}|0\rangle) .$$

Together, these operators define $\mathcal{A} = \mathcal{R}(\mathcal{P} \otimes 1)$. This is the state preparation

$$|\Psi\rangle = \mathcal{A}|0\rangle_{q+1} = \sin\vartheta|\Psi_1\rangle|1\rangle + \cos\vartheta|\Psi_0\rangle|0\rangle ,$$

which is split into a good state Ψ_1 and a bad state Ψ_0. Upon measurement of the last qubit, the probability to find $|1\rangle$ is

$$\sin^2\vartheta = \sum_{x=0}^{2^q-1} p(x)f(x) .$$

In order to generate a quantum speed-up, an amplification operator is needed. This operator is defined [25] as $Q = -\mathcal{A}S_0\mathcal{A}^{-1}S_{\Psi_1}$. The operators

$$S_0 : |0\rangle_{q+1} \mapsto -|0\rangle_{q+1} \quad \text{and} \quad S_{\Psi_1} : |\Psi_1\rangle|1\rangle \mapsto -|\Psi_1\rangle|1\rangle$$

and act trivially on other states. It follows that the operator $\mathcal{A}$ is a rotation, so that

$$Q^k|\Psi\rangle = \sin((2k+1)\vartheta)|\Psi_1\rangle|1\rangle + \cos((2k+1)\vartheta)|\Psi_0\rangle|0\rangle .$$

The amplification operator Q contains both $\mathcal{A}$ and $\mathcal{A}^{-1}$, so that operating Q once counts as two function evaluations. After k amplification steps there are $2k+1$ function evaluations performed. Because the amplified angles provide more information about the value of ϑ [6,25], amplification increases the efficiency and constitutes the potential quantum advantage of QAE.

After the circuits have been run for a selection of amplification powers, the angle ϑ must be derived from the outcomes. For each amplification power $k \in \mathcal{M}$, N_k shots are performed with h_k measurements of 1. The best choice for ϑ is the one that maximizes the likelihood

$$L(\vartheta) = \prod_{k\in\mathcal{M}} \left(\sin^2((2k+1)\vartheta) \right)^{h_k} \times \left(\cos^2((2k+1)\vartheta) \right)^{N_k-h_k} .$$

2.3   QAE on Noisy Processors

On a noiseless QPU more amplification rounds result in more information. For noisy processors the situation is more complex [28]. Larger amplification powers yield larger circuits and more noise, reducing the amount of extractable information. The challenge is to optimize the amount of generated information based on the error parameters of the QPU.

Before the noise model (2) can be used, it should be validated whether it can be used to explain the observed noisy measurements. This is done by using small dummy QAE circuits, because complete QAE circuits are too large and the outcomes would be dominated by noise. These dummy QAE circuits can be simulated without errors to obtain the noiseless outcomes p_i of measuring a state i. Repeating the experiments on a noisy QPU provides the noisy outcomes p_i'. Afterwards, the best fitting probability a is found that minimizes

$$V_1 = \sum_{i=1}^{2} \left(p_i' - (ap_i + \frac{1-a}{2}) \right)^2 = -\frac{(a-1)^2}{2} + \sum_i (p_i' - ap_i)^2 ,$$

which is solved by

$$a = \frac{2p_1' - 1}{2p_1 - 1} .$$
(5)

This means that the error probability, the noisy and the noiseless probabilities are related by

$$p_i' = ap_i + \frac{1-a}{n} .$$

The ratio of V_1 against $V_0 = \sum_{i=1}^{2}(p_i' - p_i)^2$ can be used to describe the quality of the noise model. It is given by $\mathcal{R}^2 = 1 - \frac{V_1}{V_0}$.

The interpretation as a probability dictates that $a \in [0,1]$, but fluctuations in the measurement results may lead to solutions of (5) outside this interval. Both $p_0' > p_0 > 0.5$ and $p_0' < p_0 < 0.5$ imply $a > 1$. Negative values of a are observed for $p_0' < 0.5 < p_0$ and $p_0 < 0.5 < p_0'$. Noise should drive the observed noisy probabilities p_i' from p_i towards 0.5, but statistical fluctuations may lead to negative values for single experiments.

2.4   Dummy QAE Circuits

Standard QAE circuits are too deep to run on available noisy hardware and are thus unsuited to validate the noise model. Therefore, dummy QAE circuits composed of the same gates are used. Such circuits consist of R rounds of not-gates and one multi-controlled y-rotation. The rotation acts on the last qubit and is controlled by randomly selected qubits, flipped to 1 and back after the rotation. Only the last qubit is measured. The tests are run on IBM-perth (7 qubits) and the results are used to fit the noise model. The results are shown in Sect. 3.2. Tests have also been run on IBM-guadalupe (16 qubits), but these circuits are so large that all obtained values of a will be effectively zero and prevent fitting.

Besides IBM-perth and IBM-guadalupe, it is useful to run the circuits on quantum processors of intermediate sizes. Therefore, simulations with a noise model have been performed on a series of topologies for 8 until 12 qubits that interpolate between IBM-perth and IBM-guadalupe. The idea behind the choices is that the topology should have a similar influence on the observed noise parameters in all the experiments. Since these are not available as hardware, the noise effects are simulated with a thermal noise model that mimics the noise characteristics of IBM-perth.

2.5   Fisher Information

The amount of information generated by a numerical integration method is given by the Fisher information $\mathcal{I}$. Its relevance follows from the Cramer-Rao bound

$$\text{Var}(\vartheta) \geq \frac{1}{\mathcal{I}(\vartheta)}$$

which is an equality for this approach of QAE [25]. For QAE with depolarizing noise it can be derived [6,28] from the noiseless case. Writing the noiseless and noisy probabilities for k amplification rounds as $p_{(k)} = \cos^2(2k+1)\vartheta$ and $p'_{(k)} = a_k p_{(k)} + \frac{1-a_k}{2}$, the Fisher information is given by

$$\mathcal{I}(\vartheta) = \sum_{k \in \mathcal{M}} \frac{a_k^2 N_k (2k+1)^2}{p_{(0)}(1-p_{(0)})} \frac{p_{(k)}(1-p_{(k)})}{p'_{(k)}(1-p'_{(k)})} . \tag{6}$$

The angle ϑ is obtained through maximization of the noisy likelihood function

$$\mathcal{L}'(\vartheta) = \prod_{k \in \mathcal{M}} (p'_{(k)})^{h_k} (1-p'_{(k)})^{N_k - h_k} .$$

The formulas above show that the value $\vartheta = \pi/4$ is of particular interest. In this case $p_{(k)} = p'_{(k)} = 1/2$, so that the error parameters are irrelevant and the Fisher information is maximal. QAE problems are ideally formulated in a way that the expected angle is close to $\pi/4$, so that

$$\mathcal{I}(\vartheta) = 4b \sum_k (2k+1)c^{2k} \cdot \left(N_k(2k+1)\right) . \tag{7}$$

Maximizing this against the available number of function evaluations $M = \sum_k N_k(2k+1)$ yields $N_{k_1} = \frac{M}{2k_1+1}$, where k_1 maximizes $(2k+1)c^{2k}$. To determine precisely when to switch from $k-1 \to k$ amplification steps, solve

$$(2k+1)c^{2k} \geq (2k-1)c^{2k-2} \qquad \Rightarrow \qquad k = \left\lfloor \frac{1}{1-c^2} - \frac{1}{2} \right\rfloor . \tag{8}$$

2.6   A Monte Carlo Example

The interest in QAE comes from its ability to converge faster than MCI, so a simple estimate of the reference speed of MCI is useful to choose the parameters of the quantum algorithm. For this purpose a 1D integral is determined numerically with $N = 10^7$ samples in 47.4 seconds on a legacy processor, estimated at 0.2 Tflops. In an operational setting, a computing cluster of 100 Tflops could thus evaluate approximately 10^8 samples per second. MCI works without amplification, so the Fisher information per second is $4 \cdot 10^8$, according to (7).

3   Results

3.1   Success in a Quantum Volume Experiment

The probability to find a heavy output on an ideal QPU is $r = 0.8466$ [3]. In a QV experiment on a noisy QPU with q qubits a heavy output will be found with probability $r' \leq r$ with variance $r'(1-r')$. After N measurements, the experiment has passed the test, if

$$\frac{2N}{3} + 2\sigma \leq Nr' \qquad \Rightarrow \qquad r'(1-r') \leq \frac{N}{4}(r' - \frac{2}{3})^2 ,$$

so a QV experiment can succeed as long as $r' > \frac{2}{3}$. Since half of the states are heavy outputs and we assume to measure a random state if an error occurs,

$$r' = ar + \frac{1-a}{2} .$$

A QV experiment will fail, if the probability of an error-free run drops below

$$a_Q = \frac{r' - 1/2}{r - 1/2} = 0.4809 . \tag{9}$$

From the values r' and r the value of the error parameter a can be derived, which we attempt to model by (2) with $m = q$.

3.2   Results for the Error Model

To test the assumptions behind the noise model dummy QAE circuits are run on real and simulated hardware. The results are used to fit the parameters of the noise model (2) with $m = 1$. The quality of the resulting model can then be expressed by the R^2-score. The obtained results are presented in Table 1. To test whether this noise model also works for more qubits, the procedure has been repeated for larger simulated machines with a modified thermal noise model. This causes the difference between the values for c in Fig. 1 and the value found in (the caption of) Table 1.

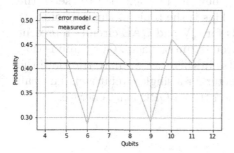

Fig. 1. The error per 100 gates c versus the number of qubits in the circuit. The average value of $b = 0.9958$. The dummy QAE circuits used here are discussed in paragraph 2.4. All experiments are simulated using a modified thermal noise model.

Table 1. The results for dummy QAE circuits on IBM-perth (7 qubits). The obtained parameters for (2) are $b = 0.938$ and $c = 0.374$. The average probability for an error-free run is given by a, its prediction by (2) is $\tilde{a}$. This noise model can explain $R^2 = 0.75$ of the variance in the data.

Rounds	Tests	Depth	Size	a	$\tilde{a}$
0	20	1.0	1.0	0.938	0.923
1	20	74.2	99.3	0.278	0.351
2	20	144.5	202.8	0.233	0.127
3	20	206.7	294.8	0.010	0.051
4	20	290.6	408.4	0.044	0.017
5	20	348.5	495.8	0.093	0.007
6	20	401.4	570.5	−0.044	0.003
7	20	497.5	725.5	0.056	0.001

3.3   Speed of the QPU

The unit of speed of IBM's quantum computers is the circuit layer operations per second [29] or CLOPS. This gives the number of layers of a QV circuit it can process per second. The size in gates of the transpiled and decomposed QV circuits can be determined to translate this speed metric into one applicable to other quantum circuits. This shows the $2.9K$ clops of IBM-perth corresponds to

$$S = 1.0 \times 10^6 \pm 0.6 \times 10^6 \text{ basis gate operations per second.} \qquad (10)$$

3.4 Estimations for the Size of Circuits

Within the frame of the noise model of paragraph Sect. 2.1, the size of the circuit is the dominant parameter. We need, therefore, the scaling of the size of the circuits for more qubits. The size S_1 of a QAE circuit as a function of the number of qubits q and the number of amplification rounds R is

$$S_1 = \alpha_1 \cdot (2R+1) \cdot 10^{\beta_1 \cdot q} \quad \text{with } \alpha_1 = 17.35 \text{ and } \beta_1 = 0.46 . \qquad (11)$$

The size of the QV circuit depends on the number of qubits only, since it is squared. The depth in layers is equal to the number of qubits. Because the circuit need to be transpiled, a quadratic relation between the number of qubits and the depth of the circuit is obtained, which implies a cubic relation between the number of qubits and the size of the circuit. This relation is given by

$$S_2 = \alpha_2 \cdot q^3 + \beta_2 \cdot q^2 \quad \text{with } \alpha_2 = 0.24 \text{ and } \beta_2 = 8.13. \qquad (12)$$

4 Application of the Noise Model to QAE

Now that all elements required are present, they can be combined to sketch the prospect of the integration power of QAE. This is the topic of this section.

The parameters b and c of the noise models for QV and QAE circuits represent properties of the QPU rather than of the circuits run on them. Our claim is that they are the same for both types of circuits. In this way we may translate the development of the QV into an error parameter for a single circuit run. This can then be used to model the Fisher information that a QPU can generate.

4.1 General Functions

The number of qubits that can pass a QV experiment (1) is expected to grow as $q(T) = 2 + \frac{T}{6}$. The expected probability $r'(T)$ to run a QV circuit without errors at future time T can be inferred from (1) and (9). In terms of the (gate) size s of a circuit this probability can be modelled as

$$a = b^q c^{\frac{s}{100}} \approx c^{\frac{s}{100}} .$$

To simplify the computations it is assumed that $b = 1$. This is justified by the values of b seen in (the captions of) Fig. 1 and Table 1.

Expressing the error probability per depth c deduced from a QV experiment with size (12) as a function of $q(T)$ yields

$$c(T) = a_Q^{\frac{100}{\alpha_2 q(T)^3 + \beta_2 q(T)^2}} .$$

Combined with the size of a QAE circuit (11) this yields the probability

$$a_{OP}(T) = a_Q^{\frac{\alpha_1(2R+1)\cdot 10^{\beta_1}Q}{\alpha_2 q(T)^3 + \beta_2 q(T)^2}} \tag{13}$$

to run a QAE circuit with R amplification rounds on Q qubits without errors.

Each function evaluation takes the same amount of time. The number of amplification rounds that yields the most Fisher information per unit of time is the same as maximizing it per number of function evaluations. The optimal number of amplification rounds (8) then is

$$R(T) = \left\lfloor \left(1 - a_Q^{\frac{2\cdot\alpha_1\cdot 10^{\beta_1}Q}{\alpha_2 q(T) + \beta_2 q(T)^2}}\right)^{-1} - \frac{1}{2} \right\rfloor.$$

This leads to an optimal number of amplification rounds shown in Fig. 2a and an optimal operational success probability of (13). This probability decreases after the number of amplifications rounds has just increased, but this is compensated by the extra information gained from the experiments.

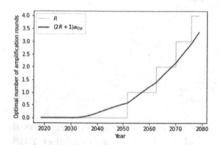 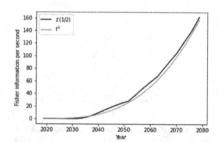

(a) The optimal number of amplification rounds (magenta) and the scaled optimal operational error rate (purple).

(b) The Fisher information per second (red) and the time November 2018 cubed with some prefactor (orange).

Fig. 2. The performance characteristics of QAE on 8 qubits. The length of the x-axis is chosen to show the trends and general development clearly, not to suggest the lifespan of our predictions.

Assuming that the speed of the QPU's remains constant and is given by (10), the achievable Fisher information per second (7) is given by

$$\mathcal{I}(\frac{1}{2}) = 4\cdot 1.0\cdot 10^6 \cdot a_{OP}^2(T) \cdot \frac{(2R(T)+1)}{\alpha_1 10^{\beta_1}Q}.$$

Figure 2b shows that the Fisher information will increase cubically in time as a function of the exponentially increasing quantum volume. This shows that this approach for QAE will not be competitive in the coming years.

4.2   Shallow Functions

Figure 2b shows that the Fisher information increases cubically with time, whereas the QV increases exponentially in time. The implementation of these functions is too deep to yield reliable QAE circuits. Significant improvements are needed to make QAE faster than classical methods. Switching to shallower functions could be an option. And to approximate the 10^8 samples a computing cluster may evaluate, the QAE algorithm should be executed on at least $Q = 27$ qubits. To see the effects of these changes, the analysis of the previous section is repeated here with a hypothetical shallow implementation size of

$$s_3 = 1.0 \cdot (2R + 1) \cdot q^3 \, . \tag{14}$$

The results are summarized in Fig. 3. This shows that the amount of Fisher information generated per second for shallow circuit implementations of the integrand increases cubically and will remain much lower than for MCI.

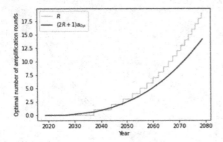

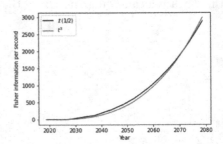

(a) The optimal number of amplification rounds (magenta) and the scaled optimal operational error rate (purple).

(b) The Fisher information per second (red) and the time since November 2018 cubed with some prefactor (orange).

Fig. 3. The performance characteristics of QAE with a shallow function implementation (14) on 27 qubits. The length of the x-axis is chosen to show the trends and general development clearly, not to suggest the lifespan of our predictions.

5   Conclusions

We have used a phenomenological noise model of measurement and depolarizing noise to model the error probability for a single run of a quantum volume experiment. This same model with the same parameters can be used in the context of other quantum algorithms, such as quantum amplitude estimation. The main parameter of the noise model is the size of the circuit decomposed into basis gates and transpiled for the topology of the quantum processor. Using information on the circuit size, the computational limits of other algorithms can be related directly to the corresponding quantum volume. This provides a novel view on the applicability of such algorithms in the near future.

Applying this method to quantum amplitude estimation, it can be used to estimate to achievable Fisher information per second in the upcoming years. This shows that both for general functions and for (hypothetical) shallow functions the implementation size of the circuit will be prohibitive to achieve an actual quantum advantages. This shows that significant improvements in the QPU are needed to achieve a quantum advantage for quantum amplitude estimation in the NISQ era.

Acknowledgements. We would like to thank Niels Neumann, Esteban Aguilera, Robert Wezeman and Ward van der Schoot for their contributions to this project.

References

1. https://metriq.info/
2. (Jan 2024). https://www.ibm.com/roadmaps/quantum.pdf
3. Aaronson, S., Chen, L.: Complexity-Theoretic Foundations of Quantum Supremacy Experiments. In: Proceedings of the 32nd Computational Complexity Conference. CCC '17, Schloss Dagstuhl–Leibniz-Zentrum fuer Informatik, Dagstuhl, DEU (2017). https://doi.org/10.5555/3135595.3135617
4. Baldwin, C.H., Mayer, K., Brown, N.C., Ryan-Anderson, C., Hayes, D.: Re-examining the quantum volume test: Ideal distributions, compiler optimizations, confidence intervals, and scalable resource estimations. Quantum **6**, 707 (2022). https://doi.org/10.22331/q-2022-05-09-707
5. Brassard, G., Hoyer, P., Mosca, M., Tapp, A.: Quantum Amplitude Amplification and Estimation. AMS Contemporary Math. Ser. **305**, June 2000. https://doi.org/10.1090/conm/305/05215
6. Brown, E.G., Goktas, O., Tham, W.K.: Quantum Amplitude Estimation in the Presence of Noise. arXiv: Quantum Physics (2020). https://doi.org/10.48550/arXiv.2006.14145
7. Cross, A.W., Bishop, L.S., Sheldon, S., Nation, P.D., Gambetta, J.M.: Validating quantum computers using randomized model circuits. Phys. Rev. A **100**, 032328 (2019). https://doi.org/10.1103/PhysRevA.100.032328
8. D'Ariano, G.M., Macchiavello, C., Sacchi, M.F.: On the general problem of quantum phase estimation. Phys. Lett. A **248**(2-4), 103–108 (1998). https://doi.org/10.1016/S0375-9601(98)00702-6
9. Ezratty, O.: Understanding Quantum Technologies 2022 (2022). https://doi.org/10.48550/arXiv.2111.15352
10. Gambetta, J.: (2022). https://twitter.com/jaygambetta/status/1529489786242744320
11. Giurgica-Tiron, T., Kerenidis, I., Labib, F., Prakash, A., Zeng, W.: Low depth algorithms for quantum amplitude estimation. Quantum **6**, 745 (2022). https://doi.org/10.22331/q-2022-06-27-745
12. Grinko, D., Gacon, J., Zoufal, C., Woerner, S.: Iterative quantum amplitude estimation. npj Quantum Inf. **7**, 52 (03 2021). https://doi.org/10.1038/s41534-021-00379-1
13. Herbert, S.: Quantum Monte Carlo integration: the full advantage in minimal circuit depth. Quantum **6**, 823 (2022). https://doi.org/10.22331/q-2022-09-29-823, https://doi.org/10.22331/q-2022-09-29-823

14. Jurcevic, P., Zajac, D., Stehlik, J., Lauer, I., Mandelbaum, R.: (Apr 2022). https://research.ibm.com/blog/quantum-volume-256

15. de Lejarza, J.J.M., Grossi, M., Cieri, L., Rodrigo, G.: Quantum fourier iterative amplitude estimation. In: 2023 IEEE International Conference on Quantum Computing and Engineering (QCE) **01**, 571–579 (2023). https://doi.org/10.1109/QCE57702.2023.00071

16. Lubinski, T., et al.: Application-oriented performance benchmarks for quantum computing. IEEE Trans. Quantum Eng. **4**, 1–32 (2023). https://doi.org/10.1109/TQE.2023.3253761

17. Manzano, A., Musso, D., Leitao, A.: Real Quantum Amplitude Estimation. EPJ Quantum, February 2023. https://doi.org/10.1140/epjqt/s40507-023-00159-0

18. Michielsen, K., Nocon, M., Willsch, D., Jin, F., Lippert, T., De Raedt, H.: Benchmarking gate-based quantum computers. Comput. Phys. Commun. **220**, 44–55 (2017). https://doi.org/10.1016/j.cpc.2017.06.011. https://www.sciencedirect.com/science/article/pii/S0010465517301935

19. Mills, D., Sivarajah, S., Scholten, T.L., Duncan, R.: Application-motivated, holistic benchmarking of a full quantum computing stack. Quantum **5**, 415 (2021). https://doi.org/10.22331/q-2021-03-22-415

20. Pelofske, E., Bärtschi, A., Eidenbenz, S.: Quantum volume in practice: what users can expect from NISQ devices. IEEE Trans. Quantum Eng. **3**, 1–19 (2022). https://doi.org/10.1109/TQE.2022.3184764

21. Plekhanov, K., Rosenkranz, M., Fiorentini, M., Lubasch, M.: Variational quantum amplitude estimation. Quantum **6**, 670 (2022). https://doi.org/10.22331/q-2022-03-17-670

22. Preskill, J.: Quantum computing in the nisq era and beyond. Quantum **2**, August 2018. https://doi.org/10.22331/q-2018-08-06-79

23. Proctor, T.J., Rudinger, K.M., Young, K.C., Nielsen, E., Blume-Kohout, R.: Measuring the capabilities of quantum computers. Nat. Phys. **18**, 75–79 (2020). https://doi.org/10.1038/s41567-021-01409-7

24. Robert, C.P., Casella, G.: Monte Carlo Integration, pp. 71–138. Springer New York, New York, NY (1999). https://doi.org/10.1007/978-1-4757-3071-5_3

25. Suzuki, Y., Uno, S., Raymond, R., Tanaka, T., Onodera, T., Yamamoto, N.: Amplitude estimation without phase estimation. Quantum Inf. Process. **19**, January 2020. https://doi.org/10.1007/s11128-019-2565-2

26. Tanaka, T., Suzuki, Y., Uno, S., Raymond, R., Onodera, T., Yamamoto, N.: Amplitude estimation via maximum likelihood on noisy quantum computer. Quantum Inf. Process. **20**, September 2021. https://doi.org/10.1007/s11128-021-03215-9

27. Tanaka, T., Uno, S., Onodera, T., Yamamoto, N., Suzuki, Y.: Noisy quantum amplitude estimation without noise estimation. Phys. Rev. A **105**, 012411 (2022). https://doi.org/10.1103/PhysRevA.105.012411

28. Uno, S., et al.: Modified grover operator for quantum amplitude estimation. New J. Phys. **23**(8) (2021). https://doi.org/10.1088/1367-2630/ac19da, https://dx.doi.org/10.1088/1367-2630/ac19da

29. Wack, A., et al.: Quality, speed, and scale: three key attributes to measure the performance of near-term quantum computers, October 2021. arXiv preprint arXiv:2110.14108

30. Wack, A., McKay, D.: (November 2023). https://research.ibm.com/blog/quantum-metric-layer-fidelity

KetGPT – Dataset Augmentation of Quantum Circuits Using Transformers

Boran Apak⬤, Medina Bandic(✉)⬤, Aritra Sarkar⬤, and Sebastian Feld⬤

Quantum Machine Learning group, QuTech, Department of Quantum and Computer Engineering, Delft University of Technology, Delft, The Netherlands
{m.bandic,a.sarkar-3,s.feld}@tudelft.nl

Abstract. Quantum algorithms, represented as quantum circuits, can be used as benchmarks for assessing the performance of quantum systems. Existing datasets, widely utilized in the field, suffer from limitations in size and versatility, leading researchers to employ randomly generated circuits. Random circuits are, however, not representative benchmarks as they lack the inherent properties of real quantum algorithms for which the quantum systems are manufactured. This shortage of 'useful' quantum benchmarks poses a challenge to advancing the development and comparison of quantum compilers and hardware.

This research aims to enhance the existing quantum circuit datasets by generating what we refer to as 'realistic-looking' circuits by employing the Transformer machine learning architecture. For this purpose, we introduce KetGPT, a tool that generates synthetic circuits in Open-QASM language, whose structure is based on quantum circuits derived from existing quantum algorithms and follows the typical patterns of human-written algorithm-based code (e.g., order of gates and qubits). Our three-fold verification process, involving manual inspection and Qiskit framework execution, transformer-based classification, and structural analysis, demonstrates the efficacy of KetGPT in producing large amounts of additional circuits that closely align with algorithm-based structures. Beyond benchmarking, we envision KetGPT contributing substantially to AI-driven quantum compilers and systems.

Keywords: quantum circuits · generative AI · dataset augmentation · Quantum Assembly · quantum compilation

1 Introduction

The journey from knowledge and rule-based artificial intelligence to the contemporary era of data-driven deep neural networks-based machine learning (ML) has marked significant milestones in artificial intelligence (AI). This type of AI, termed deep learning (DL), focuses on recognizing and extracting patterns from vast datasets. A proliferation of popular DL models and architectures contributed to use cases such as image and speech recognition, sequence prediction, and reinforcement learning. However, the application landscape changed

© The Author(s), under exclusive license to Springer Nature Switzerland AG 2024
L. Franco et al. (Eds.): ICCS 2024, LNCS 14837, pp. 235–251, 2024.
https://doi.org/10.1007/978-3-031-63778-0_17

dramatically with the emergence of generative models [16], such as generative adversarial networks (GAN) and variational autoencoders (VAE). These models marked a profound shift in the capabilities of DL, allowing machines not only to recognize patterns in the data but also to generate new, coherent data that closely resembles the patterns learned from the training data.

Amid this diversity, the model that stands out in recent advances is the generative pre-trained transformer (GPT) [34] based on the transformer architecture [42]. Transformers achieve impressive performance on tasks like realistic text and code generation [29,30] by capturing important information about the structure of sequences of data. GPT's ability to leverage massive scale with billions of parameters and self-supervised learning makes it the model of choice for natural language understanding and generation. A wide spectrum of AI applications can be formulated as a language modeling and generation task, like chatbots, text summarization, question answering, code generation, medical diagnosis, and legal document review.

Simultaneously, another groundbreaking technology is being developed: quantum computers. Quantum computers can solve certain problems faster than classical computers [28] by employing information processing capabilities governed by the laws of quantum mechanics. To solve such problems, quantum algorithms, typically expressed as quantum circuits, need to be executed on quantum computers. Besides serving the target use case, these circuits, defined in quantum assembly languages (QASM) [8], are often also used to characterize, evaluate, and benchmark the quantum processors and related system software. Moreover, system software, like the quantum compiler, often employs DL-based approaches to tackle the complexity of controlling large quantum processors. This presents the need for large datasets of quantum circuits [11,27] for the training of the ML-based quantum compilation passes, such as routing and mapping the circuits to a quantum processor. However, at the moment, only a handful of quantum algorithms [22] are known to provide quantum computational benefits. Due to the lack of large quantum circuit databases, these ML-based compilation techniques resort to randomly generated quantum circuits to train the model. This use of unrepresentative training data can critically affect the performance of the quantum computer when deployed for pragmatic use cases.

In an attempt to address this problem in quantum computing and inspired by the paradigm shift in language generation, in this work, we *employ transformer models to generate realistic-looking quantum circuits to augment quantum circuit datasets.*

This paper's contribution is threefold:

1. Introducing KetGPT, a transformer model capable of generating realistic-looking quantum circuits in the QASM language;
2. Developing a method to determine the quality of the generated QASM code using a different transformer model specifically designed for this task; and
3. Analyzing the generated circuits by extracting their structural parameters and comparing them to those of previously existing circuits.

KetGPT can immediately be applied to the following use cases:

- **Extending quantum circuit benchmarks datasets**: KetGPT circuits offer a valuable expansion to existing circuit suites, such as those in [5,33], commonly employed for benchmarking and comparing quantum compilers and systems. Unlike typical synthetic circuits that consist of random gates on random qubits, KetGPT circuits emulate the behavior of real quantum algorithms, enhancing their relevance as benchmarks. Moreover, compared to the current practice of employing entirely random circuits with consistent width and depth, they present a compelling alternative for evaluating success metrics like quantum volume [7]. Given that quantum computers are designed to accelerate specific algorithms challenging for classical computers, assessing them using circuits that closely resemble these algorithmic structures is imperative. A dataset of KetGPT-generated quantum circuits is available as part of this software in Sect. 6.
- **Automating quantum system software**: Recent research uses machine learning models to enhance quantum compilation and error correction [1,11,27,31]. The substantial data required for training these models often leads researchers to resort to generating random circuits. However, a system that solves a certain problem should be trained on representative problem instances. Therefore, training a compiler to handle realistic circuits is more beneficial than training it on a random sample of gates, which makes KetGPT ideal for such a purpose [5]. In an ongoing project, KetGPT is being used to train a reinforcement learning agent for quantum circuit mapping on noisy quantum processors.

The remainder of this paper is structured as follows: The transformer models are introduced in Sect. 2. Section 3 introduces the main contribution of this work, KetGPT, a transformer model specifically designed to generate QASM files useful for benchmarking quantum system software. Additionally, a method is proposed to quantify how realistic these QASM files are. In Sect. 4, the generated code is examined and results are presented and discussed. Ultimately, Sect. 5 contains the conclusion of this work and presents suggestions for future work.

2   Evolution and Structure of Transformers

Transformer models, as introduced in the groundbreaking work [42], have changed the landscape of natural language processing. Their applications extend to code generation [29,40] and music generation [2]. Renowned for their proficiency in capturing dependencies within sequential data, these widely adopted machine-learning models have proven effective in various domains.

Before the advent of transformers, conventional models for natural language processing tasks, such as text generation, primarily relied on Convolutional Neural Networks (CNN) [24], Recurrent Neural Networks (RNN) [37], and Long Short-Term Memory networks (LSTM) [18]. However, these models encountered several challenges, including difficulties in handling long-range dependencies and a lack of parallelizability [42]. A transformer, on the other hand, is a highly parallelizable model, well-suited for training on extensive datasets, that excels at

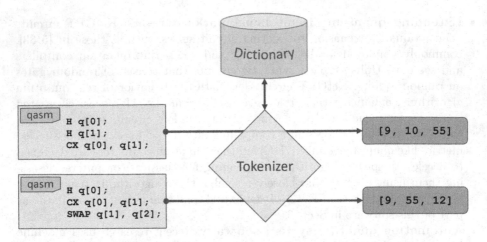

Fig. 1. Tokenization Example. A sequence of QASM operations (in text file form) is provided as input, and each statement (a line of QASM code) is assigned to a number. The number assigned to each statement does not have an intuitive meaning; rather, it just depends on how the tokenization algorithm orders its vocabulary. Consequently, tokenizing a sequence of statements will create a list of numbers. It is important to note that both gate and qubit(s), we apply the gate on, matter for the assigned token. For instance, h q[0]; and h q[1]; would have different numbers assigned as shown.

capturing longer-range dependencies and, therefore offers a significant improvement over earlier models.

In what follows, we review the three main components of the transformer model with quantum assembly language as the data.

2.1 Tokenizer

It is well known that performing any kind of computations on strings necessitates converting them to numerical *tokens* through a process called *tokenization*. While this tokenization step is not explicitly outlined in the transformer architecture defined in [42] (as it falls under the domain of dataset preparation), it plays a crucial role in comprehending how information flows through a transformer model. A tokenizer plays a significant role in our case as using QASM code as input requires a different preprocessing type than with standard text. An example of the QASM code tokenization process is presented in Fig. 1.

To fully describe a tokenization process, it is required to have a system for segmenting a sequence and a 'dictionary' to establish the numerical association for each possible segment encountered using this segmentation system. There are different types of tokenization algorithms available. For instance, instead of the scheme shown in Fig. 1, every character can be converted to a number. Thus, h q[0]; would be tokenized into 7 integers, one for each character and whitespace, instead of just a single token.

2.2   Feed-Forward Neural Network

Neural networks [12] play a key role in various machine-learning approaches and are one of the fundamental segments of transformer models. They consist of a series of layers that each perform a linear operation on the input followed by a (non-linear) activation function.

To be precise, the value of each node in the network will be a linear combination of the values of the nodes in the previous layer weighted by the corresponding weights, passed through an activation function. Then a non-linear activation function (such as softmax [6] or ReLu [17]) is applied so that the network can capture complex non-linear patterns.

A Feed-forward neural network is fully defined by specifying the number of layers, the number of nodes in each layer, the weights of every connection between nodes of a layer and a previous layer, a bias per node and the activation function per layer. To train a network, the desired architecture is initialized with (random) weights and biases. During training, the inputs are iteratively presented to the network and the weights and biases are adjusted to progressively align the network's output with the expected output for each specific input. This adjustment is typically done using a method called Stochastic Gradient Descent [36]. In this paper we are not focusing on the details of the neural networks, even though it represents the core of the transformer model, as it is widely and generally used as a base of most machine learning models. Instead, we will focus on the segments of the transformer that are specifically significant for our model, like *self-attention*.

2.3   Self-attention

Self-attention is a mechanism that helps a transformer understand the relation between words and represents the main innovation in transformer models. Consider the sentence, "The computer executes the program because it is told to." Humans effortlessly discern that "it" refers to the computer, not the program, but making automated systems distinguish this difference is very challenging. The inclusion of a self-attention component empowers transformers to establish such connections.

The input to the attention mechanism consists of queries, keys, and values. Each token in the input sequence corresponds to one query and key vector with dimension d_k and a value vector with dimension d_v, but for computational purposes, the queries, keys and values for all tokens are packed into, respectively, matrices Q, K and V. Thereafter, the main equation [42] describing the attention process is:

$$\text{Attention}(Q, K, V) = \text{softmax}\left(\frac{QK^T}{\sqrt{d_k}}\right) V, \qquad (1)$$

where softmax is the softmax function [6] and K^T is the transpose of the K matrix.

The underlying idea of this equation is in the QK^T term, representing the dot product between queries and keys to discern their "inter-relation." Subsequently, this information forms an attention matrix akin to a correlation matrix. However, unlike a correlation matrix with values between -1 and 1, the attention matrix adopts the form of a probability distribution, with values ranging from 0 to 1. The $\sqrt{d_k}$ scaling factor is there to obtain a more dimension-independent dot product, which helps train the network easier [42]. Multiplying this attention matrix with V produces the final result, enriching the original matrix V with insights into the inter-relations between queries and keys. For instance, elements with low scores in the attention matrix, close to 0, are drowned out. To illustrate, in the context of encoding the sentence "The computer executes the program because it is told to." represented by matrices Q, K, and V, the operation $\text{Attention}(Q, K, V)$ returns a matrix that embodies this sentence with information about the inter-relations between the words (e.g., clarifying that "it" refers to the computer and not the program).

3   KetGPT - Transformers for Quantum Circuit Generation

This section presents KetGPT, a novel software tool designed to generate quantum algorithm-based circuits. These circuits can serve as essential benchmarks for evaluating the performance of both existing and forthcoming quantum systems. Within this section, we delve into the technical intricacies of KetGPT, offering a comprehensive understanding of its architecture and methodology. Figure 2 shows an overview of the KetGPT design and overall workflow.

3.1   Input Dataset and Data Preprocessing

Several datasets of quantum circuits suitable for benchmarking are available [5, 26, 43], including MQT Bench [33], which is utilized in this study. QASM files were generated to depict circuits implementing algorithms spanning 2 to 100 qubits, employing OpenQASM 2.0 [8]. In cases where algorithms were incompatible with a specific qubit count, such as those requiring an uneven number of qubits, all valid circuits within the feasible range were generated. The full dataset and additional details can be found in Sect. 6.

The files taken from the dataset require preprocessing in order to comply with the transformer model. This involves making minor adjustments to the QASM files in the dataset (e.g., removing comments). Due to technical constraints – specifically, the model's incapacity to process large files – a maximum circuit length of 1024 QASM statements is enforced. This limitation is specific to the hardware's RAM constraints and not a general technical restriction. Following the preprocessing step, the final dataset comprises 713 QASM files.

3.2   Generator: Architecture and Tokenizer

When it comes to generating text and code, a decoder-only transformer architecture [40] is a popular choice. Accordingly, for the generation of QASM files, we

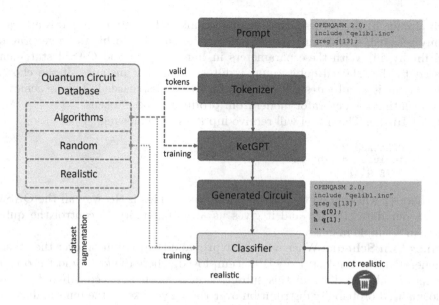

Fig. 2. KetGPT Workflow: Firstly, a given text prompt is tokenized. These tokens are fed into the KetGPT model, which was trained with quantum circuits from an existing quantum circuit database. KetGPT then generates text to continue the given prompt, yielding a synthetic circuit. A separate transformer classifier model, trained to distinguish real from random quantum circuits, tests if the generated circuit is realistic. If the test is positive, it can be used to augment the quantum circuit database.

have opted for the GPT-2 model architecture [35], known for its use of a decoder-only transformer. The Python code to construct this architecture is openly accessible through the GPT-2 implementation in the Hugging Face "Transformer" python library [21,44].

As discussed in Sect. 2.1, we employ a *tokenization* approach to transform the dataset text into tokens. The original implementation of GPT-2 relies on a form of tokenization known as Byte Pair Encoding (BPE). To comprehend this method intuitively, it dissects text into components (e.g., 'training' into 'train' and 'ing'), facilitating a better grasp of the full word's meaning. However, a drawback is that it may allow the generation of QASM code that is not syntactically correct, such as the potential generation of the line "hh q0q1;". To address this, we modified the tokenization method for the generator to only permit syntactically correct QASM code as tokens. This modification was implemented by adjusting the GPT2Tokenizer class. By compiling a list of all valid QASM statements in the dataset and using it as our vocabulary, we ensure that any token generated by the model will be a valid QASM statement. The generator model workflow consists of the following four parts:

Preparation: The process of generating tokens using the generator model unfolds as follows: i) A list is compiled containing the qubit count for every circuit in the dataset, along with another list containing the number of gates for each circuit; ii) From these lists, a qubit count and a number of gates are randomly selected, establishing the parameters for the QASM file to be generated; and finally, iii) With these parameters in hand, any invalid QASM statement related to the selected qubit count is filtered out. For instance, if the chosen qubit count is 5, all gates involving qubit 13 are disregarded. This is achieved by preventing the generator model from producing these tokens.

Model Input: The model will receive input as the following:

```
OPENQASM 2.0;
include "qelib1.inc"
qreg q[{}];
```

where {} will contain the chosen qubit count. This is the way all the QASM files in our dataset start, and it gives us an opportunity to control the qubit count in a simple manner.

Generation Scheme: Whenever a new probability distribution over the tokens is generated, the top-k strategy [10] is employed, where the $k = 5$ most probable tokens are identified. From this subset, a new token is selected based on the renormalized probability distribution over these five tokens (the renormalization ensures that all probabilities add up to one). This approach introduces additional randomness into the QASM file generation process while maintaining the realism of the generated tokens, as the five most probable tokens are typically viable candidates. Furthermore, it is specified that sequences of 15 tokens should not repeat within the file. While this constraint may not align perfectly with QASM code generation, in which algorithms often contain repetitive sequences, it serves to prevent instances where the transformer model becomes stuck in a loop, repeatedly predicting the same sequence. The top-k generation process iterates until the desired number of gates is reached.

Post-processing: Finally, to guarantee the validity of all generated files, all quantum and classical registers utilized in the generated file are instantiated at the beginning of the QASM file. This ensures every file, including its header, is syntactically correct.

3.3   Verification Method: KetGPT Classifier

Once the generator produces the QASM files, the next step is to assess their authenticity. To determine whether the generated QASM files exhibit a "realistic" quality, we employed a binary classifier. This classifier's task is to distinguish whether a generated QASM file bears a closer resemblance to files from our algorithm-based circuit dataset or aligns more with a randomly generated QASM file [5].

The classifier adopts an encoder-only transformer model, specifically the architecture of the DistilBERT model [38], leveraging the implementation from the Huggingface transformers library [21]. This model is a smaller version of the highly influential encoder-only BERT model [9] and is chosen for quicker training and inference.

Unlike the generator, which required a customized tokenization method to ensure the generation of valid QASM code, the classifier employs the *tokenization* method used to train the original DistilBERT model, known as WordPiece [45]. This method, similar to the BPE tokenizer briefly mentioned in Sect. 3.2, breaks down words into sub-words. It is important to note that the choice of how these sub-words are determined distinguishes WordPiece from BPE, but this is not pertinent to this work. To adapt the QASM sequences for the classifier, the tokenization truncates them after 512 tokens. Since these tokens represent sub-words instead of complete QASM lines, the 512-token limit corresponds to approximately 50 lines of QASM code, dependent on the sequence. This adjustment ensures compatibility with the maximum input size of the classifier model used. While this approach has the drawback of only considering the initial portion of the QASM file in determining its authenticity, it offers the advantage of expedited training and inference, necessitating a less technically intricate model. Moreover, the initial segment of a QASM file typically provides sufficient cues to discern its nature as random or structured.

During the *training* phase of the classifier, a dataset is prepared in which all real quantum circuits are assigned the label '0' (total of 1112 QASM files). Correspondingly, an equal number of QASM files are randomly generated, comprising gates randomly selected from a list of all unique QASM statements in the dataset, and labeled '1'. To ensure fairness in the classification process, akin to the methodology employed for generating KetGPT QASM files, the randomly generated QASM files are structured to encompass the same distribution of qubit counts and number of gates as the original dataset. Subsequently, the model is trained on the labeled dataset, and upon completion of training, the trained model is employed to predict whether the KetGPT-generated circuits are classified as '0' or '1', indicating their proximity to genuine algorithms or random circuits, respectively.

3.4 Implementation Details

Our experiments were conducted using a Jupyter notebook [23] executed on the *Google Colab* environment [13]. This Notebook is provided in Sect. 6. The Google Colab GPU has 16Gb of GDDR6 memory, 320 Turing tensor cores and 2560 CUDA cores. At the time of writing, Google Colab uses Python version 3.10.12. Relevant packages for the code used to obtain the results of this work are the transformers [44] (version 4.34.0) and datasets [25] (version 2.14.5) libraries from Huggingface, PyTorch [32] (version 2.0.1+cu118) and NumPy [15] (version 1.23.5).

Table 1 contains the parameters that define the structure of our generator model. Default values correspond to those used in the original GPT-2 implementation [35]. The training settings are specified in Table 2. On the other hand, Table 3 specifies the settings that were used to define the *classifier model*. The training settings for the classifier model are in Table 4. All the parameters' detailed definitions can be found in [20].

Table 1. Generator model settings

Name	Value
n_embd	768 (default)
n_layer	3
n_head	4
n_positions	1024 (default)
vocab_size	48291

Table 2. Generator training settings

Name	Value
Epochs	5
Learning Rate	5e-5 (default)
Batch Size	4
Optimiser	AdamW (default)
Loss function	Cross-entropy (default)

Table 3. Classifier model settings

Name	Value
n_embd	768 (default)
n_layer	6 (default)
n_head	12 (default)
n_positions	512 (default)
vocab_size	30522

Table 4. Classifier training settings

Name	Value
Epochs	3
Learning Rate	5e-5 (default)
Batch Size	4
Optimiser	AdamW (default)
Loss function	Cross-entropy (default)

It is worth noting that KetGPT training time was 240 s, and generating 1000 QASM files took 8818 s (147 min), or 8.8 s per generated file on average. However, the QASM files are of varying size (as explained in Sect. 3.1), and the amount of time needed to generate one file is dependent on its size, so this number should be taken as a rough estimate.

4   Results and Discussion

In this section, we unveil outcomes of this work by showing the results of the three verification steps: manual inspection and Qiskit execution, transformer-based classifier and structural analysis of the circuits. Note that the usage of the term 'realistic' or 'real' when describing the circuits generated by KetGPT is not meant to be interpreted as describing circuits that implement useful quantum algorithms. The circuits might describe some undiscovered quantum algorithms, but it is nearly impossible to reverse engineer an explainable description.

4.1   Manual Inspection

First, we manually examine the QASM lines of a circuit produced by KetGPT. We juxtapose this with the initial lines of both a genuine and a completely random circuit to establish a comparative analysis. One can observe some patterns shown in the files of Fig. 3: The lines within the KetGPT file and the real file exhibit structured patterns, such as the repetition of Hadamard and 2-qubit gates (CX and CZ), whereas the fully random circuit lacks such repetitive sequences. Additionally, it is noteworthy that the order in which the Hadamard gates are applied in the KetGPT and the real circuit follows an ascending order based on qubit numbers, whereas in the fully random circuit, as expected, there is no logical order of operations. Importantly, the random circuit includes invalid statements, such as operations on nodes that were never defined (e.g., an operation on node 4 is instructed, but node 4 was never defined). However, this error is also occasionally present in files generated by KetGPT, albeit seemingly less frequently. The fact that it is not specifically forbidden for KetGPT to generate invalid statements, but it still generates such statements considerably less often than random files, can also be seen as a realistic feature of KetGPT-generated data. Note that we also ran all our circuits within the Qiskit framework [3] where 96% of the circuits passed the compilation process successfully.

Based on the provided examples and the illustration in Fig. 3, a visual examination strongly indicates that KetGPT-generated circuits exhibit characteristics reminiscent of real quantum circuits. This observation underscores the promise of employing transformers to generate quantum circuit data.

4.2 Classifier-Based Evaluation

As a second measure of verification, we developed and trained a classifier model to determine whether KetGPT circuits resemble more real algorithm-based or random quantum circuits. As input, we created a dataset with the same amount of real and random circuits (1112 each) and used 85% of the data for training and 15% for testing the classifier.

To assess the model's performance, a confusion matrix is employed to ascertain the alignment between the model's predictions and the actual labels of the data. The corresponding confusion matrix for this evaluation is depicted in Fig. 4.

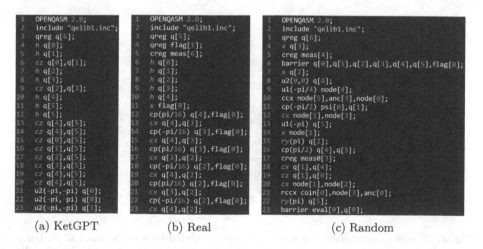

(a) KetGPT   (b) Real   (c) Random

Fig. 3. Side-by-side comparison between the lines of a 6 qubit QASM file generated by KetGPT (a), algorithm-based circuit (b) and a random circuit(c).

A total of 328 out of 334 test dataset values are predicted correctly, which means that the classifier model achieved an accuracy of 98.2%.

Subsequently, the classifier was tasked to classify 1000 KetGPT QASM files as either more similar to its training dataset (real algorithm-based) or to completely random qauantum circuits. Among the 1000 circuits evaluated, 999 were classified as authentic, indicating a classification accuracy of 99.9%.

It is difficult to evaluate the reliability of the model. The high accuracy could potentially be explained by the fact that the test dataset consists of a random subset of the total data. It is possible, for instance, that the Deutsch-Jozsa algorithm on 6 qubits is part of the training dataset, and Deutsch-Jozsa on 5 qubits is in the test dataset. The similarity between the training and testing data may influence the accuracy metric calculation. Nonetheless, using different instances of the same algorithms for the datasets was inevitable due to the limited availability of diverse algorithms. The random QASM files in the test set, however, are not similar to the random files in the training dataset, and are still predicted correctly every time.

Taking all of these considerations into account, including the classifier's accuracy when evaluated, it appears that the classifier is capable of discerning realistic features within the data. However, determining whether this proficiency results from the model overfitting to specific features of QASM files or genuinely learning relevant aspects of realistic circuits presents a challenge.

4.3   Analysis Based on Circuit Structure

Another approach to quantifying and validating KetGPT involves extracting structural parameters from circuits. Within this approach, a circuit is transformed into interaction and gate dependency graphs [5] and then analyzed based

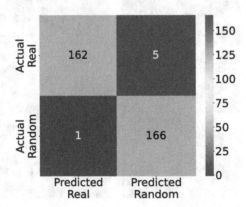

Fig. 4. Classifier performance on a test dataset illustrated by a confusion matrix. Diagonal values of the matrix are correctly predicted: only 5 QASM files that are actually "Real" are predicted as being "Random", and 1 QASM file that is 'Random' is predicted as being 'Real'.

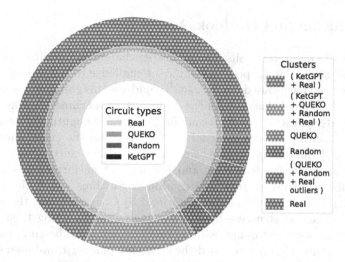

Fig. 5. The distribution of clusters obtained through structural parameters analysis is depicted. Each segment in the outer circle represents clusters characterized by the same types of circuits (e.g., the dark green segment encompasses all clusters that consist of KetGPT and real circuits). The inner circles display the quantity of each circuit type within the respective outer circle segment. (Color figure online)

on quantum compilation-related, graph theory-based (e.g., degree of nodes) parameters. Following this methodology, we extracted the suggested 23 metrics [4] from our KetGPT circuits in order to compare them with existing circuit dataset. For comparison, we followed another method suggested in [5] and clustered the circuits (KetGPT and qbench [5] circuits) based on the extracted parameters to discover groups of ultimately structurally similar circuits. The benchmark set *qbench* consists of real algorithm-based circuits, random circuits, and QUEKO circuits (synthetic circuits with predefined depth and gate count) [41], so by doing clustering, we could see where KetGPT would belong within these groups, or if it would form its own. Notably, we refrained from utilizing this benchmark set for creating KetGPT circuits, ensuring an unbiased evaluation.

Clustering is done in a two-level manner: first based on size and then sub-clusters based on the structure of the quantum circuits, resulting in a final tally of 18 clusters. For clarity, we consolidated clusters sharing identical circuit structures (in terms of circuit types) into one and illustrated the distribution in Fig. 5. The depicted clustering reveals that KetGPT circuits consistently align with real circuits and never with completely random ones. Additionally, a smaller portion of QUEKO circuits exhibit a similar association with both KetGPT and real circuits. Given that QUEKO circuits aim to mimic realistic behaviors more closely than classical random circuits [41], this observation is logical. Figure 5 also suggests how much KetGPT contributes to having more realistic circuits in the whole set (green segments of the inner circle).

5   Conclusion and Outlook

The scarcity of quantum circuits 'useful' for benchmarking, stemming from limitations in existing datasets, poses a significant challenge to the progress of quantum compiler and hardware development. To address this gap, our research introduces KetGPT, a tool that utilizes the Transformer machine learning architecture to generate synthetic circuits resembling real-world quantum algorithms. We verified our resulting circuits three-fold by: 1) Running the circuits with Qiskit framework and manual inspection, we achieved a 96% success rate (without error or warning); 2) Implementing and training a transformer-based classifier for distinguishing between 'real' and random algorithms which classified Ket-GPT circuits as real in 99% of the cases; and 3) Characterizing the generated circuits by extracting structure-based properties and clustering them together with another dataset containing real and random circuits. The analysis revealed that all our circuits closely resembled the structure of algorithm-based ones, and showcased the expansion of the dataset. In conclusion, this three-step, extensive verification shows that KetGPT can augment realistic and executable quantum circuit dataset(s).

Our future steps in expanding and improving KetGPT include: i) Exploring alternative generation schemes, such as top-p [19], beam search [14], or contrastive search [39], to compare their effectiveness in generating QASM files or, development of a generation scheme tailored specifically for QASM file generation; ii) Reconsidering the representation of QASM statements as discrete tokens: Introducing an arbitrary gate token to accommodate QASM files with arbitrary angles, using a transformer trained for this purpose in post-processing; and iii) Modifying the tokenization scheme by separating gates and target qubits into distinct tokens (e.g., treating 'Hadamard gate' and 'on qubit 1' as separate tokens) and ensuring that the adjusted scheme generates only valid QASM expressions and exploring its scalability for higher qubit counts.

In summary, we are confident that KetGPT holds the promise to not only significantly influence the benchmarking of quantum systems, but also to serve as a valuable input for data-intensive, AI-based solutions in the development of innovative quantum compilers and systems.

6   Software Availability

The code that was used for this work is provided as a Jupyter notebook [23], which was executed in the Google Colab environment [13], available at: https://colab.research.google.com/drive/1dbtJX6q8sED4yrb1I09KUuXWYH0-AVN8r.

The data that was used for this work, comprising of the training dataset, and a KetGPT folder that contains: the pre-trained Ket-GPT model, the KetGPT tokenizer, the pre-trained classifier model, all KetGPT generated circuits and all random circuits, is available at: https://www.kaggle.com/datasets/boranapak/ketgpt-data.

Acknowledgments. MB and SF acknowledge funding from Intel Corporation. AS acknowledges funding from the Dutch Research Council (NWO).

References

1. Acampora, G., Schiattarella, R.: Deep neural networks for quantum circuit mapping. Neural Comput. Appl. **33**(20), 13723–13743 (2021)
2. Agostinelli, A., et al.: Musiclm: Generating music from text. arXiv preprint arXiv:2301.11325 (2023)
3. Anis, M.S., et al.: Qiskit: An open-source framework for quantum computing (2021). https://doi.org/10.5281/zenodo.2573505
4. Bandic, M., et al.: Qauntum benchmarks structural analysis for improvement of quantum circuit mapping for single- and multi-core quantum computation (2024), (work in progress)
5. Bandic, M., Almudever, C.G., Feld, S.: Interaction graph-based characterization of quantum benchmarks for improving quantum circuit mapping techniques. Quantum Mach. Intell. **5**(2) (2023)
6. Bridle, J.: Training stochastic model recognition algorithms as networks can lead to maximum mutual information estimation of parameters. Advances in neural information processing systems 2 (1989)
7. Cross, A.W., Bishop, L.S., Sheldon, S., Nation, P.D., Gambetta, J.M.: Validating quantum computers using randomized model circuits. Phys. Rev. A **100**(3) (2019)
8. Cross, A.W., Bishop, L.S., Smolin, J.A., Gambetta, J.M.: Open quantum assembly language. arXiv preprint arXiv:1707.03429 (2017)
9. Devlin, J., Chang, M.W., Lee, K., Toutanova, K.: BERT: Pre-training of deep bidirectional transformers for language understanding. In: Proceedings of the 2019 Conference of the North American Chapter of the Association for Computational Linguistics: Human Language Technologies, vol. 1 (2019)
10. Fan, A., Lewis, M., Dauphin, Y.: Hierarchical neural story generation. In: Proceedings of the 56th Annual Meeting of the Association for Computational Linguistics (Volume 1: Long Papers), pp. 889–898. Association for Computational Linguistics, Melbourne, Australia (2018)
11. Fösel, T., Niu, M.Y., Marquardt, F., Li, L.: Quantum circuit optimization with deep reinforcement learning. arXiv preprint arXiv:2103.07585 (2021)
12. Goodfellow, I., Bengio, Y., Courville, A.: Deep learning. MIT press (2016)
13. Google, LLC: Google colaboratory (2023). https://colab.research.google.com
14. Graves, A.: Sequence transduction with recurrent neural networks. arXiv preprint arXiv:1211.3711 (2012)
15. Harris, C.R., et al.: Array programming with NumPy. Nature **585**, 357–362 (2020)
16. Harshvardhan, G., Gourisaria, M.K., Pandey, M., Rautaray, S.S.: A comprehensive survey and analysis of generative models in machine learning. Comput. Sci. Rev. **38**, 100285 (2020)
17. Hendrycks, D., Gimpel, K.: Gaussian error linear units (gelus). arXiv preprint arXiv:1606.08415 (2016)
18. Hochreiter, S., Schmidhuber, J.: Long short-term memory. Neural Comput. **9**(8), 1735–1780 (1997)
19. Holtzman, A., Buys, J., Du, L., Forbes, M., Choi, Y.: The curious case of neural text degeneration. arXiv preprint arXiv:1904.09751 (2019)
20. HuggingFaceInc.: Openai gpt2 (2020). https://huggingface.co/transformers/v3.5.1/model_doc/gpt2.html

21. HuggingFaceInc.: Transformers: State-of-the-art natural language processing (2021). https://github.com/huggingface/transformers
22. Jordan, S.: Quantum algorithm zoo. https://quantumalgorithmzoo.org. Accessed 25 Sept 2023
23. Kluyver, T., et al.: Jupyter notebooks-a publishing format for reproducible computational workflows. Elpub **2016**, 87–90 (2016)
24. LeCun, Y., Bottou, L., Bengio, Y., Haffner, P.: Gradient-based learning applied to document recognition. Proc. IEEE **86**(11), 2278–2324 (1998)
25. Lhoest, Q., et al.: Datasets: a community library for natural language processing. In: Proceedings of the 2021 Conference on Empirical Methods in Natural Language Processing: System Demonstrations, pp. 175–184. Association for Computational Linguistics, Online and Punta Cana, Dominican Republic, November 2021
26. Li, A., Stein, S., Krishnamoorthy, S., Ang, J.: Qasmbench: a low-level quantum benchmark suite for nisq evaluation and simulation. ACM Trans. Quantum Comput. **4**(2), 1–26 (2023)
27. van der Linde, S., de Kok, W., Bontekoe, T., Feld, S.: qgym: A gym for training and benchmarking rl-based quantum compilation (2023). arXiv preprint arXiv:2308.02536
28. Montanaro, A.: Quantum algorithms: an overview. npj Quantum Information **2**(1), 1–8 (2016)
29. Nijkamp, E., et al.: Codegen: an open large language model for code with multi-turn program synthesis. arXiv preprint arXiv:2203.13474 (2022)
30. OpenAI: Gpt-4 technical report (2023). arXiv preprint arXiv:2303.08774
31. Overwater, R.W., Babaie, M., Sebastiano, F.: Neural-network decoders for quantum error correction using surface codes: a space exploration of the hardware cost-performance tradeoffs. IEEE Trans. Quantum Eng. **3**, 1–19 (2022)
32. Paszke, A., et al.: Pytorch: an imperative style, high-performance deep learning library. In: Advances in Neural Information Processing Systems 32, pp. 8024–8035. Curran Associates, Inc. (2019)
33. Quetschlich, N., Burgholzer, L., Wille, R.: Mqt bench: Benchmarking software and design automation tools for quantum computing. Quantum **7**, 1062 (2023). mQTbench is available at https://www.cda.cit.tum.de/mqtbench/
34. Radford, A., Narasimhan, K., Salimans, T., Sutskever, I., et al.: Improving language understanding by generative pre-training (2018)
35. Radford, A., Wu, J., Child, R., Luan, D., Amodei, D., Sutskever, I., et al.: Language models are unsupervised multitask learners. OpenAI blog **1**(8), 9 (2019)
36. Robbins, H., Monro, S.: A stochastic approximation method. The annals of mathematical statistics, pp. 400–407 (1951)
37. Rumelhart, D.E., Hinton, G.E., Williams, R.J., et al.: Learning internal representations by error propagation (1985)
38. Sanh, V., Debut, L., Chaumond, J., Wolf, T.: Distilbert, a distilled version of bert: smaller, faster, cheaper and lighter. ArXiv abs/1910.01108 (2019)
39. Su, Y., Lan, T., Wang, Y., Yogatama, D., Kong, L., Collier, N.: A contrastive framework for neural text generation. Adv. Neural. Inf. Process. Syst. **35**, 21548–21561 (2022)
40. Svyatkovskiy, A., Deng, S.K., Fu, S., Sundaresan, N.: Intellicode compose: code generation using transformer. In: Proceedings of the 28th ACM Joint Meeting on European Software Engineering Conference and Symposium on the Foundations of Software Engineering, pp. 1433–1443 (2020)
41. UCLA: Queko benchmark (2020). https://github.com/UCLA-VAST/QUEKO-benchmark

42. Vaswani, A., et al.: Attention is all you need. Advances in neural information processing systems **30** (2017)
43. Wille, R., Große, D., Teuber, L., Dueck, G.W., Drechsler, R.: Revlib: an online resource for reversible functions and reversible circuits. In: 38th International Symposium on Multiple Valued Logic (ismvl 2008), pp. 220–225. IEEE (2008)
44. Wolf, T., et al.: Transformers: State-of-the-Art Natural Language Processing. pp. 38–45. Association for Computational Linguistics (Oct 2020)
45. Wu, Y., et al.: Google's neural machine translation system: Bridging the gap between human and machine translation (2016). arXiv preprint arXiv:1609.08144

Design Considerations for Denoising Quantum Time Series Autoencoder

Jacob L. Cybulski[1(✉)] and Sebastian Zając[2]

[1] Deakin University, Melbourne, Australia
jacob.cybulski@deakin.edu.au
[2] SGH Warsaw School of Economics, Warsaw, Poland
szajac2@sgh.waw.pl
https://jacobcybulski.com/ , https://sebastianzajac.pl/

Abstract. This paper explains the main design decisions in the development of variational quantum time series models and denoising quantum time series autoencoders. Although we cover a specific type of quantum model, the problems and solutions are generally applicable to many other methods of time series analysis. The paper highlights the benefits and weaknesses of alternative approaches to designing a model, its data encoding and decoding, ansatz and its parameters, measurements and their interpretation, and quantum model optimization. Practical issues in training and execution of quantum time series models on simulators, including those that are CPU and GPU based, as well as their deployment on quantum machines, are also explored. All experimental results are evaluated, and the final recommendations are provided for the developers of quantum models focused on time series analysis.

Keywords: Quantum machine learning · Autoencoder · Quantum encoding · Quantum measurement

1 Introduction

The area of quantum time series analysis finds its foundations in a mature and well-published field of classical time series analysis [14], as well as in the new, dynamic, and not yet tested area of quantum machine learning [27].

Our exploration of quantum machine learning approaches to time series analysis has been motivated by our long-term interest in quantum methods for the detection and removal of irregularities in temporal data [4].

In general, there are two categories of irregularities found in time series [4]: (1) noise and errors, or unwanted data that need to be removed from data; and (2) anomalies, representing events of interest, which need further analysis and understanding, often with great urgency.

The presence of abnormal data in time series may be due to poor data entry practices, the use of substandard recording devices, or the impacts of environmental factors. Eliminating noise could improve the quality of analytical models

L. Franco et al. (Eds.): ICCS 2024, LNCS 14837, pp. 252–267, 2024.
https://doi.org/10.1007/978-3-031-63778-0_18

trained with these data [1]. For example, the removal of noise and errors from time series could improve the accuracy of systems responsible for financial forecasting [28], monitoring the condition of industrial machines [17], or helping physicians perform medical diagnoses [30]. However, identification and understanding of anomalies discovered in financial transactions could reveal fraudulent customer behavior, anomalies in machine vibration could represent the onset of its catastrophic failure, while anomalies in ECG recordings could signify early signs of heart attacks.

The handling of simple noise (such as Gaussian) and simple anomalies (such as jumps in amplitude or shifts in time) can be handled by using simple filters and algorithmic solutions [19]. More complex temporal patterns involve the development of machine learning models [20]. In more demanding cases, methods to eradicate time series problems require massive computing resources, leading researchers to explore quantum solutions [15,25].

Much of this area of research is exploratory and includes a lot of fragmented and highly specialized "proof-of-concept" projects. These explorations include quantum signal processing [11], evaluating similarity measures to support temporal data analysis [21], methods of quantum time series classification [32], quantum time series forecasting [12], the adoption of deep learning models, such as RNN and LSTM, for quantum modeling [2,8,29], and some work on noise removal and anomaly detection in time series [18].

The machine learning model adopted in this project aims to analyze time series using a neural network autoencoder (AE) [13, ch14], and more specifically its quantum counterpart, the quantum autoencoder (QAE). In general, autoencoders feature lossy data compression [10], allowing data sequences to preserve their most important or recurrent features, while removing their less significant, noisy, or infrequently occurring data [9, sect. 2.2.2].

Although there are many applications of classical AEs [3,16], there are very few examples of QAE use. The best received QAE studies include efficient data compression [25], image processing [5,15], analysis of marketing media [24], and anomaly detection in signals, although only in the frequency domain [26].

Other research explored the ability of QAE to deal with data in the time domain. They investigated the impact of data encoding on QAE performance [5], performed data sequence compression and reconstruction with high accuracy [25], and efficiently managed resources in hybrid quantum-classical computational settings [15,26].

Applications of quantum autoencoders for time series analysis, compression of temporal data, noise elimination, representation learning, and anomaly detection are all in uncharted territory and hence worth further investigation.

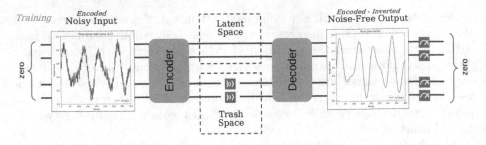

Fig. 1. Full QAE for circuit training (all qubits measured).

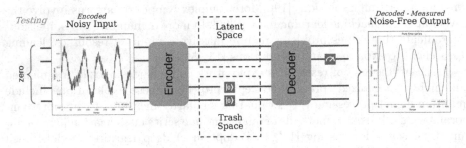

Fig. 2. Full QAE for circuit testing.

2   QuTSAE Design

There have been few prior attempts at applying quantum autoencoders to time series analysis. The design of the well-known *quantum time series autoencoders* (QuTSAE) commonly utilizes the variational quantum algorithms [7], which manipulate quantum circuits with parameterised operations (or gates) (see Figs. 1 and 2). QuTSAE operations are arranged into blocks that include the following:

- *Input block* coding classical data as the state of a quantum circuit.
- *Trainable encoder block* compressing the quantum state into a latent space.
- *Latent space* defining an active subset of qubits and discarding (reinitializing) the state of the remaining "trash" qubits.
- *Trainable decoder block* decompressing the quantum state of the latent space.
- *Output block* reversing the circuit state to the initial state (zero).
- *Measurements* converting the final quantum state into classical data.

2.1   Design Choices for QuTSAE Architecture

When designing the QuTSAE model, the following well-known architectural choices have been considered.

Replicating QAE
Initially, consider the QAE training model (see Fig. 1), of which the latent

space includes all qubits, the output block inverts the gate sequence of the input block, and a decoder is the inverse of an encoder with both sharing their parameters. The training circuit works as an identity that reverses the quantum state around the latent space, to reproduce the initial quantum state $|0\rangle^n$, regardless of the parameter values. As the QAE testing model (see Fig. 2) drops the training circuit's output block, its decoder returns a quantum state encoded on input. The final state can be measured in a variety of ways - all qubits at once or one at a time (as shown in the figure).

Approximating QAE

When the size of the latent space is reduced (see Fig. 1), the training QAE could no longer act as an identity for all sets of parameter values. However, the QAE encoder/decoder parameters can be optimized to minimize error in measured circuit values. When the output block is removed (see Fig. 2), the QAE circuit returns an approximation of the input state on the output. The smaller the size of the latent space, the more information is lost by the circuit and the more imprecise the reconstruction of the input on the output.

Denoising QAE

Autoencoders are known for their ability to denoise data. However, since the input/output and encoder/decoder blocks are their respective mirror twins, when QAE is given some noisy input, the same noise would be reconstructed on output. To prevent this from happening, the input/output and encoder/decoder pairs need to be decoupled by making their parameter sets distinct. In this case, in QAE training (see Fig. 1), input and output can be assigned noisy and pure data, respectively, and encoder/decoder parameters can be optimized independently, thus allowing the reconstruction of pure data from noisy data.

Custom QAE

Other types of QAEs have been developed, and in particular hybrid neural networks, with some interface (called dressing) between classical and quantum components [24].

2.2 Design Choices for QuTSAE Input Encoding

As the above QAEs make no assumptions about data on input and output, it is possible to apply them to time series and implement a functional QuTSAE.

Time series have to be represented in a way that allows the model to continuously fit the series and its subsequences. A sliding-window protocol was therefore selected to represent and process time series subsequences. The time series was also preprocessed by differencing and scaling its values, making the series (partially) stationary and its values manageable (as per ARIMA), [14, ch 9.1, 13.3].

As the window moves along the series during training or testing, its values need to be encoded as the QuTSAE quantum states of the input and output blocks. It is preferable to adopt a single encoding scheme for these two purposes.

There are several ways of quantum encoding schemes that are applicable to time series, i.e., basis, amplitude, QRAM, angle, and others [31]. Each of these schemes has its own strengths and weaknesses. For example:

Basis encoding
   adopts the binary representation of numbers on input; it is simple, but limits the circuit to handle single values only or multiples of imprecise values.

Amplitude encoding
   represents a window as a distribution of expectation values, as in circuit measurement; great for results interpretation; however, encoding of different input values leads to circuits of different structures, making very inefficient execution on QPUs or GPUs.

QRAM encoding
   precodes all window (or time series) values in the circuit and allows "referring" to them as needed; very flexible in processing; however, leads to large circuits, making their execution on quantum devices prone to errors.

Angle encoding
   encodes time series values as qubit state rotations, making it flexible, efficient, circuit size friendly, and easy to manipulate and measure. Its main weakness is its susceptibility to quantum noise, which could affect the accuracy of the results obtained from noisy NISQ-term quantum machines.

   Angle encoding, which was adopted for QuTSAE (see Fig. 3), can be realized as a series of $(Ry(x_0), Ry(x_1), ..., Ry(x_n))$ qubit rotations around the y axis. To facilitate later measurement and interpretation of the encoded values, we selected the state $|+\rangle$ to represent the value 0, $|1\rangle$ as -1, and $|0\rangle$ as $+1$. Any adjustment in a qubit $value \in [-1, +1]$ is a Ry rotation of $value \times 0.5\pi \times (1 - 2 \times \epsilon)$ (with ϵ as an error range). Due to time series differencing, the values handled by QuTSAE are small, resulting in qubit states very close to $|+\rangle$.

   It is worth mentioning that early experiments with the Z and ZZ feature maps, which are commonly used to encode input data, were found to be difficult to use along the QAE decoder for output reversal and result interpretation.

2.3   Design Choices for QuTSAE Output Decoding/Cost Function

Measurement and interpretation of the QuTSAE model are vital to the successful optimization of its parameters through a cost function; and, to gaining the ability to extract the reconstructed input from the output block.

Half-QAE with swap test
   When dealing with approximating QAEs, where the input-encoder pair shares its parameters with the decoder-output pair, the QAE circuit can be trained on the input-encoder half of the circuit alone [25].

   This can be accomplished with a cost function using a *swap test* circuit, which aims for the measurement of the trash state to approach zero, and thus for the state of the latent space to represent the maximum of information present on input. Unfortunately, when the trash size is large (say, greater than 5 qubits), the swap test becomes slow and inefficient.

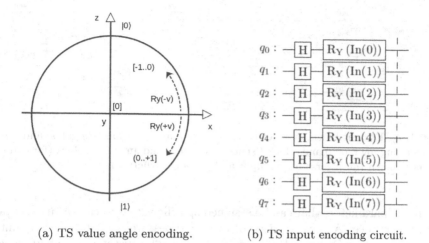

(a) TS value angle encoding.   (b) TS input encoding circuit.

Fig. 3. QuTSAE input encoding

Full-QAE and interpretation of measurements

When data are highly redundant, the model's latent space can encode input without any loss of information with just a few qubits, in which case the swap test is less effective working with large trash area.

Instead, in model training, the optimizer could measure the state of the entire input-encoder-latent-decoder-output sequence, which should approach the initial state of zero, when the probability of all qubits to be simultaneously zero is $p(|0\rangle^n) \approx 1$ (n is the number of qubits, see Fig. 1). This is achieved by minimizing the cost function of the form $Cost = 1 - p(|0\rangle^n)$, whose implementation requires measuring all qubits at once and testing a single distribution outcome of $|0\rangle^n$.

In model testing (with the output block removed), the output value can be derived from the angular state of individual qubits. This can be calculated from the probability distribution of measuring all qubits at once. However, as the number of qubits increases, the space of expectation values grows exponentially, resulting in the calculation to be of exponential complexity. To overcome this problem, it is possible to measure every qubit state individually (see Fig. 2 as an example), which provides single-qubit expectation values, easily interpreted as qubit amplitudes, cast back to qubit Ry rotations, and decoded as part of the time series output. Although the latter process has linear complexity, with the small number of qubits and a long execution time of quantum simulation, in practice the process is much slower.

2.4   Design Choices for QuTSAE Encoder/Decoder Ansatze

Another factor that affects the QuTSAE model is the selection of an *ansatz* implementing the QAE encoder and decoder. An anzatz is a template with

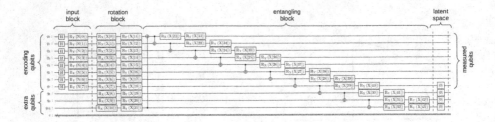

Fig. 4. QAE input block, encoder (with 3 extra qubits, Rx and Ry gates, and 1 rotation/entangling block) and a latent space. A decoder and an output block (for training) are not shown (an inverse of the input and encoder structure).

trainable parameters that can be assigned specific values to create an executable circuit instance. A great variety of anzatz types are available from quantum libraries (such as Qiskit or PennyLane). They can be customized to have specific structural characteristics, the type and number of rotation and entanglement blocks, and other properties (see Fig. 4).

The following are four important aspects of the QuTSAE ansatz design.

Ansatz structural symmetry

The structural symmetry between the encoder and decoder ansatze is necessary to ensure the reversibility of the QAE components on both sides of the latent space and the ensuing effectiveness of the adopted cost function. This decision led to the rejection of ansatz structures that interweave input and encoding blocks to facilitate reuploading of input data [23]. While data reuploading improves circuit trainability, it breaks the QAE symmetry and prevents easy interpretation of the output.

Ansatz rotation blocks

To keep the structure of the QuTSAE ansatz simple and consistent with the Ry input encoding and interpretation of the output, the ansatz was initially designed using only the Ry rotations. However, this imposed a severe limitation on the possible circuit states, which impeded the model learning. Subsequently, to take advantage of the entire space of possible qubit states, rotation blocks consisting of Rx and Ry gates were adopted.

Ansatz size (width and depth)

The QuTSAE performance is influenced by the ansatz size, i.e. its width (the number of qubits) and depth (the longest gate path). In QuTSAE, the ansatz width is controlled by extra qubits that offer additional degree of freedom but do not participate directly in input/output activities (option *aw*), and its depth by the number of rotation/entangling blocks (option *reps*).

Optimization of the ansatz parameters

Typically, the model optimizer cannot be selected without performing a preliminary investigation of the model parameter space and the effectiveness of the optimization algorithm. For example, this project reviewed gradient-based optimizers (such as ADAM and SPSA), as well as linear and nonlinear

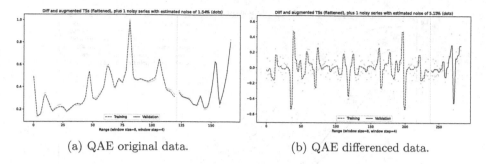

(a) QAE original data.   (b) QAE differenced data.

Fig. 5. Fragment of the original and differenced beer sales data (training set as a dashed line, validation as a solid line, noise shown as dots).

approximation methods (such as COBYLA and BFGS), with COBYLA found to be the quickest, most effective and producing the best results.

In our design deliberation of QuTSAE time series processing, we initially adopted the approximating half-QAE model developed by Romero et al. [25] and the enhancements recommended by Bravo-Prieto [5]. The resulting models produced a high level of accuracy in replicating inputs into outputs, with or without noise - the behavior undesirable in sequence denoising. Subsequently, the full-QAE model was proposed, leading to slower model optimization, but more successful noise reduction in time series.

3   Experiments

A series of experiments were conducted to test the ability of QuTSAE to remove "simple noise" from temporal data, and investigate the influence of model design decisions on its performance in training and validation. To manage the complexity of noise presence in the data and noise generated by quantum machines, it was decided to develop QuTSAE models using noise-free quantum simulators.

Data used for QuTSAE training related to beer sales in the USA, which was sampled from the IRI Marketing Data Set [6]. We selected a small sample of time series data consisting of 160 data points, which was split into two partitions (0.75/0.25) - 120 data points for model training and 40 data points for its validation (see Fig. 5). We refer to the original data as *pure*. A copy of the pure data was injected with 3% of uniformly distributed noise, and consequently we refer to these data as *noisy*. Differencing was applied to all pure and noisy data sets, resulting in data sets with noise exceeding 5.8%.

Subsequently, all data sets were segmented into windows of 8 data points each, sliding with a step of 4. This resulted in 29 training windows and 10 validation windows. Note that the window size of 8 (and various step sizes) was established in earlier experiments with half- and full-QAEs and synthetic data. After some investigation, it was decided that a window of size 8 and a step of 4 was suitable for the experiments with full-QAEs and beer sales data.

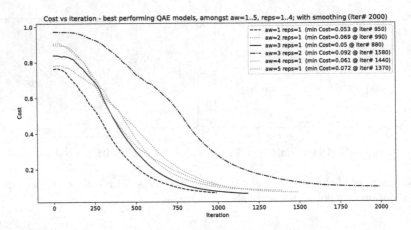

Fig. 6. Cost vs iterations in training models of the same latent space (lat=7) but different width and depth. Based on training cost, 6 best models (out of 14).

The experiments were carried out on a Qiskit quantum simulator running on Ubuntu 22.04.4 LTS, on a workstation equipped with an i9 CPU (24 cores and 32 threads), 64GB RAM, and a NVIDIA GPU GeForce RTX 4090. Initially, all experiments used GPU; however, later code improvements resulted in significantly faster CPU-based runs.

3.1   Determining the Optimum Ansatz Size

The first group of experiments aimed to decide on the optimal size of the QuT-SAE ansatz. Fourteen separate denoising models were created in Qiskit using a *TwoLocal* anzatz. All models shared the same number of input/output qubits (8, same as window size), the same size of the latent space (of 7 latent qubits, plus 1 trash qubit to test the QuTSAE data compressing behavior), same type of entanglement (shifted circular-alternating "sca" entanglement), and parameter training blocks (using Rx and Ry rotations). However, the models differed in their width (aw = 1..5), rotation/entanglement block repetitions (reps = 1..4), and the resulting number of trainable parameters (varying from 64 to 180).

In training, models with narrower and shallower circuits, and consequently smaller numbers of parameters, converged the quickest (within 1000 iterations). Models with wider and deeper circuits, featuring a large number of parameters, needed more training time (up to 2000 iterations; see Fig. 6).

In each optimization step (see Table 1, Ex. 1), the training cost and model parameters were saved for later analysis. The parameters were used to derive the training and validation scores R^2 (R-squared), $RMSE$, MAE, and $MAPE$ metrics for all intermediate stages of each model evolution. The $MAPE$ score was not used in the comparison of the performance of the models. It was, however, produced to provide information on the distribution and variance of the output (with respect to input).

Table 1. Cost and scoring results (top 4 models in Ex. 1, 2a and 2b experiments)

Experiment			Min	Training				Validation			
Lat	Aw	Reps	Cost	R2	RMSE	MAE	MAPE	R2	RMSE	MAE	MAPE
Ex. 1 (Qiskit/QNN)											
7	3	2	0.089	0.767	0.081	0.059	3.592	0.803	0.074	0.058	1.421
7	3	1	0.054	0.742	0.085	0.067	4.471	0.782	0.078	0.064	1.946
7	5	1	0.078	0.738	0.086	0.071	3.416	0.767	0.081	0.068	2.455
7	4	1	0.065	0.658	0.097	0.074	3.880	0.723	0.088	0.072	2.191
Ex. 2a (Qiskit/QNN)											
7	3	1	0.054	0.742	0.085	0.067	4.471	0.782	0.078	0.064	1.946
6	3	1	0.070	0.399	0.130	0.087	2.745	0.442	0.116	0.088	2.367
4	3	1	0.063	0.428	0.126	0.085	3.346	0.331	0.137	0.094	1.568
5	3	1	0.066	0.455	0.123	0.095	4.603	0.403	0.129	0.097	2.654
Ex. 2b (Qiskit/QNN)											
7	3	2	0.089	0.767	0.081	0.059	3.592	0.803	0.074	0.058	1.421
8	3	2	0.086	0.761	0.083	0.066	4.743	0.741	0.085	0.068	1.952
2	3	2	0.086	0.752	0.085	0.068	4.668	0.690	0.090	0.072	2.490
5	3	2	0.105	0.396	0.132	0.088	3.122	0.461	0.115	0.079	1.473
Ex. 3 (PyTorch/MLP, lat=7)											
Enc=3/Dec=3			0.012	0.996	0.010	0.006	0.225	0.751	0.081	0.059	1.310

Based on the MAE validation scores, among the models with a latent space of 7 qubits, two were selected for further experiments (see Table 1, Ex. 1 models with additional depth $aw = 3$, and repeating blocks $lat = 1$ and $lat = 2$).

3.2 Impact of Latent Space on Performance

As the size of the latent space determines the quality of time series reconstruction, the second group of experiments was conducted to identify the optimum ratio between latent and trash space in the QuTSAE circuits.

We used the two models selected previously and varied the size of their latent space, from 1 to 8 qubits (with size 0, interaction between the encoder and the decoder was not possible, thus generating errors). This resulted in 14 additional models to be investigated. Each model was optimized, its training and validation scores plotted (e.g., see Fig. 7a), and those with suboptimal performance were rejected. Subsequently, the models within each size group were compared (e.g. see Fig. 7b for models of the same size, given as $aw = 3$ and $reps = 2$).

As indicated by MAE validation scores, the best 4 models were then selected from each model size group (see Table 1, Ex. 2a and Ex. 2b). By analyzing their performance scores, the two best performing models overall were found to have a ratio of 7:1 latent to trash space (the most desirable ratio for this data set), 3 additional qubits (beyond the 8 input/output qubits), and 1 to 2 rotational/entangling blocks.

To assess the performance of QAE models against the equivalent classical autoencoders, a number of such models were developed using the PyTorch package. Each model had a differently sized latent space (from 1 to 8 variables) and consisted of two multilayer perceptrons (MLP) acting as the AE encoder and decoder, featuring 3 hidden layers each, and a total of 19,741 parameters (see Table 1, Ex. 3 PyTorch model with $lat = 7$). Training of PyTorch and quantum models used the same windowed and differenced data, and identical approaches to data coding. Although PyTorch models excelled in their training performance, their validation performance was on par with that of quantum models.

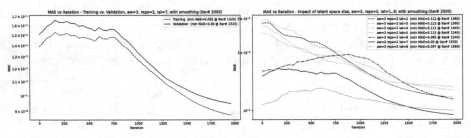

(a) MAE scores for training and validation (aw=3, reps=2, lat=7, y linear scale).

(b) Latent space impact on validation with MAE scores (aw=3, reps=2, y log scale).

Fig. 7. QuTSAE model performance.

3.3 Time Series Denoising

The final experiment was to investigate whether the QuTSAE models are capable of denoising time series.

To this end, for each model developed so far, we created its two separate instances: the first by instantiating its parameters with values found to be optimal during training, and the second by instantiating its parameters with values determined to be optimal for model validation. We then re-examined all model instances by applying the previously used scoring metrics to assess differences between reconstructed vs. pure sequences, as well as differences between noisy vs. pure sequences. Should the reconstructions be closer to the pure sequences than noise, we would then regard such models as capable of denoising time series.

Significantly, the final model scoring was carried out in two stages. As the time series windows had some significant overlaps due to their stepwise creation, in the first scoring process, we identified data points of high score variance commonly present at window edges, which were removed to improve the model scores produced in the second stage. We then reintegrated the remaining window overlaps by averaging to produce the adjusted sequences suitable for the presentation of the pure, noisy and reconstructed series (see Fig. 8).

As an example, let us take the best performing QuTSAE model, characterized by its hyperparameters $lat = 7$, $aw = 3$, and $reps = 2$. Table 2 provides its training and validation scores for pairs of pure, noisy, and reconstructed sequences. In the left column, we find the scores comparing the original pure time series vs. the measurement of noisy time series. In the right column, the scores compare pure time series vs. the reconstruction of pure time series from noisy input.

In validation, the scores of $R2$, $RMSE$, and MAE indicate that the reconstructed sequences fall between pure and noisy sequences. Hence, the model is capable of removing a modest level of noise from previously unseen time series. Unfortunately, the training scores tell a different story. Although the MAE scores are still indicative of the models' denoising abilities, the $R2$ and $RMSE$ scores, which are more sensitive to the presence of outliers, suggest that the model had difficulty with noise removal in training. In contrast, the classical models' reconstructions fitted pure data very closely, eliminating virtually all noise in the process. However, their denoising performance on previously unseen data was no better than that of the quantum models.

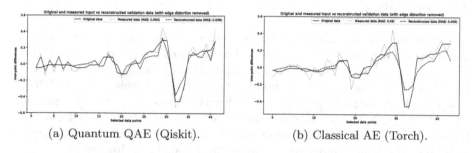

(a) Quantum QAE (Qiskit).   (b) Classical AE (Torch).

Fig. 8. Reintegration of time series for quantum and classical models, showing the original pure data (solid line), the noisy input (light dotted line), and reconstructed data with reduced noise on output (dashed line).

At the end of this discussion, it is important to take a closer look at the reintegrated time series (see Fig. 8). We can observe that the QuTSAE reconstructions (dashed line) seem to follow the shape of the noisy input (dotted line) (see Fig. 8a). However, the classical PyTorch reconstructions (dashed line) follow the shape of the pure data (solid line) (see Fig. 8b). However, the validation performance of both models is similar. The likely explanation for this phenomenon is that the QuTSAE cost function did not allow the model to fully learn, which can be confirmed by investigating the slope of the cost vs. performance scores over time (compare Figs. 6 and 7a). This may have been caused by the lack of nonlinear activations within the quantum model structure [22]. PyTorch models, in turn, converged very quickly; however, the algorithm had insufficient training data to avoid overfitting. Neither of the two models had the opportunity to reach its full potential.

Table 2. Final QAE model performance (lat = 7, aw = 3, reps = 2, ep = 2000)

Training Accuracy (no edge distortion)			
R2 (org-pure, in-noisy) =	0.782	R2 (org-pure, out-rec) =	0.767
RMSE (org-pure, in-noisy) =	0.078	RMSE (org-pure, out-rec) =	0.081
MAE (org-pure, in-noisy) =	0.064	MAE (org-pure, out-rec) =	0.059
MAPE (org-pure, in-noisy) =	3.985	MAPE (org-pure, out-rec) =	3.592
Validation Accuracy (no edge distortion)			
R2 (org-pure, in-noisy) =	0.795	R2 (org-pure, out-rec) =	0.803
RMSE (org-pure, in-noisy) =	0.076	RMSE (org-pure, out-rec) =	0.074
MAE (org-pure, in-noisy) =	0.064	MAE (org-pure, out-rec) =	0.058
MAPE (org-pure, in-noisy) =	1.962	MAPE (org-pure, out-rec) =	1.421

4 Conclusions

Quantum time series analysis is at the intersection of classical time series analysis and quantum machine learning. This article discussed a specific area of quantum time series processing, concerning the application of quantum autoencoders to the elimination of noise from temporal data. The article explained how different design decisions impact the function and performance of quantum time series autoencoders (QuTSAE). In this final section, we provide some reflections on the key design choices for quantum models in general, which could guide and assist researchers and developers of quantum machine learning models.

Requirements. In the initial steps of a quantum model development, one must consider not only the model structure and algorithms suitable for its processing (e.g. the need for the encoder and decoder), but also requirements (e.g. ability of handling temporal data) and constraints (e.g. level of noise in data not exceeding 5%) imposed on the structure and function of any acceptable solution.

Input. Input encoding schemes (such as QuTSAE *angle encoding around* $|+\rangle$) must match the strategy to measure and interpret the model output (e.g. interpretation of measurements into qubit angular states). The adopted design choices may restrict the use of certain devices for model execution, e.g. QPU or GPU.

Output. Quality design of quantum model observables, their measurement, and interpretation of results are essential for training and testing the model. Design of a suitable cost function (such as swap test or zero testing) is also of pivotal importance, and its use by an optimizer will determine the model trainability, and ultimately its capabilities (such as denoising), size (e.g. width and depth), and efficiency (due to complexity of obtaining and interpreting results) of its architectural options (e.g. half-QAE or full-QAE in QuTSAE).

Ansatz Design. Quantum toolkits offer rich libraries of ansatze suitable for the design of model components (such as the QuTSAE encoder and decoder). It is also possible to hand-craft a parameterized custom ansatz. Several important objectives in ansatz design must be considered. The ansatz structural properties must fit the model function (such as the need for QuTSAE symmetry). The rotational and entanglement blocks must match the input encoding and output decoding methods (the simplicity and compatibility of the *RealAmplitude* ansatz vs. the versatility of the *TwoLocal* ansatze in QuTSAE). The ansatz design, with its width, depth, the number of trainable parameters, as well as the required degree of freedom, will determine the suitability of its optimization strategies.

Model Training and Validation. Virtually all design choices for quantum models will have a major impact on their performance in training and validation. Typically, a series of experiments should be designed to establish which design aspects could influence model performance. To this end, an overarching experimental framework may need to be introduced, with data repositories and development tools (e.g. IRI marketing data, Qiskit and PyTorch toolkits), measuring instruments for model scoring and analysis (e.g. R^2, $RMSE$, MAE and $MAPE$), results capture and presentation facilities (e.g. MLFlow), and eventually source code and results distribution (e.g. via github).

Objectives and Success Criteria. From the outset, it is necessary to clearly define quantum project objectives (e.g. analysis of temporal data) and a set of testable success criteria (e.g. the target level of noise reduction in data). It is important to clearly state what goals are aspirational only and hence out of scope (such as the quantum model's ability to detect and remove "complex noise patterns"), but which could be investigated in the future.

As is evident from this article, quantum model development must constantly balance the quality and precision of the results with the complexity of developed models and processes, the available resources, and the practical execution time to produce useful results. It is a difficult process to manage and master. Nevertheless, the research and development process that we followed in this project provided us with some valuable learning experience.

We believe that our insights, as reported in this article, are transferable to other researchers and developers, other types of quantum time series analysis model, and quantum models in general, using different data, with distinct objectives, and a variety of quantum and machine learning tools.

Disclosure of Interests. The authors have no competing interests to declare that are relevant to the content of this article.

References

1. Bajaj, A.: Anomaly Detection in Time Series, March 2021. https://neptune.ai/blog/anomaly-detection-in-time-series
2. Bausch, J.: Recurrent quantum neural networks. Adv. Neural. Inf. Process. Syst. **33**, 1368–1379 (2020)
3. Berahmand, K., Daneshfar, F., Salehi, E.S., Li, Y., Xu, Y.: Autoencoders and their applications in machine learning: a survey. Artif. Intell. Rev. **57**(2), 28 (2024)

4. Blázquez-García, A., Conde, A., Mori, U., Lozano, J.A.: A review on out-lier/anomaly detection in time series data. ACM Comput. Surv. (CSUR) **54**(3), 1–33 (2021)
5. Bravo-Prieto, C.: Quantum autoencoders with enhanced data encoding. Mach. Learn. Sci. Technol. **2**, May 2021
6. Bronnenberg, B.J., Kruger, M.W., Mela, C.F.: Database paper —the IRI marketing data set. Mark. Sci. **27**(4), 745–748 (2008)
7. Cerezo, M., et al.: Variational quantum algorithms. Nature Rev. Phys. **3**(9), 625–644 (2021)
8. Chen, S.Y.C., Yoo, S., Fang, Y.L.L.: Quantum long short-term memory. In: ICASSP 2022-2022 IEEE Int. Conf. on Acoustics, Speech and Signal Processing (ICASSP), pp. 8622–8626. IEEE (2022)
9. Chen, S., Guo, W.: Auto-encoders in deep learning—a review with new perspectives. Mathematics **11**(8), 1777 (2023)
10. Chiarot, G., Silvestri, C.: Time series compression: a survey. ACM Comput. Surv. **55**(10), 1–32 (2023)
11. Eldar, Y., Oppenheim, A.: Quantum signal processing. IEEE Signal Process. Mag. **19**(6), 12–32 (2002)
12. Emmanoulopoulos, D., Dimoska, S.: Quantum Machine Learning in Finance: Time Series Forecasting. Tech. Rep. arXiv:2202.00599, arXiv, February 2022
13. Goodfellow, I., Bengio, Y., Courville, A.: Deep Learning. The MIT Press (2016)
14. Hyndman, R.J., Athanasopoulos, G.: Forecasting: Principles and Practice, 3rd edn. OTexts, Melbourne, Australia (2021)
15. Khoshaman, A., Vinci, W., Denis, B., Andriyash, E., Sadeghi, H., Amin, M.H.: Quantum variational autoencoder. Quantum Sci. Technol. **4**(1), 014001 (2018)
16. Li, P., Pei, Y., Li, J.: A comprehensive survey on design and application of autoencoder in deep learning. Appl. Soft Comput. **138**, 110176 (2023)
17. Liu, C.J., Liu, H.T., Bian, C., Chen, X.D., Yang, S.H., Wang, X.F.: Investigation of time-series prediction for turbine machinery condition monitoring. IOP Conf. Ser. Mater. Sci. Eng. **1081**(1), 012022 (2021)
18. Liu, N., Rebentrost, P.: Quantum machine learning for quantum anomaly detection. Phys. Rev. A **97**(4), 042315 (2018)
19. Lu, L., et al.: A survey on active noise control in the past decade–Part I: Linear systems. Signal Process. **183**, 108039 (2021)
20. Lu, L., et al.: A survey on active noise control in the past decade-Part II: Nonlinear systems. Signal Process. **181**, 107929 (2021)
21. Markov, V., Rastunkov, V., Fry, D.: Quantum Time Series Similarity Measures and Quantum Temporal Kernels, December 2023
22. Maronese, M., Destri, C., Prati, E.: Quantum activation functions for quantum neural networks. Quantum Inf. Process. **21**(4), 128 (2022)
23. Pérez-Salinas, A., Cervera-Lierta, A., Gil-Fuster, E., Latorre, J.I.: Data re-uploading for a universal quantum classifier. Quantum **4**, 226 (2020)
24. Rivas, P., Zhao, L., Orduz, J.: Hybrid quantum variational autoencoders for representation learning. In: 2021 Int. Conf. on Computational Science and Computational Intelligence (CSCI), pp. 52–57. IEEE, Las Vegas, NV, USA, December 2021
25. Romero, J., Olson, J.P., Aspuru-Guzik, A.: Quantum autoencoders for efficient compression of quantum data. Quantum Sci. Technol. **2**(4), 045001 (2017)
26. Sakhnenko, A., O'Meara, C., Ghosh, K.J., Mendl, C.B., Cortiana, G., Bernabé-Moreno, J.: Hybrid classical-quantum autoencoder for anomaly detection. Quantum Mach. Intell. **4**(2), 27 (2022)

27. Schuld, M., Petruccione, F.: Machine Learning with Quantum Computers. Springer, 2nd edn., October 2021
28. Sezer, O.B., Gudelek, M.U., Ozbayoglu, A.M.: Financial Time Series Forecasting with Deep Learning: A Systematic Literature Review: 2005-2019, November 2019
29. Takaki, Y., Mitarai, K., Negoro, M., Fujii, K., Kitagawa, M.: Learning temporal data with variational quantum recurrent neural network. Phys. Rev. A **103**(5), 052414 (2021)
30. Tripathi, P.M., Kumar, A., Komaragiri, R., Kumar, M.: A review on computational methods for denoising and detecting ECG signals to detect cardiovascular diseases. Arch. Comput. Methods Eng. **29**(3), 1875–1914 (2022)
31. Weigold, M., Barzen, J., Leymann, F., Salm, M.: Encoding patterns for quantum algorithms. IET Quantum Commun. **2**(4), 141–152 (2021)
32. Yarkoni, S., Kleshchonok, A., Dzerin, Y., Neukart, F., Hilbert, M.: Semi-supervised time series classification method for quantum computing. Quantum Mach. Intell. **3**(1), 12 (2021)

Optimizing Quantum Circuits Using Algebraic Expressions

Varun Puram[✉], Krishnageetha Karuppasamy, and Johnson P. Thomas

Oklahoma State University, Stillwater, OK 74075, USA
{vpuram,kkarupp}@okstate.edu, jpt@cs.okstate.edu

Abstract. Optimizing quantum circuits and reducing errors plays a crucial role in quantum circuit computation. Every quantum circuit can be represented using algebraic expressions, we propose an approach that directly derives algebraic expressions, ensuring that parallelism is maximized, that is, the number of circuit slices is minimized, and secondly, the computation required for obtaining the desired algebraic expression is reduced. This results in quantum circuits that are more efficient in space and computation time. The simplification of algebraic expressions offers methods to streamline optimized circuits, reducing the number of gates and depth. This reduction is aimed at minimizing the overall complexity of the expressions, resulting in more efficient quantum computations. In this paper, we also show through simulations that the optimized circuit will have less errors when compared to original circuits.

Keywords: Quantum Computing · Algebraic expressions · Quantum circuit optimization and Quantum error correction

1 Introduction

Although Quantum computing algorithms have been proven to be much faster than classical algorithms, a major obstacle to its successful realization is high error or noise rates. Due to this noise, qubits are rotated by the wrong amount that cause the final output state to not be the final correct state that is expected [7]. Unlike classical circuit optimization, quantum circuit optimization not only improves the execution time but also helps to reduce error. This is because quantum noises are linearly dependent on the size of the circuits [8]. As a quantum circuit becomes larger and more complex, they involve a higher number of qubits and quantum gates. Larger circuits tend to require longer coherence times to complete the computations successfully. Qubit coherence time is finite. Hence, the probability of errors occurring during quantum operations increases. An error in an early gate (a gate that does the computations early in execution of an algorithm) can affect the state of qubits as they propagate through subsequent gates. Thus, by reducing the circuit size the errors can also be mitigated.

Quantum algorithms are implemented as quantum circuits, which consist of quantum gates. Quantum Circuit Optimization is the process of finding a more efficient and streamlined way to represent a given quantum circuit or algorithm. This is achieved by

L. Franco et al. (Eds.): ICCS 2024, LNCS 14837, pp. 268–276, 2024.
https://doi.org/10.1007/978-3-031-63778-0_19

reducing the number of gates used and shortening the overall depth of the circuit. The optimized circuit should implement the same algorithm as the original circuit and hence yield the exact same results as the original circuit while operating more efficiently in terms of errors and execution time.

There are several ways to represent quantum circuits such as Quantum Assembly Language (QASM) [10], Quantum Gate Matrices, Directed Acyclic Graphs (DAGs) [1]. Every representation has its own advantages and disadvantages. One common technique for optimization involves representing a quantum circuit as a Directed Acyclic Graph (DAG) with gate commuting properties. This approach allows for the cancellation of gates and the merging of gates to simplify the quantum circuit. [1–5, 9] proposes various optimization and pattern matching techniques using DAG.

In this paper we propose an algorithm to obtain efficient algebraic expressions for quantum circuits. Algebraic expressions ensure that the algorithm itself does not change because of optimization. Using algebraic expressions, quantum circuits can be analyzed and optimized. Algebraic simplification leads to less complex logic expressions which results in reduced circuit complexity. This leads to improved performance, lower power consumption and fewer errors. Simulations show that the resultant optimized circuit has less errors than the original circuit.

2 Proposed Approach

Quantum algorithms are expressed using quantum circuits, which are sequences of quantum gates. These gates perform quantum computations. Each quantum gate can be mathematically represented by a unitary matrix, which describes the transformation as it applies to quantum states. A quantum circuit can also be viewed as a composite unitary matrix. If there is a n-qubit quantum circuit, the corresponding unitary matrix to the circuit will have a dimension of $2^n \times 2^n$. At each depth or layer in the quantum circuit, a quantum gate acts on its respective qubit. If the quantum circuit has n depth of layers (or "slices") of parallel computations, then there will be n slices present. Multiplication of the slices result in composite unitary matrix of the quantum circuit.

When a quantum circuit contains single-qubit gates, they are represented using their respective matrices. When the circuit includes multi-qubit gates, these quantum matrices are decomposed into elementary unitary matrices. An efficient approach for this decomposition was developed by Hutsell et. al. [6]. They proposed the use of two 2×2 matrices, D_0 and D_1, as a means of decomposing larger unitary matrices into smaller 2×2 matrices. This decomposition allows for a more efficient representation of the algebraic expressions involved.

$$D_0 = \begin{pmatrix} 1 & 0 \\ 0 & 0 \end{pmatrix} D_1 = \begin{pmatrix} 0 & 0 \\ 0 & 1 \end{pmatrix} \{D_0 + D_1 = I\}$$

where I is the identity matrix. We utilize this work [6] as a foundation for our proposed new approach.

We use Hutsell's work [6] to address the general case where a slice of a quantum circuit contains multiple target qubits acting on a control qubit or multiple multi-qubit gates. Figure 2a provides an illustration. Using the approach proposed in [6], the given slice is broken down into further slices to achieve the desired algebraic expression, as demonstrated in Fig. 1b. A slice refers to a layer of parallel computation.

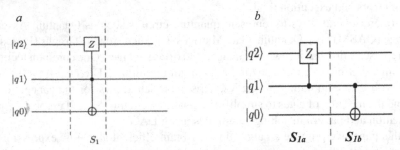

Fig. 1. a: Circuit with the slice which has 2 targets (Z and Not) with only 1 respective control qubit (|q1⟩). b: Slice S1 is divided into two slices S1a and S1b to get algebraic expressions

The disadvantage with this approach is when the number of target qubit exceeds the number of control qubits, the circuit will decompose into many slices which will result in an inefficient algebraic expression, resulting in a less optimized algebraic expression which leads to a less optimized circuit. Our proposed work introduces a novel method to handle situations where the number of target qubits exceeds the number of control qubits. Figure 2a provides an example. To get an algebraic expression for circuit S_1 the previous method [6] needs slicing into S_{1a} and S_{1b} (Fig. 1b). The disadvantage of slicing is that it increases the depth of the circuit (Fig. 1b). This means more slices which results in more inefficiency and errors. In quantum circuit computation, each slice of n qubits is represented by $2^n \times 2^n$ Unitary matrix, Although the final matrix remains unchanged, obtaining the final matrix involves more matrix multiplications as the number of slices increases. There is a computation for each slice followed by a computation to merge the results from each slice. The primary objective of the proposed work is to eliminate the need for further dividing slices so as not to increase the depth of the circuit. Instead, our method directly derives algebraic expressions, ensuring that parallelism is maximized (or the depth is minimized) and thereby the computation required for obtaining the desired expression is reduced.

In this paper, we propose a method that directly yields algorithmic expressions without the necessity of dividing the circuit into additional slices. This approach maximizes parallelism while effectively reducing the computational burden associated with obtaining algebraic expressions. Secondly, we optimize the resulting algebraic expressions by using circuit identities and cancellation rules to get optimized circuits. The proposed approach is outlined in Algorithm_1.

Algorithm_1: Optimization Outline

Input: - Original Quantum circuit C
Output: - Optimized Quantum circuit C^1
1) Represent Quantum Circuit C as algebraic Expression.
2) Optimize algebraic Expressions by using circuit identities and gate cancellation rules.
3) Convert optimized algebraic expression to optimized Quantum Circuit C^1

Algorithm_1: Outline of optimization procedure using algebraic expressions.

3 Representing Quantum Circuits as Algebraic Expressions

The proposed approach to represent quantum circuits as Algebraic expressions while maximizing parallelism is outlined next.

Step 1: - Identifying Number of terms: Given a slice with N control nodes, the number of terms in the algebraic expression will be 2^N terms. If T_i represents term i, then the slice will have the algebraic form.

$$U = \sum_{i=0}^{2^n} T_i$$

Step 2: - Representing each term: If we have a m qubit quantum circuit (q_{m-1}, q_{m-2} .... q_0), each term T_i represents the Tensor products of qubits from left to right. In each term, there will be m characters represented by C_i representing each quantum bit in the quantum circuit.

$$T_i = \otimes_{j=m=1}^{0} C_j = C_{m-1} C_{m-2} \cdots \cdots C_0$$

Step 3: - Finalizing the expression: If the control bit is in state $|1\rangle$, then the target bit undergoes the corresponding target gate computation. On the other hand, if the control bit is in state $|0\rangle$, the target gate acts as an identity gate, leaving the target bit unchanged. We term that D_0 as the inactive control matrix and D_1 as the active control matrix.

If in the slice, there are k control bits represented as $a_1, a_2....,a_k$ qubits respectively ({for all a_i} $\in$ {0, 1,... m-1}), then we have 2^k terms as mentioned in step 1 and each term in the final expression will be represented in binary notation as shown below.

$$U = \sum_{i=0}^{2^k-1} T_k, (k \in \{0, 1\} \text{ n})$$

For example, If there are three control nodes acting on a_1, a_2, a_3 qubits where each a_i is a control qubit, the final expression can be represented as

$U = T_0 + T_1 + T_2 + T_3 + T_4 + T_5 + T_6 + T_7 = T_{\{000\}} + T_{\{001\}} + T_{\{010\}} + \cdots \cdots + T_{\{111\}}$

1.1. In each of the terms generated from the k control bits, in the term represent each bit which equal to '0' as the inactive control qubit and '1' as the active control qubit respectively.

1.2. Step 2 presents how each term T_i can be represented as a tensor product $\otimes \, {}^{0}_{j=m-1}$ $C_j = C_{m-1} \, C_{m-2} \ldots \ldots C_0$. In the binary representation of each term $T_{\{k\}}$, each k_p which is a control qubit represents a character in the algebraic expression. Where $p \in (0 \ldots n\text{-}1)$ and n is the number of control nodes in the respective slice. Each k_p represents a character representing the a_i qubit. Therefore, the character k_p is in the $C_{ai}{}^{th}$ position. In each term, if k_p is 0 then D_0 replaces the respective $C_{ai}{}^{th}$ position in the term. If k_p is 1 then D_1 replaces the respective $C_{ai}{}^{th}$ position in the term.

1.3. In the quantum circuit, each control node may be associated with one or more target bits. Similar to the previous step described, when considering the algebraic representation, if a target bit is associated with an inactive control node, an identity gate will be substituted at the respective target qubit position within the term. If a target bit is associated with an active control node, which is replaced by the matrix D_1, the target gate matrix will be substituted at the respective qubit position within the term. This allows for the incorporation of control gates into algebraic expression. The behavior of the control node determines whether the target gate is applied or replaced with an identity gate. By considering the status of the control node, the corresponding gate operation can be appropriately represented within the term, resulting in an accurate and comprehensive algebraic expression for the quantum circuit. When there are multiple targets, similar steps will take place, that is, multiple targets which are associated to the same control node will be replaced at once. When dealing with multiple targets in a quantum algorithm, similar steps will occur where multiple targets associated with the same control node are replaced

Example: -

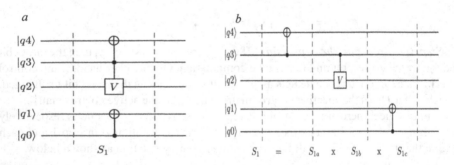

Fig. 2. a: Example for proposed algorithm. b: Existing algorithm breaking into slices.

Consider the quantum circuit with one slice S_1 and multiple quantum gates as shown in Fig. 2a. This example illustrates how the proposed approach works and also its efficiency in generating optimized algebraic expressions.

The algebraic expression using our proposed approach is:

1. There are two control nodes in the slice, so the number of terms in the algebraic expression is $2^2 = 4$. By step 1, $S_1 = T_0 + T_1 + T_2 + T_3$
2. Each term can be represented as a tensor product of characters (= number of qubits). Here there are five qubits. By step 2,

$$T_0 + T_1 + T_2 + T_3 = C_4 C_3 C_2 C_1 C_0 + C_4 C_3 C_2 C_1 C_0 + C_4 C_3 C_2 C_1 C_0 + C_4 C_3 C_2 C_1 C_0$$

3. There are two control nodes acting on the q3 and q0 qubits, Binary indices of term indices are represented as $S_1 = T_{\{00\}} + T_{\{01\}} + T_{\{10\}} + T_{\{11\}}$
4. In the Binary representation of $T_{\{ij\}}$ where i represents qubit q3 and j represents qubit q0. For example, the 2nd term $T_{\{01\}}$, $ij = 01$; here $i = 0$ representing the q3 qubit and $j = 1$ representing the q0 qubit.
5. In step 1.3 replace C_3 with D_0 and C_0 with D_1.
6. In $T_{\{01\}}$, q3 is inactive and q0 is active, the replacement process involves replacing the target nodes that are associated with q3, namely q2 and q4, with the identity matrix. At the same time, substitute the target matrices associated with q0, that is the q1 qubit, at their respective positions within the algebraic expression. The final expression is given in Eq. 1 and is represented in Fig. 2a.

$$S_1 = ID_0 IID_0 + ID_0 IND_1 + ND_1 VID_0 + ND_1 VND_1 \tag{1}$$

Comparing Eq. (1) with the algebraic expression generated with the existing method [6] as shown in Fig. 2b.

$$S_1 = S_{1a} S_{1b} S_{1c}$$
$$= [(ID_0 III + ND_1 III)(ID_0 III + ID_1 VII)(IIIID_0 + IIIND_1)] \tag{2}$$

We observe Eq. (1) is a simplified version when compared to Eq. 2. Hence, the proposed methodology not only offers a correct method (that is, the outcome remains the same), but also a more efficient approach by directly providing the equation to be solved.

4 Optimizing the Algebraic Expressions

The next step is to optimize the algebraic expressions. Algebraic expression of quantum circuits provides a concise and simple way to represent complex quantum circuits. It uses elementary quantum operators such as X, Y, Z, H, phase shift, CX, CCX, SWAP to express the relationships between inputs and outputs. The use of these universal unitary operators ensures consistency in representing quantum circuits across different designs. This readable representation of a quantum circuit allows circuit behavior to be expressed in a natural and intuitive way, making it easier to understand and communicate quantum circuit functionality.

The quantum circuit is represented as a list of gates that are applied sequentially. The following transformation rules are then applied to optimize the quantum circuits.

To simplify or eliminate gates in a quantum circuit, remove gates that are directly next to their inverse. By rearranging the gates in the circuit, it becomes possible to transform or eliminate them. Specifically, when there is a U gate in the circuit, the optimization algorithm looks for a corresponding U† gate where U† is the adjoint. If a U† gate is found, the U gate is successfully canceled out. For some elementary gates such as X,

Y, Z, H, and CNOT, U† is equal to U. In such cases, if two instances of those gates are adjacent to each other, they get cancelled out.

For two rotation gates RZ(θi) and RZ(θj) that have a shared control line or adjacent to each other, merge the two rotations to make it a single gate. This we call gate fusion.

By using techniques such as the cancellation rule and gate fusion the depth is reduced by rearranging for possible commuting quantum gates. The quantum circuit can be simplified, leading to improved performance in terms of time and size of the circuits. Until now as shown in the previous section, we have represented the quantum circuit using the most optimal algebraic expressions. In this section we extend this work in the optimization of quantum circuits by building on the algebraic expressions discussed in Sect. 3. The proposed method uses the following steps, to begin the optimization process, we first analyze the circuit shown in Fig. 3a and express it using algebraic expression notation. This algebraic expression represents the overall behavior of the circuit in terms of mathematical operations and variables. By manipulating and simplifying this expression using basic identities [1] and distributive properties of algebra, we arrive at an optimized form. Once the optimized algebraic expression is derived, the next step involves converting it back into a quantum circuit. This conversion process translates the algebraic operations and variables into corresponding quantum gates and qubits which results in Fig. 3b.

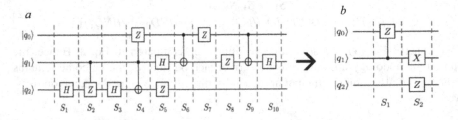

Fig. 3. a: Circuit C_2. b: Simplified version of circuit C_2

The process of optimizing the given quantum circuit C_2 involves the following steps: The unitary matrix U representing the circuit C_2 is formed by,

$U = S_1 \times S_2 \times S_3 \times S_4 \times S_5 \times S_6 \times S_7 \times S_8 \times S_9 \times S_{10}$ where each slice is given by algorithm mentioned in above section

$S_1 = IIH; S_2 = ID_0I + ID_1Z; S_3 = IIH; S_4 = ID_0I + ZD_1X; S_5 = IHZ;$
$S_6 = ID_0I + ID_1X; S_7 = ZII; S_8 = IZI; S_9 = ID_0I + ID_1X; S_{10} = IHI;$
$S_1 \times S_2 \times S_3 \times S_4 \times S5 \times S6 \times S7 \times S8 \times S9 \times S10 =$
$= (IIH)(ID_0I + ID_1Z)(IIH)(ID_0I + ZD_1X)(IHZ)(D_0II + D_1XI)(ZII)(IZI)(D_0II + D_1XI)(IHI)$
$= (ID_0H + ID_1H.Z)(IIH)(ID_0I + ZD_1X)(D_0HZ + D_1H.XZ)(ZII)(IZI)(D_0HI + D_1X.HI)$
$= (ID_0I + ID_1H.Z.H)(ID_0I + ZD_1X)((D_0HZ + D_1H.X Z)(ZII)(D_0Z.HI + D_1Z.H.ZI)$
$= (ID_0I + ID_1X)(ID_0I + ZD_1X)(D_0HZ + D_1H.X Z)(D_0Z.HI - D_1XI)$
$= (ID_0I + ZD_1I)IXZ$

the resultant unitary matrix for circuit C2; $U = (ID0\ I + ZD1I)\ (IXZ)$

Table 1. Optimization results

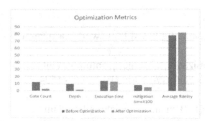

Fig. 4. Optimization Results

Optimization Metrics	Circuit before Optimization	Circuit after Optimization
Gate Count	12	3
Depth	10	2
Mitigation Overheads	88.16	87.35
Mitigation time	0.082	0.051
Average fidelity	77.87	81.6
Execution time	13.5	13

This algebraic expression simplification process can be automated/implemented by creating an algorithm that parses algebraic expressions into Abstract Syntax Trees (AST), applying predefined simplification rules recursively until we get the same expression as the outcome and converts the simplified AST back into algebraic expression format. The expression is then converted into a quantum circuit. As a result of simplifying using our approach, the circuit is reduced by 75% in terms of the number of gates in C_2. The simplified optimized circuit is shown in Fig. 3b. Instead of 10 levels there are only 2 levels resulting in improved performance and reduced error.

5   Results

Quantum circuit optimization can be measured by gate count, depth, Mitigation Overheads, Mitigation time, Average fidelity, and Execution time. Average fidelity quantifies how closely the actual gates on your device match the ideal gates. The ideal process matrix is subtracted from estimated calibration matrices (readout, Decoherence, cross talk). Error mitigation is a process for enhancing the reliability of quantum computations. The average resultant state of a given circuit is obtained by sampling. This will be done by the processor. Obviously if the circuit size is large the mitigation time and mitigation overheads increase. The circuits 3a and optimized 3b are executed on the ibm_perth processor [11]. Both circuits return the same resultant state |010 > with different optimization metrics values shown in Fig. 4 and Table 1.

The results show that the proposed algebraic simplification method results in reduced Gate count, depth, execution time, mitigation time, while increasing the average fidelity. This is for a small circuit. The improvement using our proposed approach on larger circuits will be more substantial and significant.

6   Conclusion

This paper proposes a novel approach to represent quantum circuits using algebraic expressions. These expressions serve as identities to analyze and simplify quantum circuits. By applying algebraic simplification techniques, we can transform complex logical structures within circuits into more straightforward and efficient forms. This simplification process reduces circuit complexity, which in turn improves the overall performance of quantum circuits and lowers their power consumption.

References

1. Jiang, H.-J., Li, D.L., Deng, Y., Xu, M.: A pattern matching-based framework for quantum circuit rewriting (2022)
2. Nam, Y., Ross, N.J., Su, Y., et al.: Automated optimization of large quantum circuits with continuous parameters. NPJ Quant. Inf. **4**, 23 (2018). https://doi.org/10.1038/s41534-018-0072-4
3. Chen, M., Zhang, Y., Li, Y., Wang, Z., Li, J., Li, X.: QCIR: pattern matching based universal quantum circuit rewriting framework. In: Proceedings of the 41st IEEE/ACM International Conference on Computer-Aided Design (ICCAD 2022), vol. 55, pp. 1–8 (2022). https://doi.org/10.1145/3508352.3549405
4. Iten, R., Moyard, R., Metger, T., Sutter, D., Woerner, S.: Exact and practical pattern matching for quantum circuit optimization. ACM Trans. Quant. Comput. **3**(1), 1–41 (2022). https://doi.org/10.1145/3498325
5. Mazder Rahman, M., Dueck, G.W., Horton, J.D.: An algorithm for quantum template matching. J. Emerg. Technol. Comput. Syst. **11**, 3 (2015). https://doi.org/10.1145/2629537
6. Hutsell, S.R., Greenwood, G.W.: Efficient algebraic representation of quantum circuits. J. Disc. Math. Sci. Cryptogr. **12**(4), 429–449 (2009). https://doi.org/10.1080/09720529.2009.10698246
7. França, D., Garcia-Patron, R.: Limitations of optimization algorithms on noisy quantum devices. Nat. Phys. **17**, 1–7 (2021). https://doi.org/10.1038/s41567-021-01356-3
8. Wang, S., Fontana, E., Cerezo, M., et al.: Noise-induced barren plateaus in variational quantum algorithms. Nat. Commun. **12**, 6961 (2021). https://doi.org/10.1038/s41467-021-27045-6
9. Abdessaied, N., Soeken, M., Drechsler, R.: Quantum circuit optimization by hadamard gate reduction. In: Yamashita, S., Minato, S. (eds.) Reversible Computation, pp. 149–162. Springer, Cham (2014). https://doi.org/10.1007/978-3-319-08494-7_12
10. Cross, A.W., et al.: Open quantum assembly language. arXiv preprint arXiv:1707.03429 (2017)
11. IBM quantum Composer. https://quantum-computing.ibm.com/composer. Accessed 5 Dec 2023

Implementing 3-SAT Gadgets for Quantum Annealers with Random Instances

Pol Rodríguez-Farrés[1]([✉]), Rocco Ballester[1,2], Carlos Ansótegui[3], Jordi Levy[1], and Jesus Cerquides[1]

[1] IIIA-CSIC, Cerdanyola, Spain
pol.rofa@gmail.com, {levy,cerquide}@iiia.csic.es
[2] Enzyme Advising Group, Barcelona, Spain
rocco.ballester@enzyme.biz
[3] Universitat de Lleida (DIEI, UdL), Lleida, Spain
carlos@diei.udl.cat

Abstract. The Maximum Boolean Satisfiability Problem (also known as the Max-SAT problem) is the problem of determining the maximum number of disjunctive clauses that can be satisfied (i.e., made true) by an assignment of truth values to the formula's variables. This is a generalization of the well-known Boolean Satisfiability Problem (also known as the SAT problem), the first problem that was proven to be NP-complete. With the proliferation of quantum computing, a current approach to tackle this optimization problem is Quantum Annealing (QA). In this work, we compare several gadgets that translate 3-SAT problems into Quadratic Unconstrained Binary Optimization (QUBO) problems to be able to solve them in a quantum annealer. We show the performance superiority of the not-yet-considered gadgets in comparison to state-of-the-art approaches when solving random instances in D-Wave's quantum annealer.

Keywords: Quantum Annealing · Satisfiability · Optimization

1 Introduction

As the field of Quantum Computing (QC) continues to grow, researchers have been seeking to use quantum properties, such as superposition and entanglement, to help or even surpass the performance of classical computers [2]. One of the main focuses of this field has been to tackle intractable problems that cannot be solved by classical algorithms with running times that grow only polynomially in relation to the input length [9], i.e., problems that are not (or have not been proved to be) in P. A clear example is large number factorization, a problem that Shor's quantum algorithm has theoretically proven to solve in polynomial time, something far away from achieving with classical computers

---

P. R. Farrés and R. Ballester—These authors contributed equally to this work.

[17]. Another well-known example is the so-called Boolean Satisfiability Problem (SAT, for short). The SAT problem, and in particular, the 3-SAT problem, is of paramount importance in computer science since it is the first problem to be proven NP-complete, meaning that any other NP problem can be translated, i.e., mapped, into a 3-SAT problem [8]. This problem has been tackled recently using Adiabatic Quantic Optimization (AQO) [14], which addresses it by encoding the solution into the state of a Hamiltonian and letting the system evolve, relying on the adiabatic theorem of quantum mechanics [9]. More specifically, these optimization tasks have been attended to solve using Quantum Annealing (QA), a technique which can be understood as a heuristic to adiabatic quantum computing [16]. In addition, experimental annealing such as D-Wave experiments, performed in this paper, will not have provable scaling as other algorithms do, and the best which could be shown is an empirically observed scaling, which is unlikely to resolve complexity theory questions since NP-hardness is a worst-case rather than typical-case statement. More on SAT, 3-SAT, and QA will be explained in detail in Sect. 2.

Our work focuses on implementing and comparing different translations, hereinafter referred to as *gadgets*, to map random 3-SAT instances to Quadratic Unconstrained Binary Optimization (QUBO) problems to be able to solve them using a quantum computer.

The main contributions of our work are:

- We implement new gadgets for translating 3-SAT instances into QUBO formulations and experimentally test them in D-Wave's commercial quantum annealer[1].
- We analyze and compare the characteristics of each gadget, such as the required number of logical and physical qubits, when implementing them in D-Wave's quantum annealer.
- We compare the performance of the new-implemented gadgets in a quantum computer with state-of-the-art gadgets and show their superiority.
- We compare their performance against a classical SAT solver.

This paper is organized as follows. First, in Sect. 2, we introduce the main concepts relevant to this work. In Sect. 3, we provide an overview of the relevant research that has contributed to advancements in this domain, highlighting the state-of-the-art gadgets that map 3-SAT to QUBO. Then, in Sect. 4, we delve into a few gadgets that have been theoretically proposed but have not been implemented yet. Later, in Sect. 5, we show, compare, and analyze the results obtained from the conducted experiments. Finally, in Sect. 6, we provide conclusive remarks along with avenues for future exploration.

2   Fundamental Concepts

In this section, we introduce the concepts we will work with, namely the 3-SAT, Max-3-SAT and QUBO problems, and we provide a brief introduction to QA.

---

[1] https://www.dwavesys.com.

2.1   The 3-SAT and Max-3-SAT Problems

The Boolean Satisfiability Problem, also known as SAT for short, is a computational problem in which the goal is to verify if a propositional logic statement formed by boolean variables can be satisfied.

In this work, we focus on 3-SAT, i.e., instances characterized by n boolean variables $\{v_i\}_{i=1}^n$ and a conjunction of m clauses, each one containing the disjunction of 3 literals:

$$F = (l_1 \vee l_2 \vee l_3) \wedge (l_4 \vee l_5 \vee l_6) \wedge \ldots \wedge (l_{3m-2} \vee l_{3m-1} \vee l_{3m}), \qquad (1)$$

where every literal l_i refers to a variable v_k or its negation $\neg v_k$.

As stated in Sect. 1, 3-SAT is of utmost importance since it can provide insights into inherent challenges and strategies involved in solving broader classes of combinatorial optimization and decision problems [8].

On the other hand, Max-SAT (or Max-3-SAT, in our case) is an optimization problem. In this case, the goal is not to verify if the problem is satisfiable but to find an assignment for the variables that satisfy the maximum number of clauses. For 3-SAT, there exists a ratio between the number of variables n and the number of clauses m, $\frac{m}{n} \approx 4.2$, where the generated instance is or is not satisfiable with approximately the same probability. This inflection point is an interesting scenario since, in this situation, the required computational time for classical solvers increases exponentially as the number of variables grows [10]. This work focuses on the number of clauses that are solved for a given instance and not if the instance is itself satisfiable, being the latter a special case of the former. Moreover, we do not expect to get the optimal solution (exactly the maximum number of satisfiable clauses), but a good approximation.[2]

2.2   The QUBO Problem and Its Equivalence in Ising Models

In order to solve a SAT problem with a quantum computer, such as a quantum annealer, it is common for the input problem to be formulated in terms of a Binary Quadratic Model (BQM). A typical representation for BQMs are QUBO models, which are characterized by an upper triangular matrix $Q \in \mathbb{R}^{K \times K}$ with K binary variables that take values $x_i = \{0,1\}$, for i in $0, ..., K-1$. The objective of a QUBO problem is to optimize a function that has the following expression:

$$\min_{\mathbf{x} \in \{0,1\}^K} f(\mathbf{x}), \qquad \text{where } f(\mathbf{x}) = \sum_{0 \leq i \leq j < K} x_i Q_{ij} x_j + C, \qquad (2)$$

$Q_{ij} \in \mathbb{R}$ are the values of the Q matrix, $\mathbf{x} = (x_0, \ldots, x_{K-1})$ refer to the variables of the problem, and C is a constant that it is usually used to shift the optimal output to 0.

---

[2] Notice that 3-SAT, Max-3-SAT, and even Max-2-SAT are NP-complete problems. Therefore, a bounded-error quantum polynomial-time algorithm for any of these problems would result in a proof of $NP \subseteq BQP$, which, obviously, we do not have.

Analogously to SAT, the solution will be an assignment of $\mathbf{x}$ that minimizes the function $f(\mathbf{x})$.

An equivalent representation of a QUBO problem can be written in terms of an Ising model [11]:

$$\min_{\mathbf{z}\in\{-1,1\}^K} H(\mathbf{z}), \qquad \text{where } H(\mathbf{z}) = -\sum_{0\le i<j<K} J_{ij}z_iz_j - \sum_i h_iz_i, \quad (3)$$

$J_{ij}, h_i \in \mathbb{R}$ and $z_i = \{-1,1\}$ for i in $0, ..., K - 1$. This formulation plays a prominent role in physics since the variables z_i represent magnetic or quantum spins of a physical system [14].

Via the linear transformation, $x_i = (1 + z_i)/2$ a QUBO problem can be easily mapped to an Ising model, and vice-versa.

In this work, we will work mainly with QUBO problems. However, since the two formulations are equivalent, it is important to recognize that, essentially, we are discussing the same concept when referring to either formulation.

2.3 Quantum Annealing

QA is a quantum computational method to find the global minimum of an objective function. It is a process that occurs in a quantum computer, characterized by a Quantum Processor Unit (QPU). Thus, it relies on quantum bits, commonly referred to as qubits, which stand out for their ability to have their two possible values in superposition, i.e., to be in the ground state (i.e., 0) with probability p and in the excited state (i.e., 1) with probability $1 - p$.

The QPU of a QA has the capability to adjust both the biases of individual qubits and the coupling factors between them, meaning that it can fine-tune the behavior of each qubit and how they interact with one another in order to solve an optimization task. Nevertheless, not all qubits are interconnected via coupling factors; rather, the specific connections depend on the particular QPU architecture.

The experiments conducted in this work are based on D-Wave's *Advantage System 5.4*[3] which uses the *Pegasus* architecture, containing 5000+ qubits and 35000+ coupling factors.

All in all, in order to solve a problem with QA, an objective function needs to be submitted to the QPU, i.e., the values of the bias and coupling factors have to be set. In our case, the diagonal and the upper-diagonal terms of the QUBO matrix Q refer to the biases and coupling factors, respectively.

Once the equivalent physical system of the problem is prepared in the QPU, has a temperature close to absolute zero and it is isolated from the surrounding environment, the system is supposed to work adiabatically and can evolve in time following the quantum mechanics. Consequently, all the amplitudes of the different possible states (i.e., all potential solutions) that started equally will

---

[3] https://www.dwavesys.com/solutions-and-products/systems/.

start to change time-dependently. As a result, the states with higher amplitudes will be the most probable to be present in the system and, in turn, to obtain, once the solution is returned.

As one can expect, the states with higher amplitudes will be those that minimize the objective function submitted to the system. However, as the method is stochastic, i.e., not guaranteed to obtain the most optimum solution, we can obtain either the global minimum of the problem equation or any other state. The results depend on the efficiency of the adiabatic physical system, the defined objective function, the value assignment for the characteristic parameters of the quantum computer, and other errors that can occur during the process. At the end of the QA process, the system will return as a solution the final set of qubits' values and the energy of the system in that state, which ideally will correspond to the solution that minimizes the problem.

3   Related Work

One of the first works that mention the use of gadgets to translate NP-complete problems into Ising formulations is the one from A. Lucas [14]. In this article, the author translates many known problems, such as the knapsack problem, the traveling salesman problem, the graph coloring problem, and the SAT problem, among others, driven by the motivation to solve them via adiabatic quantum optimization algorithms, e.g., quantum annealing. Regarding the satisfiability problem, the author exposes the reduction of 3-SAT to the Maximal Independent Set (MIS) problem, proposed in [6], which is then translated to an Ising model following the procedure in [7].

Nuesslein et al. [15] compare the two main existing translations of 3-SAT to QUBO (i.e., [5,7]) and propose two novel approaches to solve instances in the critical region examined by [10]. The authors show theoretically the advantages of their proposed gadgets and achieve better performance when testing them in D-Wave's quantum annealer.

The work of Bian et al. [4] presents a rigorous study on mapping and encoding techniques to solve SAT problems in a quantum annealer and, more specifically, using D-Wave's *Chimera* architecture. However, their proposed translation from 3-SAT to Ising, summarised in Sect. 4.1, uses a large number of auxiliary variables, being inefficient in our study case. Moreover, their study is not trivially extended to Max-SAT instances.

Finally, Ansotegui & Levy [1] study and propose several new gadgets to tackle the SAT problem using quantum annealers. However, the authors do not show any experimental results nor compare the gadgets to the current state-of-the-art.

Thus, we build our work from [15], completing their comparison by including the new gadgets from [1].

In the following subsections, we expose the gadgets from [15], reviewing the current state-of-the-art and in Sect. 4, we present the intuition for the gadgets found in [1]. For the sake of simplicity, we are not reporting the results for two of the mappings from max-3-SAT compared in [15], namely the ones proposed in [5,7], since they show significantly worse results than the ones presented here. The interested reader can find the results for those methods in the experiments' code repository.

We will show in Sect. 5, that the gadgets implemented in this work achieve better results compared to the current state-of-the-art.

3.1 Nuesslein^{2n+m}

The first reduction from 3-SAT to QUBO presented in [15], while not optimal, offers valuable insights into some concepts. Following the authors' notation, we will refer to this gadget as *Nuesslein1* from now on.

This gadget requires $2n+m$ qubits, as it allocates two logical qubits for each variable in the problem: one representing the variable and another representing its negation. In addition, it uses a logical qubit for each clause. Note that we explicitly differentiate between logical and physical qubits. The first ones represent the boolean variables of the QUBO problem, and the physical qubits represent the total number of qubits needed to embed the QUBO problem into the machine's particular architecture.

Using 3 auxiliary variables is a useful strategy because the main characteristic of this transcription is to count how many times each variable appears negated or not, along with the relation between variables in each clause. In this way, the importance and weight of each auxiliary variable are easily determined. In summary, with this method, the gadget implicitly assigns the best value $\{True, False\}$ to each variable to satisfy the most number of clauses.

For example, if the variable x_j appears more times negated than not negated, the logical qubit representing the negated variable will have a higher weight than the logical qubit representing the non-negated variable. In this way, the algorithm will try to assign the value *False* to that particular variable.

However, as the authors point out, because of the necessity of using $2n+m$ logical qubits, the number of physical qubits needed to embed the problem is large, making it difficult to obtain optimal results.

3.2 Nuesslein^{n+m}

The second approach presented in [15] is a reduction from 3-SAT to QUBO using only $n+m$ logical variables, i.e., it uses a logical qubit per variable and another logical qubit per clause. Following the authors' notation, we will refer to this gadget as *Nuesslein2* from now on.

The idea behind this gadget is to update the QUBO matrix by recursively scanning each clause individually. In essence, the algorithm first sorts all clauses depending on the number of negated literals appearing in each one, obtaining an expression similar to the following:

$$(a \vee b \vee c), (a \vee b \vee \neg c), (a \vee \neg b \vee \neg c), (\neg a \vee \neg b \vee \neg c). \tag{4}$$

The goal is to construct matrices that depict the relationships between literals and their existing connections, aiming to determine how each clause pattern influences the QUBO matrix.

The usefulness of determining relations among literals and other associations allows the gadget to identify patterns and, when possible, merge two clauses into one.

For instance, if we are presented with the following pair of clauses: $(v_1 \lor v_2 \lor v_3), (v_1 \lor v_2 \lor v_4)$ the gadget can spot that the relation $(v_1 \lor v_2)$ is repeated, merging the two clauses into only one: $(v_4 \lor (v_1 \lor v_2) \lor v_3)$. As a result, and as opposed to the previous gadget, one auxiliary logic qubit is no longer needed.

4   New Approaches

In this section, we expose two additional transcriptions to reduce 3-SAT to QUBO, extracted from [1].

Both approaches require $n + m$ logical qubits. That is, they require a logical variable x_i, for each variable v_i, for $i = 1, \ldots n$, and another logical variable b_j, for each clause c_j, for $j = 1, \ldots m$. Despite using the same number of logical variables than *Nuesslein2*, superior experimental results will be shown in Sect. 5.

4.1   CJ1^{n+m}

In [4], a "divide-and-conquer" approach is presented for transforming a SAT formula into an Ising problem. The general procedure, detailed in [4], follows three steps:

1. Factorize the input formula rewriting every conjunct that is not small enough to be easily mapped to Ising. Each of these large conjuncts should be transformed, by means of Tseitin transformation [18], into an equivalently-satisfiable formula where each conjunct can be easily mapped to Ising. As a result of this step, our formula will be a conjunction of subformulas, with each subformula easily mapped to Ising.
2. Independently map each of the subformulas to Ising. This means that new variable replicas should be created for each subformula.
3. For each different variable, link their variable replicas by means of a chain of equivalences.

However, this approach requires a large number of logical variables. In particular, it requires $n + 2m$ logical variables plus those needed to fulfill the last step of the algorithm (i.e., to link each variable's replicas).

As an alternative, Ansotsegui & Levy theoretically develop a similar approach in [1], which benefits from the Tseitin encoding but requires only $n + m$ logical variables. We label this gadget as *CJ1*.

The intuition behind this method is to transcript each clause independently, modifying iteratively the values of the QUBO matrix $Q \in \mathbb{R}^{K \times K}$. Since each clause is translated independently, we will only refer to the j'th 3-SAT clause as a reference to this explanation.

The reduction from 3-SAT starts by implementing the Tseitin encoding in [18], obtaining the following expressions:

$$l_1 \vee l_2 \vee l_3 \rightarrow \begin{cases} l_1 \vee l_2 \leftrightarrow b_j \\ b_j \vee l_3, \end{cases} \tag{5}$$

where l_i represents the variable in the i'th literal of the clause, and b_j represents the auxiliary variable added in the j'th clause.

The QUBO matrix can now be computed directly by expressing (5) in terms of the contribution to the objective function (2).

As explained in [1], each of the expressions in 5 contributes to an Ising model in the following way:

$$l_1 \vee l_2 \leftrightarrow b_j \quad \Leftrightarrow \quad H_1(\mathbf{x'}, \mathbf{b'}) = \frac{5}{2} + \frac{1}{2}x_1' + \frac{1}{2}x_2' - b_j' + \frac{1}{2}x_1'x_2' - x_1'b_j' - x_2'b_j' \tag{6}$$

$$b_j \vee l_3 \quad \Leftrightarrow \quad H_2(\mathbf{x'}, \mathbf{b'}) = \frac{1}{2} - \frac{1}{2}b_j' - \frac{1}{2}x_3' + \frac{1}{2}x_3'b_j' \tag{7}$$

where $x_i' = s_i\sigma_i$, being $s_i = -1$ (+1) the (not) negation of the literal l_i, and $\sigma_i = \{-1, 1\}$ the value assigned to that literal. The final contribution for the entire gadget will be the sum of both terms.

As stated in Sect. 2, an Ising model can be easily mapped to a QUBO problem. Using a variable transformation such that $\sigma_i = 2x_i - 1$ and $b_j' = 2b_j - 1$, the final expression that contributes to the QUBO objective function is:

$$\begin{aligned} f(\mathbf{x}, \mathbf{b}) =& (3s_1 - s_1s_2)x_1 + (3s_2 - s_1s_2)x_2 - 2s_3x_3 - (3 + 2s_1 + 2s_2 - s_3)b_j \\ &+ 2s_1s_2x_1x_2 - 4s_1x_1b_j - 4s_2x_2b_j + 2s_3x_3b_j + C, \end{aligned} \tag{8}$$

where C is a constant.

The values that contribute to each term of the matrix $Q \in \mathbb{R}^{L \times L}$ are the coefficients of 8. For instance, $Q_{l_1l_1} = 3s_1 - s_1s_2$, and $Q_{l_3b_j} = 2s_3$. Thus, all contributions for each clause will be summed into the matrix.

This approach stands out for the couple between the two first literals in the clause, while the third literal only couples with the auxiliary variable. This is important when working in a QPU, in which the architecture and the available connections between qubits are determining factors in the results obtained.

4.2   CJ2^{n+m}

The second approach presented in [1] follows a similar implementation, also focusing on transcribing each clause individually. However, this gadget's first reduction is from 3-SAT to Max-2-SAT, that is, to a conjunction of clauses involving only 2 variables.

Analogously to what is done in Sect. 4.1, the expression obtained for the j'th clause using this gadget is:

$$l_1 \vee l_2 \vee l_3 \rightarrow \begin{cases} l_1 \vee l_3, \neg l_1 \vee \neg l_3 \\ l_1 \vee \neg b_j, \neg l_1 \vee b_j \\ l_3 \vee \neg b_j, \neg l_3 \vee b_j \\ l_2 \vee b_j \end{cases} \tag{9}$$

As before, the authors show the intuition to express 9 into contributions to an Ising model:

$$H(\mathbf{x'}, \mathbf{b'}) = \frac{1}{2}(-x_2' - b_j' + x_1' x_3' - x_1' b_j' - x_3' b_j' + x_2' b_j') \tag{10}$$

Applying the same transformation as in the previous section, we obtain that the contribution to the QUBO objective function is:

$$\begin{aligned} f(\mathbf{x}, \mathbf{b}) =& (s_1 - s_1 s_3)x_1 - 2s_2 x_2 + (s_3 - s_1 s_3)x_3 - (1 - s_1 + s_2 - s_3)b_j \\ & + (2s_1 s_3)x_1 x_3 - 2s_1 x_1 b_j + 2s_2 x_2 b_j - 2s_3 x_3 b_j + C, \end{aligned} \tag{11}$$

where C is a constant.

The values for $Q_{L_i L_i}$ of the QUBO matrix will be the sum of all the coefficients in 11 for each variable x_i and b_j.

Similar to the previous gadget, this approach only adds one auxiliary variable for each clause, obtaining a total of $n + m$ logical variables.

5   Experimental Results

Following the procedure used in [15], we have generated random 3-SAT instances in the critical region, i.e., where the ratio of the number of clauses to the number of variables approaches $\frac{m}{n} \approx 4.2$ [10]. In particular, we have generated 20 random instances for 5 different scenarios: $n = 5$ and $m = 21$, $n = 10$ and $m = 42$, $n = 12$ and $m = 50$, $n = 20$ and $m = 84$ and for $n = 50$ and $m = 210$.

We let D-Wave find an appropriate embedding automatically for each instance and generated 100 shots, i.e., the algorithm returned 100 possible solutions. Moreover, we set 100 μs as the annealing time in each solution, and we used the parameter *reduce_ intersample_ correlation* to reduce sample-to-sample correlations. The values for the number of shots and the annealing time have been chosen according to the results in [12], where they benchmark these two parameters for different sizes of QUBO problems using D-Wave's architecture.

Analogously to [15], we present in Fig. 1 the number of non-zero elements existent in the corresponding QUBO matrices. Although the *Nuesslein1*, *CJ1*, and *CJ2* approaches utilize the same number of logical variables, the assignment of each coupling factor (Q_{ij}, $i < j$) differs from one another. Thus, it can be seen how this results in a slight enhancement in the number of non-zero couplings for

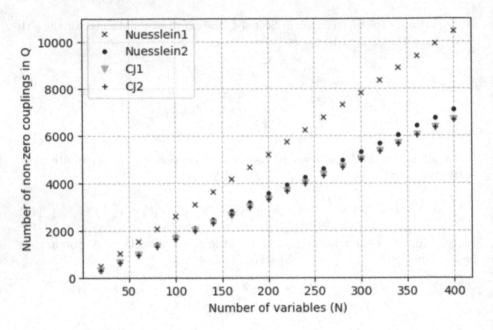

Fig. 1. Number of non-zero couplings in the QUBO matrix for 20 different 3-SAT instances varying the number of variables (N). Variance is negligibly discernible to the naked eye.

the *CJ1* and *CJ2* gadgets. In other words, these gadgets are able to construct the QUBO matrix with fewer elements.

On the other hand, the number of physical qubits depending on the instance's number of variables is essentially equivalent for *Nuesslein2*, *CJ1*, and *CJ2*. As mentioned in [15], *Nuesslein1* exhibits poorer results in this experiment. For a detailed comparison, readers are directed to consult Table 2 in Appendix A.

In Table 1, we present the mean number of satisfied clauses achieved with the best-found variable assignment for each scenario and each gadget. The last row of Table 1 displays the outcomes obtained by the *MaxSatZ* exact solver [13], which finds always the optimal solution.

We can see from Table 1 that we recover the results presented in [15]; that is, Nuesslein's second gadget performs better than their first gadget. Nevertheless, the two gadgets reviewed in this work perform better than Nuesslein's second gadget in all scenarios. For those problems with fewer variables, the *CJ1* gadget performs slightly better, while for bigger problems, the *CJ2* gadget exhibits the best performance.

Although the trait of non-zero couplings present in the QUBO matrix could suggest that the *CJ1* and the *CJ2* approaches may scale and demonstrate superior performance as the problem size expands, we think that the difference depicted in Fig. 1 is not sufficient to account for the enhanced performance observed in the newly implemented gadgets.

Table 1. Experimental results obtained i) by using the analyzed gadgets and D-Wave's quantum annealer (rows 1–4) and ii) an exact SAT solver (row 5). The numbers quoted represent the means of satisfied clauses for all different instances in each scenario.

	n=5, m=21	n=10, m=42	n=12, m=50	n=20, m=84	n=50, m=210
Nuesslein1	18.35	37.3	43.95	73.2	183.4
Nuesslein2	20.65	41.4	49.3	82.6	204.55
CJ1	**20.7**	**41.7**	**49.7**	83.05	204.25
CJ2	**20.7**	41.65	49.65	**83.25**	**207.0**
MaxSatZ	20.7	41.75	49.8	83.6	209.65

Moreover, since the number of logical qubits is the same for the three top-performing gadgets (i.e., *Nuesslein2*, *CJ1* and *CJ2*), and there is no notable difference in the use of physical qubits when embedding the problems into the QPU, we conclude that some other inner characteristic of the gadgets must play a crucial role. In this sense, we strongly believe that the energy gap of each gadget, i.e., the energy difference between the ground state of the system and the first excited state, or in other words, the difference between the optimal output value and the subsequent suboptimal output value, must be of paramount importance.

Figure 2 provides a new insight for assessing the efficiency of the analyzed gadgets. Each bar in the histogram represents the percentage of shots, i.e., the percentage of sample solutions that found the optimal or subsequent suboptimal solutions for each gadget and problem scenario. Thus, the x-axis comprises three labels: "0", "1", and "2", which denote whether the solution found by a given shot was the optimal, the first suboptimal, or the subsequent second suboptimal, compared to the exact solver.

For instance, from Fig. 2a, it can be inferred that when using the *CJ2* gadget within the instances composed of $n = 5$ variables, the optimal solution was found approximately 80% of the time. In other words, around 1600 shots found the optimal solution out of a total of $20 \cdot 100 = 2000$ shots. In addition, the subsequent suboptimal, that is, the solution that satisfied the most clauses but not as many as in the optimal case, was found around 20% of the time. As opposed to this, one can also infer that the *CJ1* gadget obtained less optimal solutions within this context. Note that *Nuesslein2* does not practically appear in the plots, being the worst approach in this case. Based on the number of samples and the results in Fig. 2 we believe that the results are statistically significant.

From Fig. 2, it can be concluded that *CJ2* seems to be the best approach when solving (Max-)3-SAT instances in D-Wave's quantum computer since it is the gadget that more frequently finds the optimal solution. Moreover, it can also be inferred that *CJ2* exhibits superior scalability relative to its counterparts, making it the best option for dealing with this type of problem.

A repository with all the analyses and experiments can be found in *https://github.com/IIIA-ML/QuantumSAT.git*.

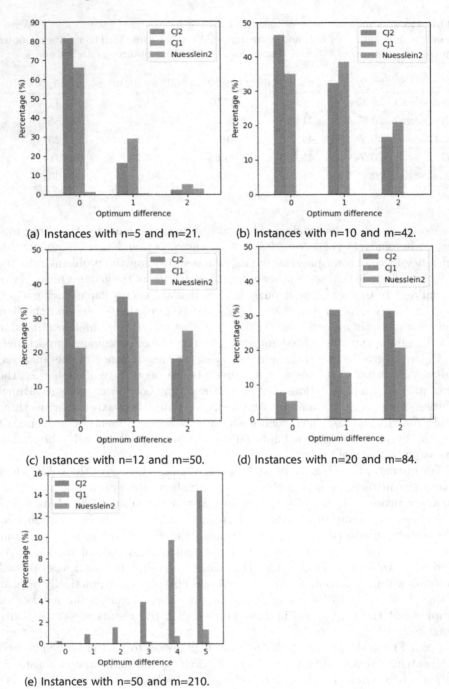

(a) Instances with n=5 and m=21.

(b) Instances with n=10 and m=42.

(c) Instances with n=12 and m=50.

(d) Instances with n=20 and m=84.

(e) Instances with n=50 and m=210.

Fig. 2. Histogram representing the percentage of shots that find the optimal or subsequent suboptimal solutions for each gadget and problem scenario. The x-axis indicates the quality of the solution found, i.e., 0 indicates the optimal solution, 1 indicates the next suboptimal solution, and so on and so forth. Each triplet of bars in each subfigure depicts, from left to right, the *CJ2*, *CJ1*, and *Nuesslein2* approach.

6   Conclusion and Future Work

Motivated to solve NP problems using quantum computers and constructing our work completing Nuesslein et al.'s prior research [15], we have implemented new gadgets to reduce 3-SAT instances to QUBO formulations. We have empirically tested the performance using D-Wave's quantum annealer and have outperformed the current state-of-the-art approaches.

The immediate subsequent task should entail comprehending the reasons behind the superior performance of these results compared to the state-of-the-art approaches. In [3] it is mentioned that, in general, larger energy gaps lead to higher success rates in the QA process. A theoretical examination of the energy gap among the gadgets could be conducted under the same parameter conditions (e.g. the range of the QUBO values when developing the gadget), ensuring a fair and equitable comparison. It is, however, non-trivial to normalize all research in a unique family of topology parameters.

It also remains a future challenge to deal with the embedding of bigger problems into the current architecture to further test the scalability and robustness of our methods and benchmark the capabilities of the current QPUs. Furthermore, we seek to extend our investigations into real-world applications by exploring industrial instances or those sourced from SAT competitions, enhancing the practical relevance of our findings.

Along these lines, it would also be promising to delve into more complex problem domains, such as k-SAT with $k > 3$, broadening the scope of our research to encompass a wider range of challenges.

Acknowledgments. This work was founded partially by the Government of Catalonia Exp. Num. 2023 DI 00071, by Enzyme Advising Group under the development of an industrial PhD on Quantum Machine Learning, by CSIC's JAE Intro JAEICU_23_00782 and by grant PID2022-136787NB-I00 funded by MCIN/AEI/ 10.13039/501100011033. Jesus Cerquides is funded by European Union KHealthInAir, GUARDEN, and Humane-AI projects with No. 101057693, 101060693, and 952026.

Author contributions. Pol Farrés and Rocco Ballester: These authors contributed equally to this work.

Disclosure of Interests.. The authors have no competing interests to declare that are relevant to the content of this article.

A   Number of Physical Qubits Required per Gadget

Table 2. Mean values and standard deviations of the required number of physical qubits for each gadget and each scenario.

	n=15, m=63	n=18, m=75	n=21, m=88	n=24, m=100	n=27, m=113
Nuesslein1	238.90 ± 13.6	308.40 ± 20.6	392.00 ± 22.2	466.90 ± 23.7	561.40 ± 24.7
Nuesslein2	150.15 ± 6.0	196.55 ± 11.7	239.75 ± 10.8	282.40 ± 11.5	342.80 ± 15.5
CJ1	147.25 ± 4.9	185.95 ± 11.6	234.05 ± 13.5	277.90 ± 14.4	332.05 ± 21.5
CJ2	145.75 ± 8.0	187.85 ± 8.6	234.50 ± 10.9	280.55 ± 8.9	331.10 ± 13.9

References

1. Ansótegui, C., Levy, J.: SAT, Gadgets, Max2XOR, and Quantum Annealers (2024). http://arxiv.org/abs/2403.00182 [quant-ph]
2. Biamonte, J., Wittek, P., Pancotti, N., Rebentrost, P., Wiebe, N., Lloyd, S.: Quantum machine learning. Nature **549**(7671), 195–202 (2017). https://doi.org/10.1038/nature23474, https://www.nature.com/articles/nature23474, number: 7671. Publisher: Nature Publishing Group
3. Bian, Z., Chudak, F., Israel, R., Lackey, B., Macready, W.G., Roy, A.: Discrete optimization using quantum annealing on sparse Ising models. Front. Phys. **2** (2014). https://www.frontiersin.org/articles/10.3389/fphy.2014.00056
4. Bian, Z., Chudak, F., Macready, W., Roy, A., Sebastiani, R., Varotti, S.: Solving SAT (and MaxSAT) with a Quantum Annealer: foundations, encodings, and preliminary results. Inf. Comput. **275**, 104609 (2020). https://doi.org/10.1016/j.ic.2020.104609, https://www.sciencedirect.com/science/article/pii/S0890540120300973
5. Chancellor, N., Zohren, S., Warburton, P.A., Benjamin, S.C., Roberts, S.: A direct mapping of Max k-SAT and high order parity checks to a chimera graph. Sci. Rep. **6**(1), 37107 (2016).https://doi.org/10.1038/srep37107, https://www.nature.com/articles/srep37107, number: 1. Publisher: Nature Publishing Group
6. Choi, V.: Adiabatic quantum algorithms for the NP-complete maximum-weight independent set, exact cover and 3SAT problems. ArXiv (2010). https://www.semanticscholar.org/paper/Adiabatic-Quantum-Algorithms-for-the-NP-Complete-Choi/1fcb4d5749074714ca2eb0b56ab986ca66fd0b95
7. Choi, V.: Minor-embedding in adiabatic quantum computation: I. The parameter setting problem. Quantum Inf. Process. **7**(5), 193–209 (2008). https://doi.org/10.1007/s11128-008-0082-9
8. Cook, S.A.: The complexity of theorem-proving procedures. In: Proceedings of the Third Annual ACM Symposium on Theory of Computing, pp. 151–158. STOC '71, Association for Computing Machinery, New York, NY, USA (1971). https://doi.org/10.1145/800157.805047
9. Farhi, E., Goldstone, J., Gutmann, S., Lapan, J., Lundgren, A., Preda, D.: A quantum adiabatic evolution algorithm applied to random instances of an NP-complete problem. Science **292**(5516), 472–475 (2001). https://doi.org/10.1126/science.1057726, http://arxiv.org/abs/quant-ph/0104129
10. Gabor, T., et al.: Assessing solution quality of 3SAT on a quantum annealing platform (2019). http://arxiv.org/abs/1902.04703 [quant-ph] version: 1
11. Glover, F., Kochenberger, G., Du, Y.: A tutorial on formulating and using QUBO models (2019). https://doi.org/10.48550/arXiv.1811.11538 [quant-ph]
12. Gonzalez Calaza, C.D., Willsch, D., Michielsen, K.: Garden optimization problems for benchmarking quantum annealers. Quantum Inf. Process. **20**, 305 (2021). https://doi.org/10.1007/s11128-021-03226-6
13. Li, C.M.: Detecting disjoint inconsistent subformulas for computing lower bounds for Max- SAT (2006)
14. Lucas, A.: Ising formulations of many NP problems. Front. Phys. **2** (2014). https://doi.org/10.3389/fphy.2014.00005, http://arxiv.org/abs/1302.5843 [cond-mat, physics:quant-ph]
15. Nüßlein, J., Zielinski, S., Gabor, T., Linnhoff-Popien, C., Feld, S.: Solving (Max) 3-SAT via quadratic unconstrained binary optimization (2023). https://doi.org/10.48550/arXiv.2303.03536 [quant-ph]

16. Schuld, M., Petruccione, F.: Supervised Learning with Quantum Computers. QST, Springer, Cham (2018). https://doi.org/10.1007/978-3-319-96424-9
17. Shor, P.W.: Polynomial-time algorithms for prime factorization and discrete logarithms on a quantum computer. SIAM J. Comput. **26**(5), 1484–1509 (1997). https://doi.org/10.1137/S0097539795293172. Publisher: Society for Industrial and Applied Mathematics
18. Tseitin, G.S.: On the complexity of derivation in propositional calculus. In: Siekmann, J.H., Wrightson, G. (eds.) Automation of Reasoning: 2: Classical Papers on Computational Logic 1967–1970, pp. 466–483. Symbolic Computation, Springer, Berlin, Heidelberg (1983). https://doi.org/10.1007/978-3-642-81955-1_28

Quantum Annealers Chain Strengths: A Simple Heuristic to Set Them All

Valentin Gilbert[✉][iD] and Stéphane Louise[iD]

Université Paris-Saclay, CEA-List, 91120 Palaiseau, France
{valentin.gilbert,stephane.louise}@cea.fr

Abstract. Quantum annealers (QA), such as D-Wave systems, become increasingly efficient and competitive at approximating combinatorial optimization problems. However, solving problems that do not directly map the chip topology remains challenging for this type of quantum computer. The creation of logical qubits as sets of interconnected physical qubits overcomes limitations imposed by the sparsity of the chip at the expense of increasing the problem size and adding new parameters to optimize. This paper explores the advantages and drawbacks provided by the structure of the logical qubits and the impact of the rescaling of coupler strengths on the minimum spectral gap of Ising models. We show that logical qubits encoded over densely connected physical qubits require a lower chain strength to maintain the ferromagnetic coupling. We also analyze the optimal chain strength variations considering different minor embeddings of the same instance. This experimental study suggests that the chain strength can be optimized for each instance. We design a heuristic that optimizes the chain strength using a very low number of shots during the preprocessing step. This heuristic outperforms the default method used to initialize the chain strength on D-Wave systems, increasing the quality of the best solution by up to 17.2% for tested instances on the max-cut problem.

Keywords: quantum annealing · Ising chain strength · minor embedding

1 Introduction

The idea of using Quantum Annealers to solve combinatorial problems is not new and was exposed by Kadowaki et al. [13]. Despite strong theoretical proofs based on the quantum adiabatic theorem, the speed-up brought up by quantum annealers still needs to be quantified for real useful applications. It comes from the fact that quantum annealers implement a noisy version of the more general Adiabatic Quantum Computation (AQC) (the reader may refer to Albash et al. [2] for an introduction to AQC). Indeed, D-Wave systems suffer from more than five different sources of Integrated Control Errors (ICE) [1] and have limited connectivity between qubits. The maximal number of qubits available on the

L. Franco et al. (Eds.): ICCS 2024, LNCS 14837, pp. 292–306, 2024.
https://doi.org/10.1007/978-3-031-63778-0_21

quantum chip limits the size of the problem that can be solved. Minor embedding algorithms transform the initial problem into a new one that fits the sparsely connected quantum chip. The transformation consists of mapping the initial problem's logical variables to a set of physical variables that can be straightly encoded on the physical qubits of the quantum annealer. As mentioned in Gilbert et al. [10], assessing the quality of an embedding is not trivial. The number of physical qubits used in the embedding can serve as a first quality indicator, but the physical qubit's topology used to encode the logical qubits can also be considered. To the authors' knowledge, a single contribution on this issue was made by Pelofske [15]. Topologies used for physical qubits are usually chains of qubits because this structure maximizes the number of potential connections of the logical qubit. E. Pelofske shows that the performance of the quantum solver is increased when logical qubits are encoded on cliques instead of chains. The minor embedding of a logical variable into a set of physical qubits requires the setting of an accurate ferromagnetic coupling. Current state-of-the-art methods often scan a static range of values to find the appropriate ferromagnetic coupler strength.

In this context, we explore the advantages and drawbacks of different logical qubit encodings. This first analysis shows that the minimum spectral gap varies depending on the qubit encoding, as well as the minimum chain strength required to maintain ferromagnetic couplings. We perform a detailed analysis of the average performance of the two main existing minor embedding methods. This analysis helps to select sets of instances on which we compute the chain break tendency w.r.t the chain strength value. In particular, we discover that the optimal value of the chain strength varies depending on the embedding choice, which suggests that a per-instance chain strength setting method is desirable. The final contribution of this article is the design of a simple algorithm used to find appropriate values for the chain strength using very few preprocessing calls to the quantum computer. This new algorithm outperforms the default method implemented by D-Wave.

The rest of the paper is organized as follows: Sect. 2 introduces background about quantum annealing and minor embedding methods. Section 3 surveys the related work used in the study. Section 4 describes the method and technical settings for the experiments. Section 5 analyzes and proves the minimum spectral gap reduction induced by coupler strength rescaling. Section 6 gathers experiments on embedding and chain breaks that are used to build the algorithm presented in Sect. 7.

2 Quantum Annealing and Minor Embedding Methods

D-Wave systems are based on the transverse field Ising model. This model can be described with a linear interpolation of two Hamiltonians: a mixing Hamiltonian H_{M}, which ground state (i.e., state of lower energy which can be degenerate) is easy to prepare, and a problem Hamiltonian H_{P}, which ground state encodes the solution to the problem. The evolution of the system at annealing fraction $s = \frac{t}{T}$ is described by the Hamiltonian:

$$H(s) = (1 - s)H_{\mathrm{M}} + sH_{\mathrm{P}} \tag{1}$$

where T denotes the total run time of the quantum evolution and $t \in [0, T]$. This evolution scheme is simplified compared to the real annealing schedules of D-Wave systems available at [1]. Considering a Hamiltonian based on a graph $G_s = (V_s, E_s)$ with $n = |V_s|$, the transverse field Hamiltonian $H_{\mathrm{M}} = \sum_{v \in V_s} \sigma_v^x$ has its ground state defined by a uniform superposition of all the quantum states of the computational basis. The problem Hamiltonian can be fully specified by the user:

$$H_{\mathrm{P}} = \sum_{v \in V_s} h_v \sigma_v^z + \sum_{(u,v) \in E_s} J_{uv} \sigma_u^z \sigma_v^z. \tag{2}$$

The ground state of the Hamiltonian H_{P} gives the solution to the Ising cost function minimization problem :

$$\min C(\mathbf{x}) = \sum_{v \in V_s} h_v x_v + \sum_{(u,v) \in E_s} J_{uv} x_u x_v \tag{3}$$

where $x_u, x_v \in \{-1, +1\}$ and $h_v, J_{uv} \in \mathbb{R}$. The optimal solution is given by $\mathbf{x^*} = (x_0, x_1, ..., x_{n-1})$. The Hamiltonian $H(s)$, at a fixed annealing fraction s, corresponds to a Hermitian matrix H_s, and can be decomposed in terms of its eigenvalues $E_i(s)$ and eigenvectors $|v_i(s)\rangle$:

$$H_s|v_i(s)\rangle = E_i(s)|v_i(s)\rangle \text{ with } E_0(s) < E_1(s) < ... < E_k(s). \tag{4}$$

This decomposition is also called eigenenergies decomposition as the eigenvalues $E_i(s)$ correspond to the energy of each eigenvector $|v_i(s)\rangle$. The minimum spectral gap of the Hamiltonian $\Delta_{\min}$ is the difference between the energy of the first excited state and the energy of the ground state at any annealing fraction s:

$$\Delta_{\min} = \min_{0 \leq s \leq 1} (E_1(s) - E_0(s)). \tag{5}$$

The adiabatic theorem guarantees that a quantum state remains in its instantaneous ground state if T is chosen large enough to smooth the quantum evolution. In the best case, the time T scales as $O(1/\Delta_{\min}^2)$ [2].

When the Ising cost function of interest cannot be straigthly mapped on the chip topology, one has to use a method to minor embed the problem into the quantum chip. This problem is well defined by the theory of graph minors developed by Robertson and Seymour [18]. The problem is defined as follows:

Given a source graph $G_s = (V_s, E_s)$ and a target graph $G_t = (V_t, E_t)$, the aim is to find a mapping function $\phi : V_s \rightarrow \mathcal{P}(V_t)$ such that :

1. *each vertex $v \in V_s$ is mapped onto a connected subgraph $\phi(v)$ of G_t.*
2. *each connected subgraph must be vertex disjoint $\phi(v) \cap \phi(v') = \emptyset$, with $v \neq v'$.*
3. *each edge $(u, v) \in E_s$ is mapped onto at least one edge in E_t : $\forall (u, v) \in E_s, \exists u' \in \phi(u), \exists v' \in \phi(v)$, such that $(u', v') \in E_t$.*

For each vertex $v \in V_s$, a ferromagnetic coupling strength $F_{\phi(v)}$ (also called chain strength) is applied to each edge of the subgraph $\phi(v)$. When this ferromagnetic coupling strength is the same for all ferromagnetic couplers, we note it F_ϕ. In the rest of the paper, we refer to V_s as the set of logical qubits and V_t as the set of physical qubits. A chain break on a logical qubit v means that at least one ferromagnetic coupling in $\phi(v)$ is corrupted. An edge of the subgraph $\phi(v)$ is a ferromagnetic coupling.

3 Related Work

Two categories of methods are used to find a minor embedding of an Ising problem. The first category comprises a set of heuristics for which the source graph G_s and target graph G_t are given as input. The state-of-the-art implementation for this category is the CMR heuristic of Cai et al. [4]. The first step of this algorithm aims to find an initial embedding of each logical qubit with possible overlaps between their associated connected subgraph of physical qubits. The second step is a refinement that tries to reduce this overlap to return a valid embedding. The second category of methods considers that the source graph G_s is a clique (i.e., a complete graph) and that the target graph G_t is static. One example is the method used to generate Clique Minor Embeddings (CME) on D-Wave systems [3], which is an iterative method that leverages the regular topology of the quantum annealer and considers inoperable qubits. Both CMR and CME methods have been extended with some pre or post-processing [21].

In [5], Choi formulates two bounds that estimate the value of the ferromagnetic coupling strength $F_{\phi(v)}$. The first bound is specific to each vertex $v \in V_s$:

$$
F_{\phi(v)} < - \left(|h_v| + \sum_{u \in \mathrm{nbr}(v)} |J_{uv}| \right) \tag{6}
$$

where $\mathrm{nbr}(v)$ gives the set of nodes connected to v. This bound is fast to compute with $O(D)$ complexity, where D is the vertex degree. V. Choi also details a more elaborated bound calculated in $O(DL)$ where L is the chain length. In the paper of Fang et al. [9], the authors derive a new tighter bound that can be computed in $O(D2^L)$. The main idea is to set the chain strength with a negative strength with its absolute value greater than the maximum potential energy gain of any configuration of the physical qubits that are not part of the ground state. Thus, this method scales exponentially w.r.t the size of the chain length. Venturelli et al. [19] provided a second approach. They study the value of the optimal coupling strength by setting a global chain strength F_ϕ. They observe that the optimal coupling value F_ϕ grows as the critical point of temperature of their embedded Sherrington-Kirkpatrick model. Beyond this critical value, the success probability decreases. In the paper of Raymond et al. [17], the authors suggest that the chain strength should be tuned as:

$$
\lambda = \lambda_0 \sqrt{std^2 n} = \lambda_0 \tau \sqrt{n} \tag{7}
$$

where $std^2 = \frac{2}{n(n-1)} \sum_{(u,v) \in E_s} J_{uv}^2$ is the variance of the coupling strength (which is 1 for clique spin glasses). Their motivation is that a spin glass transition exists at optimal λ_0. However, λ_0 remains to be set empirically. This paper led to the default method *uniform_torque_compensation* implemented on D-Wave systems [1], which sets the value of the chain strength:

$$F_\phi = -1.414 \times \sqrt{\bar{d}} \times \sqrt{\frac{1}{|E_s|} \sum_{(u,v) \in E_s} J_{uv}^2} \qquad (8)$$

where $\bar{d}$ is the average degree of the graph G_s. This formula comes from the fact that, for general Ising problems, the chain strength scales as $\tau \sqrt{n}$ where τ is the root mean square of the quadratic couplers. In practice, the value of the chain strength is mostly set with a basic chain scan method used to maximize the average expectation value of the QA. For detailed experimental studies on chain strength scanning, the reader may refer to [12,20]. The chain scan method performs well but is very expensive in the number of calls to the QA. The optimization of the chain strength is usually done based on the expectation value. A single advanced optimization method of the chain strength has been recently developed for the max-clique problem and is based on augmented lagrangian method [8]. Recent studies have benchmarked the chain break properties. Grant et al. [11] show that the chain break concentrates on specific locations of the D-Wave *2000Q* processor and design post-processing strategies to limit this bias. In another recent paper of Pelofske [16], it is seen that the approximation ratio is inversely correlated with the rate of getting chain breaks. The author shows that the optimal chain strength is also conditioned by the density of the problem and the type of quantum solver used. We reuse these conclusions as well as the first theoretical bound given in Eq. 6 to create an efficient method to set the chain strength.

4   Method

We test our methodology on two problems: the weighted Ising problem and the max-cut problem. Section 5 is based on a non-degenerate instance of the weighted Ising problem, which consists in minimizing the Ising cost function formulated in Eq. 3. The minimum spectral gap of the problem Δ_{min} is calculated using the exact diagonalization of the Hermitian matrix H_s at each step of the annealing schedule $s \in [0,1]$. The schedule used is the one corresponding to D-Wave *Advantage6.4* [1].

The max-cut problem is studied in Sects. 6 and 7. The weights J_{uv} are set to 1 for all edges, h_v weights are set to 0. All the experiments are run at a constant annealing time of $100\mu s$. We use uniform logical weight spreading, majority vote to unembed the problem and the auto-scale method for weight rescaling. We do not use spin reversal methods or annealing offsets. The figures always show the chain strength in absolute value. The heatmaps from Fig. 2 are generated from sets of 100 instances for each size and density. Erdős-Rényi graphs and random d-regular graphs are generated using the Python *networkx* library. The CMR heuristic runs until a valid embedding is found. The CME heuristic is the default method provided by D-Wave [1]. For all the experiments, we select the first valid embedding found. The plots from Fig. 3 are averaged over 30 instances of random Erdős-Rényi graphs of size $n = 80$ and density $p = 0.3$. The solver used is the *Advantage6.4* and the number of shots is set to 1024 for each chain strength value. The plots from Fig. 4 are obtained from a single Erdős-Rényi instance of size $n = 60$ and density $p = 0.4$. The solver used is the *Advantage2_prototype2.2* and the number of shots is set to 4096 for each chain strength and for the runs with *uniform_torque_compensation* method. The results obtained in Table 1 are obtained with solvers *Advantage6.4* and *Advantage2_prototype2.2*. The results are averaged over 30 instances of Erdős-Rényi graphs for each specified size and density. The number of shots in the preprocessing method is set to 128. The final run used to retrieve the statistics is set to 4096. The run with *uniform_torque_compensation* method is also set to 4096.

We compare the performance of CMR and CME methods in Fig. 2 using the average ratio of the number of qubits in mappings found by each method:

$$\overline{r_{emb}} = \frac{1}{N_p} \sum_{i=1}^{N_p} \frac{n_i^{CMR}}{n_i^{CME}} \qquad (9)$$

where N_p is the number of instances, n_i^{CMR} and n_i^{CME} is the number of physical qubits found by the embedding method CMR resp. CME on instance i. The average breaking chain rate ϵ_b of a single shot is defined by:

$$\epsilon_b = \frac{1}{n_{lq}} \sum_{i=1}^{n_{lq}} p_b(i) \qquad (10)$$

with n_{lq} the number of logical qubits. $p_b(i)$ is set to 1 if the logical qubit contains at least one broken chain and 0 otherwise. The average breaking chain rate of a serie of shots n_s is given by:

$$\bar{\epsilon}_b = \frac{1}{n_s} \sum_{i=1}^{n_s} \epsilon_b^{(i)}. \qquad (11)$$

5   Logical Qubit Structure

We study the evolution of the minimum spectral gap Δ_{min} of a single weighted Ising instance by exhaustively simulating the quantum system evolution via matrix diagonalization. While Δ_{min} is not believed to directly impact the evolution of D-Wave systems that have a long annealing time [7], recent experiments have shown that these systems can operate with much shorter annealing times [14] (in the order of the nanosecond). It makes the study of the spectral gap relevant for future experiments. For this purpose, we design a single instance of an

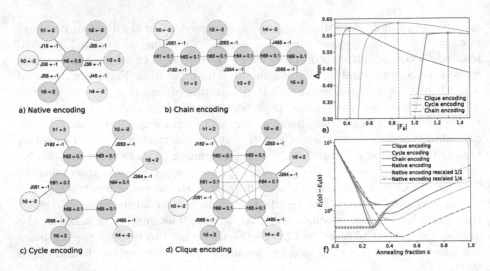

a) Native encoding

b) Chain encoding

c) Cycle encoding

d) Clique encoding

Fig. 1. Minimum spectral gap evaluations considering different types of logical qubits encoding a) Native Ising problem instance b) Same instance as in a. with the red qubit encoded as a chain of physical qubits c) Same instance as in a. with the red qubit encoded as a cycle of physical qubits d) Same instance as in a. with the red qubit encoded with a clique of physical qubits. Black edges represent logical couplers and red edges represent ferromagnetic couplers parametrized by the chain strength F_ϕ. The auto-coupler strength h_6 is uniformly spread on each physical qubit. e) Evolution of $\Delta_{\min}$ considering the whole annealing schedule for different values of the global chain strength $|F_\phi|$. f) Spectral gap evolution of each encoding type programmed with the corresponding optimal chain strength indicated by dashed lines in e. (Color figure online)

Ising problem with 6 nodes (see Fig. 1.a). We choose three possible encodings for the physical qubits $\phi(v_{\mathrm{red}})$ that replace the red logical qubit v_{red} in Fig. 1.a. The selected structures are: chain, cycle and clique. Each structure is respectively shown in Fig. 1.b, 1.c and 1.d. Figure 1.e shows the evolution of the minimum spectral gap of each encoding according to the chain strength value used to maintain the ferromagnetic coupling. For each encoding, the chain strength starts at the minimum analytical value, which is sufficient to maintain the ferromagnetic coupling of the logical qubit (0.3 for the clique, 0.45 for the cycle and 0.9 for the chain). Each embedding exhibits some sweet spots that maximize the minimum spectral gap. The size of the minimum spectral gap decreases when the chain becomes too strong. The optimal value of the chain strength decreases with the density of the logical qubit encoding. This is a desirable feature as D-Wave quantum processors have a limited working range for auto-coupler (h_range) and coupler ($extended\_j\_range$) strengths. In addition, the maximum coupling range limit imposes that the total strength of the couplers linked to each qubit remains in a specific range. These two physical limitations lead to a global rescaling of coupler strengths when the values exceed these ranges.

Reusing the work of Choi [6], it is straightforward to demonstrate that rescaling the coupler strength in the problem Hamiltonian also rescales with the same factor the minimum spectral gap of the Ising Hamiltonian. Let the α-rescaled Hamiltonian with $s_2 \in [0, 1]$ and $\alpha > 1$:

$$H^{1/\alpha}(s_2) = (1 - s_2)H_M + s_2\frac{1}{\alpha}H_P. \tag{12}$$

We take Eq. 1 as the initial Hamiltonian H with annealing fraction $s_1 \in [0, 1]$. Consider the system:

$$\begin{cases} \frac{H(s_1)}{1-s_1} = H_M + \frac{s_1}{1-s_1}H_P \\ \frac{H^{1/\alpha}(s_2)}{1-s_2} = H_M + \frac{s_2}{\alpha(1-s_2)}H_P \end{cases} \tag{13}$$

The equality $\frac{s_1}{1-s_1} = \frac{s_2}{\alpha(1-s_2)}$ can be solved with $s_1 = \frac{s_2}{(\alpha-1)(1-s_2)+1}$ and $s_2 = \frac{s_1}{\frac{1}{\alpha}(1-s_1)+s_1}$. Using this correspondence, we have:

$$\frac{H^{1/\alpha}(s_2)}{1 - s_2} = \frac{H(s_1)}{1 - s_1}. \tag{14}$$

The rescaled Hamiltonian has the form:

$$H^{1/\alpha}(s_2) = \frac{1 - s_2}{1 - s_1}H(s_1) = \left(1 + \left(\frac{1}{\alpha} - 1\right) s_2\right) H(s_1). \tag{15}$$

The eigenenergies of the Hamiltonian are then rescaled with the same factor. According to Eq. 4, we have:

$$E_i^{1/\alpha}(s_2) = \left(1 + \left(\frac{1}{\alpha} - 1\right) s_2\right) E_i(s_1). \tag{16}$$

Figure 1.f shows the effect of couplers strength rescaling on the spectral gap. This figure shows that rescaling the global coupling strength reduces the minimum spectral gap by the same factor. It also shifts the spectral gap to the right. Encoding logical qubits on a set of physical qubits has a detrimental effect on the spectral gap of the instance. The difference in spectral gap reduction between these encodings seems negligible compared to the rescaling of the weights. Logical qubit encodings that are dense, such as clique, require a lower coupling strength to maintain ferromagnetic coupling within the physical qubits. This type of encoding could be favored compared to chain encoding in specific cases to limit the effect of coupler strength rescaling imposed by D-Wave systems.

6   Embeddings and Chain Break Analysis

Figure 2 compares the embedding performance in terms of the number of qubits used by the CMR method [4] against the CME method [3]. The comparison is done on Erdős-Rényi and d-regular graphs. It shows that CMR performs better on sparse graphs of small size, while CME is almost always preferred for large

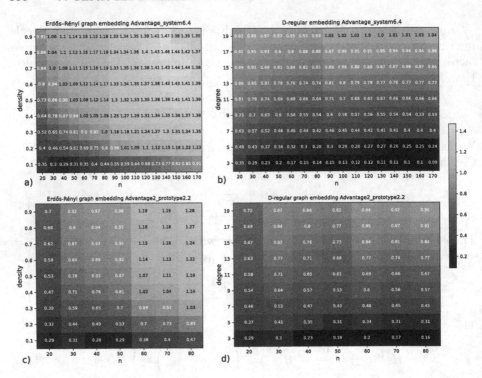

Fig. 2. Heatmaps showing the average percentage overhead of the number of qubits used to embed similar instances using either CMR or CME methods. Each score is an average done over 100 instances. The score calculation in each cell is detailed in Eq. 9. a) and b) are embeddings generated for *Advantage6.4* topology for Erdős-Rényi and d-regular graphs. b) and c) are embeddings generated for *Advantage2_prototype2.2* topology using the same instances. *Advantage6.4* and *Advantage2_prototype2.2* can embed complete graphs of maximum size 174 and 82.

graphs. We use this heatmap to select sets of instances for which CMR and CME methods produce embeddings with approximately the same number of physical qubits. We set each edge weight to 1 to build the max-cut instance associated with the graph. We choose to generate 30 random instances of the max-cut problem with size $n = 80$ and density $p = 0.3$. The aim is to analyze the impact of the embedding method on the chain strength and length. CMR and CME produce embeddings with a similar number of physical qubits for these instances. To be fair in the comparison, we force the CMR method to generate minor embeddings with $\pm 1\%$ qubits compared to the embeddings generated by CME. Figure 3.a and b. respectively show the repartition of the chain length for CMR resp. CME embeddings. While the CME method produces almost uniform chains of lengths 8 and 9, the CMR method produces chain lengths that approach a Gaussian distribution and vary from 4 to 15. The coupling strength required by the CME embeddings to obtain the same breaking chain rate is higher than for CMR embeddings. For the CMR method, the optimal

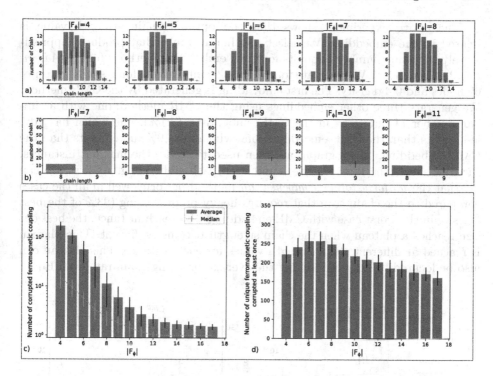

Fig. 3. Statistics on the breaking chain rate (see Eq. 11) averaged over 30 instances of Erdős-Rényi graphs of size $n = 80$ and density $p = 0.3$. a) resp. b) show the average chain length repartition of embedding obtained with CMR resp. CME methods. Blue bars show the average frequency of each chain length. Orange bars show the average breaking chain rate with a black error bar for the standard deviation. c) shows the average and median frequency of corrupted ferromagnetic couplings on CMR embeddings. d) shows the number of different ferromagnetic couplings that are corrupted at least once during the 1024 shots on CMR embeddings. (Color figure online)

solution probability occurs at chain strength 8, while this value is raised to 12 for CME embeddings. Figure 3.d shows that the number of different corrupted couplings only slightly changes when the chain strength is increased. Figure 3.c shows that when the chain strength is insufficient to maintain ferromagnetic couplings, the average number of corrupted ferromagnetic couplings is very high compared to the median. It suggests that the corruptions concentrate on the same few ferromagnetic couplings. When the chain strength is increased, the average number of corruptions decreases but the number of different corrupted couplings remains the same. It suggests that the corruption becomes sparse when the chain strength becomes sufficient. We computed statistics on the number of corrupted ferromagnetic couplings per logical qubit. At the optimal chain strength value, the most likely scenario is to have only a single corruption of couplings on the connected subgraph $\phi(v)$ (99.9% cases) and very few chances of double corruption of couplings (0.01% cases).

The above experiment shows that the optimal chain strength value is related to the embedding. We select another instance that produces approximately the same number of qubits for both embedding methods for the *Advantage2_prototype2.2*. We generate four different embeddings for a single graph of 60 nodes and 0.4 density. The first embedding is generated by the CME method. We generate three other embeddings using the CMR method: one with a similar number of qubits as in the CME embedding ($\pm1\%$), one which has 10% less qubits than the CME embedding, one which has 10% more qubits than the CME embedding. These graphs are then used to create the max-cut instances. Figure 4 shows the best cut size found with each embedding. At first, we can see that the *uniform_torque_compensation* method performs well in all the cases compared to the chain scan that requires heavy preprocessing (93% of the best cut size in the worst case with CME embedding). For each instance, the best cut size reaches a plateau when the chain strength becomes sufficient. This plateau is reached at different chain strength values for each instance. This plateau can also be located by only considering the average breaking chain rate from Eq. 11

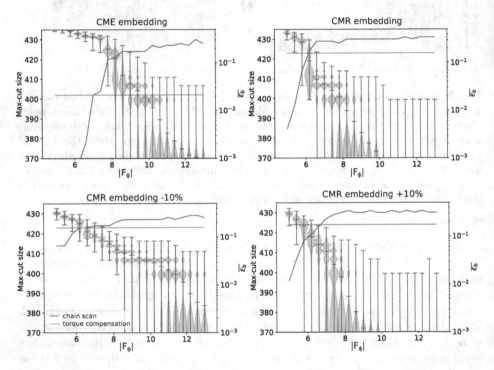

Fig. 4. Maximum cut size obtained with a chain scan on four different embeddings of the same instance. The *uniform_torque_compensation* is run once for each embedding (hence, it is independent of the chain scan). The CME and upper right CMR embedding have the same number of physical qubits ($\pm1\%$). The two other embeddings have 10% more and 10% less qubits than the CME embedding. The red violins show the average breaking chain rate $\bar{\epsilon}_b$ related to the chain scan curve. (Color figure online)

(for example, considering that the chain break probability should remain under 2×10^{-2}). As specified in [15], the chain break probability susceptible to provide the best result depends also on the quantum computer.

7   Chain Strength Setting Heuristic

The previous section detailed our motivations for designing a heuristic to set the chain strength for general instances of Ising models. Algorithm 1 describes this method in pseudo-code. The algorithm performs a binary search to find the optimal chain strength value within an initial interval of chain strengths *csInterval*. When the user does not specify it, the default interval's upper bound is set according to Choi's first bound (see Eq. 6 and line 3):

$$|F_\phi| = \max_{v \in V_s} |F_{\phi(v)}|. \tag{17}$$

At each iteration, the chain strength *cs* is chosen as the midpoint of the chain strength interval *csInterval* (line 6). The embedded instance *embInstance* is then sent to the quantum annealer with the specified chain strength *cs* (line 7). We then compute the breaking chain rate of the result according to Eq. 11 and check if the new breaking chain rate is in the interval specified by the user *cbInterval*. If this is the case, the algorithm converges (lines 9 and 10), and the loop breaks. If the chain break is higher than the upper bound of *cbInterval*, the chain strength is insufficient and requires an increase: we then rescale the lower bound of *csInterval* (lines 11 and 12). In the other case, the upper bound of the chain strength interval is rescaled (line 14). The subtlety

---

Algorithm 1. Chain strength binary search

Require: *embInstance, cbInterval, csInterval* (nullable)
Ensure: $cbInterval[0] < cbInterval[1]$
 1: $hasConverged \leftarrow False$
 2: **if** *csInterval* is None **then**
 3:     $csInterval \leftarrow [0, getUpperBound(embInstance)]$ ▷ (see Eq. 17)
 4: **end if**
 5: **while** not *hasConverged* **do**
 6:     $cs \leftarrow csInterval[0] + (csInterval[0] - csInterval[1])/2$
 7:     $res \leftarrow runQA(embInstance, cs)$
 8:     $\bar{\epsilon}_b \leftarrow getChainBreakRate(res)$ ▷ (see Eq. 11)
 9:     **if** $cbInterval[0] \leq \bar{\epsilon}_b \leq cbInterval[1]$ **then**
10:         $hasConverged \leftarrow True$
11:     **else if** $\bar{\epsilon}_b > cbInterval[1]$ **then**
12:         $csInterval[0] \leftarrow cs$
13:     **else**
14:         $csInterval[1] \leftarrow cs$
15:     **end if**
16: **end while**

---

of this algorithm relies in the setting of the chain break interval *cbInterval* which stops the algorithm when the sampling produces a breaking chain rate in this interval. This interval depends on the effective noise of the quantum computer and may also depend on the size of the instance. We test our heuristic on the max-cut problem and compare the results obtained with the default method implemented by D-Wave that relies on Eq. 8. The comparison is done both on *Advantage2_prototype2.2* on small Erdős-Rényi instances ($n = 40$ and $n = 80$) and *Advantage6.4* for large instances ($n = 100$ and $n = 170$) with density $p \in \{0.1, 0.5, 0.9\}$. For each instance set, we choose the embedding heuristic that provides the best average performance based on the size and density of the instance (see Fig. 2). We empirically set the interval *cbInterval* to $[6 \times 10^{-3}, 2 \times 10^{-2}]$ for the *Advantage2_prototype2.2* and to $[2 \times 10^{-2}, 5 \times 10^{-2}]$ for the *Advantage6.4*. This range stays untouched during the whole experiment. Each preprocessing step is done with 128 shots. The final run of our heuristic, as well as the default D-Wave method, is programmed with 4096 shots for instances run on *Advantage2_prototype2.2* and 3072 shots for instances run on *Advantage6.4*. Table 1 shows the result of this experiment. With a global setting of the value for *cbInterval*, our heuristic is able to outperform in almost every case the default *uniform_torque_compensation* method. The breaking chain rate

Table 1. Performance comparison between D-Wave's default method (*uniform_torque_compensation*) and our binary search heuristic on max-cut instances. The column Best cut size shows the average of the maximum cut size obtained for each instance with the *uniform_torque_compensation* method. The column Cut size improvement contains the minimum and maximum cut size improvement obtained with the chain strength found by Algorithm 1. The standard deviation is shown is column std. The column Step counts the average number of iterations required by our heuristic.

Advantage2_prototype2.2			Best cut size	Cut size improvement				Step
Instance size	Density	Embedding		min	max	mean	std	
$n = 40$	0.1	CMR	66.4	+0%	+0%	+0%	0%	5.4
	0.5	CMR	243	+0%	+2%	+0.2%	0%	5.8
	0.9	CMR	362.8	+2.1%	+8.2%	+5%	1.6%	4.6
$n = 80$	0.1	CMR	235.5	+0%	+0%	+0%	0%	4.7
	0.5	CME	804	+9.8%	+17.2%	+12.5%	0.2%	4.2
	0.9	CME	1435	+2%	+4.7%	+3.2%	0.6%	4.2
Advantage6.4			Best Cut size	Cut size improvement				Step
Instance size	Density	Embedding		min	max	mean	std	
$n = 100$	0.1	CMR	355.9	+0%	+0.3%	+0%	0%	4.5
	0.5	CME	1271.4	+5.6%	+14.5%	+8.8%	1.8%	2.7
	0.9	CME	2243	+1.4%	+3.7%	+2.5%	0.5%	3.7
$n = 170$	0.1	CMR	950.8	−2.1%	+0.6%	−0.5%	0.5%	2.1
	0.5	CME	3631.4	+2.8%	+6.2%	+4.5%	0.7%	2.1
	0.9	CME	6519.4	+0.4%	+1.4%	+0.8%	0.2%	3.2

can be obtained with relatively high fidelity in few shots. It gives a considerable advantage to our method compared to the basic chain strength scans that always use the expectation value to find optimal values of the chain strength and, hence, require a very high number of shots. Our optimization method does, on average, between 2 and 6 iterations to find a suitable value for the chain strength, which corresponds to approximately 750 extra shots per instance in the worst case. This overhead is almost negligible compared to the 4096 and 3072 shots used to evaluate the final chain strength and the *uniform_torque_compensation* method. The default *uniform_torque_compensation* performs well on embeddings generated by CME for dense graphs and on embeddings generated by CMR on sparse graphs. Our heuristic seems to perform the best with mid-density instances with the increase of the cut size by up to 17.2% for the *Advantage2_prototype2.2* and 14.5% for the *Advantage6.4*.

8   Conclusion

This paper presented a detailed analysis of the minimum spectral gap evolution considering different topologies for the set of physical qubits representing a single logical qubit. Each encoding required a specific chain strength value to maintain ferromagnetic couplings between the physical qubits. The denser the logical qubit encoding was, the lower the chain strength had to be. This feature is desirable as the coupler strength rescaling, usually driven by chain strength values in combinatorial problems, reduces the minimum spectral gap of the problem, which may have an impact on very short annealing time evolutions. This consideration could be included in future embedding heuristics designs to enhance the quality of the generated minor embeddings.

The analysis of the breaking chain rate considering different embeddings of the same instance has shown that the optimal chain strength varies depending on the embedding used. This experiment led to the design of a simple but fast heuristic used to optimize the chain strength for each instance. The heuristic converged in most cases in less than 5 preprocessing steps. This number has to be considered cautiously as it strongly depends on the chain break interval, which acts as the breaking condition of the heuristic. We used large intervals in this experiment. The precise identification of the breaking chain rate that produces the best results for each quantum computer could help to refine these bounds. Even if this new method does not provide better solutions than the basic chain scan, the original use of the breaking chain rate in the optimization process drastically reduces the preprocessing time required by a simple chain scan and does not require any assumption over the scanning range of chain strengths.

The performance of our heuristic can be questioned due to the relatively low gain on the max-cut size (17% in the best case). However, even a small percentage gain in the cut size can reduce the Time To Solution of several orders of magnitude, which can justify the use of our heuristic.

A detailed study on the determination of refined breaking chain intervals for each solver is a relevant perspective. This heuristic could also be used as a fast preprocessing routine to find relatively good global chain strengths, followed by a refinement step optimizing each ferromagnetic coupling strength individually.

References

1. D-wave system. Solver properties and parameters (2024). https://docs.dwavesys.com/docs/latest/doc_solver_ref.html
2. Albash, T., Lidar, D.A.: Adiabatic quantum computation. Rev. Mod. Phys. **90**(1), 015002 (2018)
3. Boothby, T., King, A.D., Roy, A.: Fast clique minor generation in chimera qubit connectivity graphs. Quantum Inf. Process. **15**, 495–508 (2016)
4. Cai, J., Macready, W.G., Roy, A.: A practical heuristic for finding graph minors. arXiv preprint arXiv:1406.2741 (2014)
5. Choi, V.: Minor-embedding in adiabatic quantum computation: I. The parameter setting problem. Quantum Inf. Process. **7**, 193–209 (2008)
6. Choi, V.: The effects of the problem Hamiltonian parameters on the minimum spectral gap in adiabatic quantum optimization. QIP **19**(3), 90 (2020)
7. Dickson, N.G., et al.: Thermally assisted quantum annealing of a 16-qubit problem. Nat. Commun. **4**(1) (2013)
8. Djidjev, H.N.: Logical qubit implementation for quantum annealing: augmented Lagrangian approach. Quantum Sci. Technol. **8**(3), 035013 (2023)
9. Fang, Y.L., Warburton, P.: Minimizing minor embedding energy: an application in quantum annealing. Quantum Inf. Process. **19**(7), 191 (2020)
10. Gilbert, V., Rodriguez, J.: Discussions about high-quality embeddings on Quantum Annealers. In: EU/ME meeting. Troyes, France (2023)
11. Grant, E., Humble, T.S.: Benchmarking embedded chain breaking in quantum annealing. Quantum Sci. Technol. **7**(2), 025029 (2022)
12. Hamerly, R., et al.: Experimental investigation of performance differences between coherent ising machines and a quantum annealer. Science Advances **5**(5) (2019)
13. Kadowaki, T., Nishimori, H.: Quantum annealing in the transverse ising model. Phys. Rev. E **58**, 5355–5363 (1998)
14. King, A.D., et al.: Computational supremacy in quantum simulation. arXiv preprint arXiv:2403.00910 (2024)
15. Pelofske, E.: 4-clique network minor embedding for quantum annealers. arXiv preprint arXiv:2301.08807 (2023)
16. Pelofske, E.: Comparing three generations of d-wave quantum annealers for minor embedded combinatorial optimization problems. arXiv:2301.03009 (2023)
17. Raymond, J., et al.: Improving performance of logical qubits by parameter tuning and topology compensation. In: 2020 IEEE International Conference on Quantum Computing and Engineering (QCE), pp. 295–305. IEEE (2020)
18. Robertson, N., Seymour, P.: Graph minors .XIII. the disjoint paths problem. J. Comb. Theory, Ser. B **63**(1), 65–110 (1995)
19. Venturelli, D., Mandrà, S., et al.: Quantum optimization of fully connected spin glasses. Phys. Rev. X **5**(3), 031040 (2015)
20. Willsch, D., et al.: Benchmarking advantage and D-wave 2000q quantum annealers with exact cover problems. QIP **21**(4) (2022)
21. Zbinden, S., Bärtschi, A., Djidjev, H., Eidenbenz, S.: Embedding algorithms for quantum annealers with chimera and Pegasus connection topologies. In: Sadayappan, P., Chamberlain, B.L., Juckeland, G., Ltaief, H. (eds.) ISC High Performance 2020. LNCS, vol. 12151, pp. 187–206. Springer, Cham (2020). https://doi.org/10.1007/978-3-030-50743-5_10

Quantum Variational Algorithms for the Aircraft Deconfliction Problem

Tomasz Pecyna[1]([✉]) [iD], Krzysztof Kurowski[1] [iD], Rafal Rózycki[2] [iD],
Grzegorz Waligóra[2] [iD], and Jan Węglarz[2] [iD]

[1] Poznań Supercomputing and Networking Center, IBCH PAS, Poznan, Poland
{tpecyna,krzysztof.kurowski}@man.poznan.pl
[2] Poznań University of Technology Institute of Computing Science, Poznan, Poland
{rafal.rozycki,grzegorz.waligora,jan.weglarz}@cs.put.poznan.pl

Abstract. Tactical deconfliction problem involves resolving conflicts between aircraft to ensure safety while maintaining efficient trajectories. Several techniques exist to safely adjust aircraft parameters such as speed, heading angle, or flight level, with many relying on mixed-integer linear or nonlinear programming. These techniques, however, often encounter challenges in real-world applications due to computational complexity and scalability issues. This paper proposes a new quantum approach that applies the Quantum Approximate Optimization Algorithm (QAOA) and the Quantum Alternating Operator Ansatz (QAOAnsatz) to address the aircraft deconfliction problem. We present a formula for designing quantum Hamiltonians capable of handling a broad range of discretized maneuvers, with the aim of minimizing changes to original flight schedules while safely resolving conflicts. Our experiments show that a higher number of aircraft poses fewer challenges than a larger number of maneuvers. Additionally, we benchmark the newest IBM quantum processor and show that it successfully solves four out of five instances considered. Finally, we demonstrate that incorporating hard constraints into the mixer Hamiltonian makes QAOAnsatz superior to QAOA. These findings suggest quantum algorithms could be a valuable algorithmic candidate for addressing complex optimization problems in various domains, with implications for enhancing operational efficiency and safety in aviation and other sectors.

Keywords: Tactical Aircraft Deconfliction · Quantum Approximate Optimization Algorithm · Quantum Alternating Operator Ansatz

1 Introduction

The global COVID-19 pandemic was not enough to stop the long-term trend of increasing demand for aviation services. According to Airports Council International, in 2023 the number of passengers reached almost 95% of the levels from 2019, and projections indicate a surpassing of the 2019 level in 2024 [1]. Along with this trend, problems with airspace congestion are returning, and

L. Franco et al. (Eds.): ICCS 2024, LNCS 14837, pp. 307–320, 2024.
https://doi.org/10.1007/978-3-031-63778-0_22

the demand for specialized algorithms dealing with airspace management comes back, one of the problems being the tactical aircraft deconfliction.

In literature, aircraft deconfliction, also known as a conflict detection and resolution problem, refers to the natural and common challenge of ensuring appropriate and safe separation among aircraft operating in the same controlled airspace. The problem arises due to the limited airspace and the need to accommodate multiple aircraft at different directions, altitudes, speeds, and planned maneuvers. The problem has been a subject of interest among many researchers within the community. Despite extensive exploration of conflict detection and resolution, numerous models struggled to sufficiently address the challenges of considered problem, as noted in a seminal work by Kuchar and Yang [16]. Then, the work by Pallottino et al. [21] gained much community attention by introducing the velocity change model, which utilizes mixed-integer linear programming (MILP) to allow real-time maneuvering to resolve aircraft conflicts. This approach was further refined by Alonso-Ayuso et al. [3], who incorporated altitude changes, weather conditions and trajectory recovery into the model while maintaining real-time capabilities.

In a separate study [27], Vela et al. concentrated on addressing the problem of future conflicts, which could occur within a time frame ranging from 15 to 45 min, to minimize fuel costs. They reported achieving near-optimal solutions using the MILP approach, incorporating control over both velocity and altitude. Furthermore, Omer [20] observed that air traffic controllers and aircraft pilots do not favor all velocity, heading, and altitude changes. Consequently, he suggested a discretization approach to facilitate easier handling by human operators, resulting in a minor increase in fuel consumption, amounting to a few kilograms.

Instead of employing MILP, some researchers have proposed using nonlinear programming to address the issue of aircraft deconfliction. In their study [7], Cafieri and Durand utilized Mixed Integer Nonlinear Programming (MINLP) as a natural choice to model separation conditions, addressing the problem using only velocity change. The study conducted by Alonso-Ayuso et al. [4] also applied MINLP formulation to solve the deconfliction problem via turn changes. One notable work that builds upon these two approaches and combines them was conducted by Cafieri and Omheni [8]. They suggest initially resolving the problem by adjusting heading angles and subsequently using this solution as a preprocessing step for modifying velocities.

Various other studies have explored the deconfliction problem, considering factors such as stochasticity and three-dimensional space [17], or employing a new method such as bi level programming [9]. For an in-depth review of research on deconfliction over the past two decades, one should refer to [22].

Given the recent advancements in quantum computing and still persistent challenges in the broad domain of air traffic management, it is not surprising that researchers have been exploring alternative approaches. The initial study that focused on the application of quantum computers in aviation was conducted by Stollenwerk et al. [25], who proposed a method to solve flight-gate assignment problem using the D-Wave 2000Q quantum annealer. Using the same device, Stollenwerk et al. [26] addressed the strategic aircraft deconfliction problem by incorporating takeoff delays into wind-optimal trajectories. Additionally, they outlined a simplified model for trajectory modifications proposing pairwise exclusive avoidance or introducing delays between two consecutive conflicts. The D-Wave 2000Q quantum annealer was also used to solve the Tail Assignment Problem [12] in a study presented by Martins et al. [18]. The problem had been addressed also thanks to classical simulation of a universal quantum computer in [28]. Real gate-based quantum hardware, however, was employed to successfully solve only toy instances of flight-gate assignment in [10,19].

In this paper, we introduce a novel approach to address the tactical aircraft deconfliction problem using gate-base quantum computers. Inspired by the ideas presented in [20], we advocate for conflict resolution through discretized maneuvers. Our main contributions include designing a proper cost Hamiltonian for the Quantum Approximate Optimization Algorithm coupled with the effective relocation of a subset of hard constraints into the mixer Hamiltonian of the Quantum Alternating Operator Ansatz. Furthermore, we establish a connection with our previous research by benchmarking our approach against a widely-used circle problem dataset published by Rey and Hijazi [23], which has been downscaled to align with the capacity of current quantum machines.

The paper is organized as follows. In Sect. 2, we formulate the problem, both classically and in quantum terms. In Sect. 3, we show how to use our formulation with existing quantum algorithms. In Sect. 4, we describe the results, and conclude the paper with future work in Sect. 5.

2   Problem Representation and Assumptions

Let us assume that during the flight, an aircraft must maintain a minimum separation of 5 nautical miles horizontally and 1000 ft vertically from other aircraft, where a nautical mile equals 1852 m and a foot equals 30.48 cm. A conflict between two aircraft arises when a pair of aircraft violates at least one of these constraints. If a particular conflict is detected and resolved within five to thirty minutes, then we consider the tactical deconfliction. [22]. We further assume that aircraft motion can be described by a sequence of line segments, maintaining a constant speed within each segment and allowing instantaneous speed changes at the beginning of each segment.

2.1   Classical Formulation

We present a graphical summary of our approach to the deconfliction problem in Fig. 1. The diagram illustrates the key components of our methodology, including the set of proposed maneuvers and the conflict matrix, which is introduced mathematically later in this subsection.

Given a set of n aircraft with their respective positions, heading angles, speeds, and flight levels, our approach begins by proposing a set of discretized

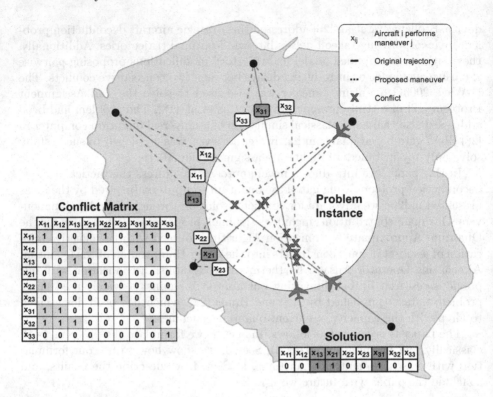

Fig. 1. Diagram summarizing our approach to the deconfliction problem. Initially, three aircraft are in conflict. After proposing 2 additional maneuvers (totaling 3 maneuvers), one feasible solution is proposed: aircraft 1 maneuver 3, aircraft 2 maneuver 1, aircraft 3 maneuver 1. After conflicts are resolved, aircraft may return to their original destinations, which, however, is beyond the scope of our approach.

maneuvers for each aircraft. Maneuvers could be of various kinds, including heading angle change, speed change, or flight level change. For simplicity, we assume that each aircraft can perform m maneuvers, although the actual number may vary for an aircraft depending on specific flight requirements. To keep track of these maneuvers let us introduce a set of the following binary variables:

$$X = \{x_{ij} : \ i = 1 \ldots, n, \ j = 1, \ldots, m, \ x_{ij} \in \{0,1\}\}. \tag{1}$$

If the variable x_{ij} is assigned the value 1 it indicates that the aircraft i performs maneuver j, whereas a value of 0 indicates the opposite. In this work, we assume that maneuvers are disjoint for an aircraft, i.e., an aircraft must perform one and only one maneuver:

$$\sum_{j=1}^{m} x_{ij} = 1 \quad \forall i, \ i = 1 \ldots, n. \tag{2}$$

After proposing the set of maneuvers for each aircraft, we can then fill a 4-dimensional Conflict Matrix (CM) of size $n \times m \times n \times m$ with binary values indicating presence or absence of a conflict between two aircraft,

$$CM(i, j, i', j') = \begin{cases} 1 \text{ if aircraft } i \text{ performing maneuver } j \text{ conflicts} \\ \quad \text{with aircraft } i' \text{ performing maneuver } j' \\ 0 \text{ otherwise.} \end{cases} \tag{3}$$

To detect the potential conflicts, we use a subroutine proposed by Bilimoria [5] wherein we appropriately transform the coordinate system and calculate the relative aircraft speed. Naturally, the entire matrix is redundant due to its symmetry, i.e., $CM(i, j, i', j') = CM(i', j', i, j)$.

The primary focus of the tactical deconfliction problem is to modify aircraft trajectories to resolve all conflicts. This objective can be achieved by satisfying the following constraint:

$$\sum_{i=1}^{n} \sum_{j=1}^{m} \sum_{i'=1}^{n} \sum_{j'=1}^{m} x_{ij} x_{i'j'} \text{CM}(i, j, i', j') = 0. \tag{4}$$

We can clearly see that, while it is relatively efficient to check whether the solution is feasible, the number of possible solutions grows exponentially with the number of aircraft and maneuvers.

The aircraft deconfliction problem extends beyond the sole consideration of avoiding conflicts as it also encompasses the optimization of various parameters such as fuel consumption or average delay. Typically, such criteria can be aggregated into a cost function to minimize, comprising partial costs for each aircraft:

$$C = \sum_{i=1}^{n} \sum_{j=1}^{m} C_{ij}. \tag{5}$$

In this work, we simplify the optimization process by focusing solely on minimizing the total number of changes to the original trajectory. Nevertheless, the objective can be easily expanded to incorporate more sophisticated criteria as needed.

2.2   Quantum Formulation and Encoding

When addressing optimization challenges, quantum computing offers a variety of approaches to choose from [2]. In this study, our emphasis is on two different optimization algorithms, namely the Quantum Approximate Optimization Algorithm (QAOA) [11] and the Quantum Alternating Operator Ansatz (QAOAnsatz) [14]. These two algorithms are rooted in the Adiabatic Theorem [6], which states that a quantum system in an eigenstate undergoing slow enough changes will remain in that eigenstate. The mathematical connection between these algorithms and the Adiabatic Theorem is not rigid. In practice, the process begins with an arbitrary state, preferably an easy-to-prepare ground state [15]. This initial state then evolves into the ground state that corresponds to the solution of the problem described by the problem Hamiltonian. The subsequent discussion outlines how to construct such a Hamiltonian.

For the translation of the formulas derived in Sect. 2.1 to quantum Hamiltonians we employ the composition rules described in [13]. In this process, we make use of the Pauli matrices: $I = \left(\begin{smallmatrix} 1 & 0 \\ 0 & 1 \end{smallmatrix} \right)$, $Z = \left(\begin{smallmatrix} 1 & 0 \\ 0 & -1 \end{smallmatrix} \right)$. The first constraint, ensuring that an aircraft can perform one and only one maneuver, can be described in the following way:

$$H_1 = \sum_{i=1}^{n} I - \sum_{j=1}^{m} \left(H_x(x_{ij}) \prod_{j'=1, j' \neq j}^{m} (H_{\text{not}}(x_{ij'})) \right). \tag{6}$$

The Hamiltonian term $H_{\text{not}}(x_{ij'}) = \frac{1}{2}(I + Z_{ij'})$ represents a boolean clause that has a value of 1 if aircraft i does not perform maneuver j'. The product represents a clause with a 1 if any other maneuver, except j, is not performed. We specify the clause that has a value of 1 if aircraft i performs maneuver j by the Hamiltonian term $H_x(x_{ij}) = \frac{1}{2}(I - Z_{ij})$. We repeat the process for every possible maneuver j to achieve a boolean clause that has a value of 1 if we have a correct one-hot encoding. Note that we want the ground state to represent the desired solution, so we must negate the Hamiltonian. Afterwards, we sum over all possible aircraft.

The second constraint, ensuring that no two aircraft are in conflict, is represented as follows:

$$H_2 = \sum_{i,j,i',j' : \text{CM}(i,j,i',j')=1} H_{\text{and}}(x_{ij}, x_{i'j'}). \tag{7}$$

The Hamiltonian term $H_{\text{and}}(x_{ij}, x_{i'j'}) = \frac{1}{4}I - \frac{1}{4}(Z_{ij} + Z_{i'j'} - Z_{ij}Z_{i'j'})$ represents a boolean clause that evaluates to 1 only if aircraft i performs maneuver j and aircraft i' performs maneuver j'. Summing these situations gives us the total number of conflicts. Naturally, our objective is to minimize the number of conflicts, aiming for a value of 0.

The optimization criterion is determined by a Hamiltonian that assigns appropriate weights to the chosen maneuvers of each aircraft:

$$H_{\text{opt}} = \sum_{i=1}^{n} \sum_{j=1}^{m} w_{ij} H_x(x_{ij}). \tag{8}$$

Here, w_{ij} represents the cost associated with aircraft i performing maneuver j. When aiming to minimize the number of changes from the original trajectories, the weights for the original trajectories are set to 0, while a positive value is assigned to the weights corresponding to modified trajectories.

These partial Hamiltonians have been crafted to be combined into a final Hamiltonian, where the ground state aligns with our desired deconflicted solution:

$$H = \theta_1 H_1 + \theta_2 H_2 + \theta_{\text{opt}} H_{\text{opt}}. \tag{9}$$

In the final Hamiltonian, we introduced additional multipliers to ensure that the ground state consistently corresponds to a feasible solution, regardless of the number of changes needed in the original trajectory. A simple valid assignment can be made as follows: $\theta_1 = 1$, $\theta_2 = 1$, $\theta_{\text{opt}} = \text{sum}(CM)$, where $\text{sum}(CM)$ is the number of all conflicts (all 1s) in the CM.

3   Application

The two algorithms, QAOA and its enhancement, QAOAnsatz, are both hybrid quantum-classical variational algorithms. In these approaches, a parametrized quantum circuit is designed, and the variational parameters are iteratively adjusted using a classical optimizer to minimize the cost function defined by the expectation value of a chosen observable. We provide a brief overview of the foundations of each of these algorithms and their application in solving the tactical deconfliction problem.

3.1   Quantum Approximate Optimization Algorithm

Given R qubits, QAOA initializes by preparing the quantum register in the state $|+\rangle^{\otimes R}$, which is the ground state of a mixing Hamiltonian composed of Pauli-X gates, $H_M = \sum_{i=1}^{R} X_i$. It then alternately applies the problem Hamiltonian (also known as the cost Hamiltonian) and the mixer Hamiltonian to the initial state, p times, where p is a positive integer. The number p is also referred to as the depth of QAOA. The evolution of Hamiltonians is parameterized by two sequences of variational parameters, namely $\overrightarrow{\gamma}$ and $\overrightarrow{\beta}$. The former controls H_c, while the latter controls H_m. Combining these elements, the final state $|\psi\rangle$ after evolution is expressed as follows

$$|\psi_p(\overrightarrow{\gamma}, \overrightarrow{\beta})\rangle = e^{-i\beta_p H_M} e^{-i\gamma_p H_C} \dots e^{-i\beta_1 H_M} e^{-i\gamma_1 H_C} |+\rangle^{\otimes R}. \tag{10}$$

The role of H_c is to distinguish our desired problem solution by applying a change in phase to it. In the context of the tactical deconfliction problem, we simply need to set $H_C = H$, see Eq. 9. The H_M, on the other hand, aims to amplify the phase increasing the probability of measuring the desired solution. This is achieved by adjusting the variational parameters using a classical optimizer which minimizes the expectation value:

$$\min_{\overrightarrow{\gamma}, \overrightarrow{\beta}} \langle \psi_p(\overrightarrow{\gamma}, \overrightarrow{\beta}) | H_C | \psi_p(\overrightarrow{\gamma}, \overrightarrow{\beta}) \rangle. \tag{11}$$

The expectation value of the circuit measurement is also commonly known as energy. Minimizing the energy is equivalent to increasing the probability of measuring solution to our problem. It is noteworthy that H_c serves a dual purpose, as it also functions as a cost function in this context.

3.2 Quantum Alternating Operator Ansatz

We can modify the approach by initializing the quantum register with a state that corresponds to a feasible solution (or a semi-feasible solution, such as one that satisfies only one of several constraints). The algorithm then applies H_C as usual to distinguish our desired solution in phase, but the mixer Hamiltonian is used differently. It is designed to provide transitions from one feasible solution to another. This way we explore and search for the lowest-energy solution only within a feasible subspace constrained by the hard constraints of our problem, which is the essence of the QAOAnsatz algorithm [14].

In the context of the tactical deconfliction problem, we have chosen to encode only the one-hot constraint (Eq. 6) into H_M. To achieve this, we employ a single-qubit ring mixer defined as follows:

$$H_M = \sum_{i=1}^{n} X_{im}X_{i1} + Y_{im}Y_{i1} + \sum_{j=1}^{m} X_{ij}X_{ij+1} + Y_{ij}Y_{ij+1}. \tag{12}$$

Here, the Y symbol represents the Pauli-Y gate. The term $X_{im}X_{i1} + Y_{im}Y_{i1}$ closes the loop between the last and the first qubit, representing the one-hot encoding for each aircraft.

As we have encoded the one-hot constraint into H_M, we can remove the constraint from H_C:

$$H_C = \theta_2 H_2 + \theta_{\text{opt}} H_{\text{opt}}. \tag{13}$$

However, it's important to note that in the presence of noisy hardware, the evolution may drift away from feasible-only solutions. In such cases, having a redundant term in the cost Hamiltonian might be advantageous. In this paper, we choose to use the full cost Hamiltonian, as formulated in Eq. 9.

4 Experimental Results

In the proposed encoding, the number of qubits was equal to the product of the number of aircraft and their maneuvers. Consequently, instances with an identical number of variables could differ in the ratio of aircraft to maneuvers. We started our set of experiments by investigating how altering these two factors affects instance difficulty. For this purpose, we introduced a set of instances that require only 12 qubits but feature different numbers of aircraft and maneuvers, and these instances were artificially generated by constructing CM to ensure only one solution exists.

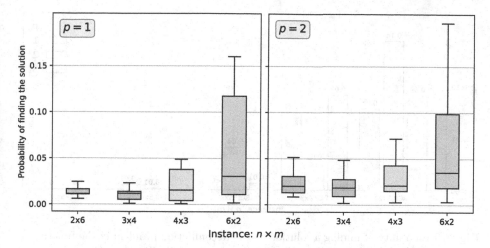

Fig. 2. Average success probability as a function of instance type. Instances are ordered based on the number of aircraft, ranging from 2 aircraft with 6 possible maneuvers to 6 aircraft with only 2 possible maneuvers. The comparison involves two different QAOA depths.

For each instance, we executed 100 QAOA circuits on a noisy simulator with varying initial variational parameters, and the results of success probability were averaged. We used SPSA [24] as the optimizer, as it has proven to perform well on noisy setups. The outcomes are presented in Fig. 2. Observing both circuit depths p, we noted that the algorithm faces increasing difficulty in finding the correct solution as the number of maneuvers grows. Conversely, increasing the number of aircraft at the expense of maneuvers tends to make the instance easier. This behavior aligns with our expectations, as ensuring that no two aircraft are in conflict requires less entanglement between qubits compared to constraining that an aircraft can perform one and only one maneuver. More entanglement naturally makes the circuit longer, introducing additional noise. Moreover, entangling gates are typically more error-prone than single-qubit gates. As a side note, we observed that increasing the circuit depth also appears to result in a slight improvement in the average success probability. After conducting initial experiments on a quantum simulator, we evaluated the capabilities of physical quantum computers.

Existing quantum hardware in the noisy intermediate-scale quantum (NISQ) era provides access to several hundred superconducting qubits. Promising qubit implementations use other quantum technologies, such as trapped ions, neutral atoms, or photons. However, the superconducting quantum architectures lack all-to-all qubit connectivity, requiring multiple swaps to make them adjacent before entanglement. Introducing extra SWAP quantum gates may cause additional errors, potentially degrading the solution quality and, in extreme cases, leading to a failure to find one. With this in mind, we decided to downscale the Random Circle Problem (RCP) instances [23] to involve 3, 4 and 5 aircraft.

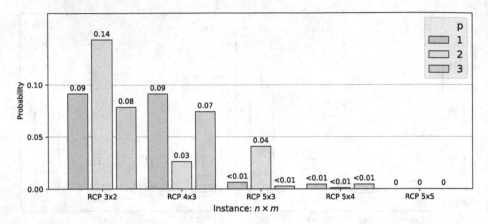

Fig. 3. Probability of finding a solution to the deconfliction problem in the function of instance difficulty and QAOA depth. The instances are the Random Circle Instances with n aircraft, each of the aircraft having m maneuvers to choose from (e.g., $n = 5$, $m = 3$ for RCP 5×3). Experiments were launched on the 133-qubit *ibm_torino*.

For instance, with 3 aircraft, we proposed 2 maneuvers and for 4 aircraft we proposed 3 maneuvers. The instance with 5 aircraft was approached with 3, 4, and 5 maneuvers. This results in a total of five RCP instances, requiring 9, 12, 15, 20, and 25 qubits, respectively.

We evaluated the performance of the latest IBM quantum computer using the superconduting 133-qubit *ibm_torino* quantum computer in solving all instances across three different QAOA depths. The results are illustrated in Fig. 3. Clearly, instances requiring fewer qubits are generally easier to solve. As we move to cases with 5 aircraft, the probabilities of measuring a correct solution drop below 0.01 (less than 1%). It is important to note that this low success probability does not indicate failure, as each circuit is typically measured several thousand times. Given the exponential complexity of the tactical deconfliction problem, achieving a correct solution for even a dozen qubits surpasses the performance of a random guess. Even a single positive outcome is sufficient to solve the considered instance. Unfortunately, the quantum computer selected for our experiments could not solve the problem instance with 5 aircraft and 5 maneuvers. Additionally, we could not identify any noticable trend within the circuit depth, largely due to the inherent randomness of a quantum device. Consequently, further experiments are necessary.

Our final set of experiments involved a comparison between QAOA and QAOAnsatz. Once more, we measured the difference on a quantum simulator and take the average of 100 runs. Given that the tactical deconfliction problem is an optimization problem, we choosen to minimize the number of changes to the original flight schedule. Consequently, we present the probabilities of finding a correct solution for the RCP 5×3 instance in a function of the number of changes required to achieve a correct solution. The results are shown in Fig. 4.

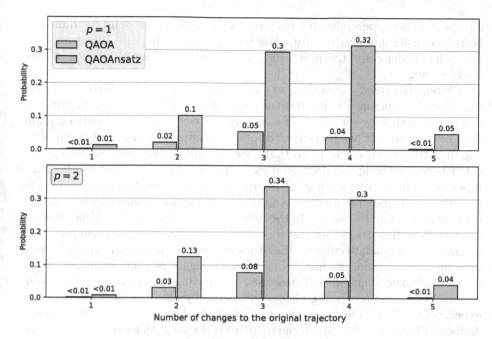

Fig. 4. Comparison between QAOA and QAOAnsatz on a noisy quantum simulator across various two depths, with a focus on the optimization criterion of minimizing changes to the original trajectory. The probabilities of finding a solution to the RCP 5×3 problem are averaged over 100 runs.

We observed that leveraging the feature of QAOAnsatz, which allows for incorporating hard constraints into the mixer Hamiltonian, provides a significant advantage over using mixers from the vanilla QAOA. The probabilities of measuring a solution to the problem are much higher for QAOAnsatz. However, QAOAnsatz still faces challenges in finding solutions that require only one change to the flight schedule to deconflict aircraft. The experiments demonstrate that QAOAnsatz might be a noteworthy algorithm candidate capable of solving deconfliction instances that QAOA could not handle. We leave this investigation for future work.

5   Conclusions and Future Work

In this paper, we have successfully shown how to formulate the aircraft deconfliction problem in a way that is applicable to solve using quantum variational algorithms. By designing a proper cost Hamiltonian for the Quantum Approximate Optimization Algorithm (QAOA) and incorporating hard constraints into the mixer Hamiltonian of the Quantum Alternating Operator Ansatz (QAOAnsatz), we have demonstrated the efficacy of quantum computing in addressing this challenge. Our experiments have validated the feasibility of quantum approaches in handling the complexity of aircraft deconfliction and shed light on the nuanced interplay between aircraft and maneuvers in determining solution difficulty. Moreover, using physical quantum machines, such as the IBM quantum computer, has underscored the practicality of our proposed methodologies in real-world settings.

We plan to extend our work in a twofold manner. Firstly, we plan to enhance the series of experiments qualitatively. One intriguing avenue for exploration involves investigating the effects of removing the constraint that limits each aircraft to one and only one maneuver. Suppose an airplane can execute more than one maneuver simultaneously. In that case, it implies that both maneuvers are conflict-free, enabling the decision-making process to be deferred to the post-processing phase. Another way of improving the solution finding would be to perform a more in-depth analysis of QAOAnsatz variants, mainly by incorporating controlled state transitions to the mixer Hamiltonian. We should not neglect the fact to address trajectory recovery, which was considered in some papers.

Secondly, we plan to enhance the series of experiments quantitatively by performing more experiments and trying to solve bigger problem instances. Some of the implemented qualitative measures, e.g. moving the one and only one constraint to the post-processing phase, will naturally allow for performing larger experiments. A notable consequence of the time-dependent three-dimensional domain of the problem is that some maneuvers do not with each other. It means that we can find such a bijection between variables and qubits so that the no-conflicting maneuvers correspond to qubits which are distant from each other on the quantum computer processor topology, which could significantly reduce the need for SWAP gates, suppressing the noise. Finally, performing more experiments on the same size instances would also improve the precision and potential findings of the experimental results.

Acknowledgments. This research was supported by Poznan Supercomputing and Networking Center, project no DOB-SZAFIR/01/B/023/01/2021 and partially funded by Poznan University of Technology, project no 0311/SBAD/0734. The access to quantum devices was possible through IBM Quantum Innovation Center established at Poznan Supercomputing and Networking Center.

References

1. Airports council international europe | aci europe - media 2024 (2024). https://www.aci-europe.org/media-room/477-passenger-traffic-reaches-nearly-95-of-pre-pandemic-levels-in-2023.html
2. Abbas, A., et al.: Quantum optimization: Potential, challenges, and the path forward. arXiv preprint arXiv:2312.02279 (2023)
3. Alonso-Ayuso, A., Escudero, L.F., Martín-Campo, F.J.: Collision avoidance in air traffic management: a mixed-integer linear optimization approach. IEEE Trans. Intell. Transp. Syst. **12**(1), 47–57 (2010)
4. Alonso-Ayuso, A., Escudero, L.F., Martín-Campo, F.J.: Exact and approximate solving of the aircraft collision resolution problem via turn changes. Transp. Sci. **50**(1), 263–274 (2016)

5. Bilimoria, K.: A geometric optimization approach to aircraft conflict resolution. In: 18th Applied Aerodynamics Conference, p. 4265 (2000)
6. Born, M., Fock, V.: Beweis des adiabatensatzes. Z. Phys. **51**(3–4), 165–180 (1928)
7. Cafieri, S., Durand, N.: Aircraft deconfliction with speed regulation: new models from mixed-integer optimization. J. Global Optim. **58**, 613–629 (2014)
8. Cafieri, S., Omheni, R.: Mixed-integer nonlinear programming for aircraft conflict avoidance by sequentially applying velocity and heading angle changes. Eur. J. Oper. Res. **260**(1), 283–290 (2017)
9. Cerulli, M., d'Ambrosio, C., Liberti, L., Pelegrín, M.: Detecting and solving aircraft conflicts using bilevel programming. J. Global Optim. **81**, 529–557 (2021)
10. Chai, Y., Epifanovsky, E., Jansen, K., Kaushik, A., Kühn, S.: Simulating the flight gate assignment problem on a trapped ion quantum computer. arXiv preprint arXiv:2309.09686 (2023)
11. Farhi, E., Goldstone, J., Gutmann, S.: A quantum approximate optimization algorithm. arXiv preprint arXiv:1411.4028 (2014)
12. Grönkvist, M.: The tail assignment problem. Citeseer (2005)
13. Hadfield, S.: On the representation of Boolean and real functions as Hamiltonians for quantum computing. ACM Trans. Quantum Comput. **2**(4), 1–21 (2021)
14. Hadfield, S., Wang, Z., O'gorman, B., Rieffel, E.G., Venturelli, D., Biswas, R.: From the quantum approximate optimization algorithm to a quantum alternating operator ansatz. Algorithms **12**(2), 34 (2019)
15. He, Z., et al.: Alignment between initial state and mixer improves QAOA performance for constrained optimization. NPJ Quantum Inf. **9**(1), 121 (2023)
16. Kuchar, J.K., Yang, L.C.: A review of conflict detection and resolution modeling methods. IEEE Trans. Intell. Transp. Syst. **1**(4), 179–189 (2000)
17. Lehouillier, T., Omer, J., Soumis, F., Desaulniers, G.: Two decomposition algorithms for solving a minimum weight maximum clique model for the air conflict resolution problem. Eur. J. Oper. Res. **256**(3), 696–712 (2017)
18. Martins, L.N., Rocha, A.P., Castro, A.J.: A QUBO model to the tail assignment problem. In: ICAART (2), pp. 899–906 (2021)
19. Mohammadbagherpoor, H., et al.: Exploring airline gate-scheduling optimization using quantum computers. arXiv preprint arXiv:2111.09472 (2021)
20. Omer, J.: A space-discretized mixed-integer linear model for air-conflict resolution with speed and heading maneuvers. Comput. Oper. Res. **58**, 75–86 (2015)
21. Pallottino, L., Feron, E.M., Bicchi, A.: Conflict resolution problems for air traffic management systems solved with mixed integer programming. IEEE Trans. Intell. Transp. Syst. **3**(1), 3–11 (2002)
22. Pelegrín, M., d'Ambrosio, C.: Aircraft deconfliction via mathematical programming: review and insights. Transp. Sci. **56**(1), 118–140 (2022)
23. Rey, D., Hijazi, H.: Complex number formulation and convex relaxations for aircraft conflict resolution. In: 2017 IEEE 56th Annual Conference on Decision and Control (CDC), pp. 88–93. IEEE (2017)
24. Spall, J.C.: An overview of the simultaneous perturbation method for efficient optimization. J. Hopkins APL Tech. Dig. **19**(4), 482–492 (1998)
25. Stollenwerk, T., Lobe, E., Jung, M.: Flight gate assignment with a quantum annealer. In: Feld, S., Linnhoff-Popien, C. (eds.) QTOP 2019. LNCS, vol. 11413, pp. 99–110. Springer, Cham (2019). https://doi.org/10.1007/978-3-030-14082-3_9
26. Stollenwerk, T., et al.: Quantum annealing applied to de-conflicting optimal trajectories for air traffic management. IEEE Trans. Intell. Transp. Syst. **21**(1), 285–297 (2019)

27. Vela, A., Solak, S., Singhose, W., Clarke, J.P.: A mixed integer program for flight-level assignment and speed control for conflict resolution. In: Proceedings of the 48h IEEE Conference on Decision and Control (CDC) Held Jointly with 2009 28th Chinese Control Conference, pp. 5219–5226. IEEE (2009)
28. Vikstål, P., Grönkvist, M., Svensson, M., Andersson, M., Johansson, G., Ferrini, G.: Applying the quantum approximate optimization algorithm to the tail-assignment problem. Phys. Rev. Appl. 14(3), 034009 (2020)

Adaptive Sampling Noise Mitigation Technique for Feedback-Based Quantum Algorithms

Salahuddin Abdul Rahman[1]($^{(\boxtimes)}$) (ID), Henrik Glavind Clausen[1] (ID),
Özkan Karabacak[2] (ID), and Rafal Wisniewski[1] (ID)

[1] Automation and Control section, Department of electronic systems,
Aalborg University, Aalborg, Denmark
{saabra,hgcl,raf}@es.aau.dk
[2] Department of Mechatronics Engineering, Kadir Has University, Istanbul, Turkey
ozkan.karabacak@khas.edu.tr

Abstract. Inspired by Lyapunov control techniques for quantum systems, feedback-based quantum algorithms have recently been proposed as alternatives to variational quantum algorithms for solving quadratic unconstrained binary optimization problems. These algorithms update the circuit parameters layer-wise through feedback from measuring the qubits in the previous layer to estimate expectations of certain observables. Therefore, the number of samples directly affects the algorithm's performance and may even cause divergence. In this work, we propose an adaptive technique to mitigate the sampling noise by adopting a switching control law in the design of the feedback-based algorithm. The proposed technique can lead to better performance and convergence properties. We show the robustness of our technique against sampling noise through an application for the maximum clique problem.

Keywords: FALQON · QLC · Sampling Noise Mitigation

1 Introduction

Solving combinatorial optimization problems is one of the leading applications where noisy intermediate scale quantum (NISQ) devices are expected to show an advantage over classical algorithms. For NISQ devices, the leading algorithms that can fulfil these devices' requirements and are expected to show quantum advantage are the variational quantum algorithms (VQAs) [3]. VQAs have applications in quantum chemistry, error correction, quantum machine learning, and combinatorial optimization [3,13,14].

In [11,12], the feedback-based algorithm for quantum optimization (FALQON) was proposed as an alternative to VQAs to solve quadratic unconstrained binary optimization (QUBO) problems. Unlike VQAs, FALQON avoids the classical optimization problem associated with VQAs. Instead, it updates the circuit parameters layer-wise through feedback from measuring the qubits in the

L. Franco et al. (Eds.): ICCS 2024, LNCS 14837, pp. 321–329, 2024.
https://doi.org/10.1007/978-3-031-63778-0_23

previous layer to estimate expectations of certain observables. FALQON was also applied to find the ground state of Hamiltonian, specifically in the Fermi-Hubbard model [10]. It can also be used as a potential initialization technique for the parameters of the Quantum Approximate Optimization Algorithm (QAOA) [12].

In FALQON, the parameters of the circuit are calculated by estimating the expected values of some observables using finite number of samples, leading to noise in the estimation of the parameters. This noise will be fed directly to the next layer through the feedback law, affecting the algorithm's performance.

In this work, leveraging tools from quantum Lyapunov control (QLC) theory, we propose an alternative controller design that can mitigate the noise caused by the finite number of samples. Inspired by the switching controller design for QLC proposed in [9], we propose a switching control law that switches between the standard Lyapunov control law and the bang-bang control law. Unlike the open loop control problem of quantum systems considered in [9], FALQON requires estimating expected values of observables via quantum measurements, the number which we strive to reduce by proposing an adaptive sampling technique with the switching bang-bang and standard Lyapunov control feedback law.

We apply FALQON to the maximum clique problem (MCP) with our proposed approach. Through simulations, we show that our proposal is robust against sampling noise and can perform better than the standard Lyapunov technique employed in FALQON with the same number of samples.

The remainder of the paper is structured as follows. In Sect. 2, we review QLC and FALQON. In Sect. 3, we introduce our adaptive sampling noise mitigation technique. In Sect. 4, we investigate the robustness of our approach against sampling noise through application to MCP. Finally, we give a conclusion in Sect. 5.

2   Preliminaries

In this section, we provide an overview of FALQON [11,12] for finding ground states and solving QUBO problems. We start by reviewing QLC and subsequently establish its connection to FALQON.

2.1   Quantum Lyapunov Control

Let us consider the Hilbert space $\mathcal{H} = \mathbb{C}^L$ with associated orthonormal basis $\mathcal{B} = \{|n\rangle\}_{n \in \{0,...,L-1\}}$ and the set of quantum states given by $\mathcal{Y} = \{|\psi\rangle \in \mathbb{C}^L : \langle\psi|\psi\rangle = \||\psi\rangle\|^2 = 1\}$. In the following, all operators will be represented on the $\mathcal{B}$ basis. Consider a quantum system whose dynamics are governed by the controlled time-dependent Schrödinger equation

$$i|\dot{\psi}(t)\rangle = (H_0 + u(t)H_1)|\psi(t)\rangle. \tag{1}$$

where the Planck constant $\hbar$ is normalized to 1, $u(t)$ is the control input and H_0 and H_1 are the drift and control Hamiltonian, respectively. In this work, both H_0 and H_1 are assumed to be time-independent and non-commuting, i.e., $[H_0, H_1] \neq 0$. The main objective here is to find a control law in a feedback form, $u(|\psi(t)\rangle)$, that guarantees the convergence of the quantum system (1), from any initial state to the ground state of the Hamiltonian H_0, i.e., the state $|\psi_g\rangle = \mathrm{argmin}_{|\psi\rangle \in \mathcal{H}} \langle\psi| H_0 |\psi\rangle$. Consider a Lyapunov function of the form $V(|\psi\rangle) = \langle\psi| H_0 |\psi\rangle$, whose derivative along the trajectories of system (1) is given by $\dot{V}(|\psi(t)\rangle) = \langle\psi(t)| \, i[H_1, H_0] \, |\psi(t)\rangle u(t)$. Hence, designing $u(t)$ as

$$u(t) = -Kf(\langle\psi(t)| \, i[H_1, H_0] \, |\psi(t)\rangle), \tag{2}$$

where $K > 0$ and f is a continuous function satisfying $f(0) = 0$ and $xf(x) > 0$, for all $x \neq 0$, will ensure that $\dot{V} \leq 0$. Applying the controller (2), under some assumptions [6,12], guarantees asymptotic convergence to the ground state $|\psi_g\rangle$.

2.2 Feedback-Based Quantum Optimization Algorithm

The quantum evolution propagator $U(t)$ associated to (1) is $U(t) = \tau e^{-i \int_0^t H(t')dt'}$, where τ is the time-ordering operator. By breaking it into p number of piecewise constant time intervals of length Δt, we get $U(T,0) \approx \prod_{k=1}^{p} e^{-iH(k\Delta t)\Delta t}$, where the time step Δt is chosen to be small enough such that $H(t)$ is approximately constant over the interval Δt. This can be simplified using Trotterization as $U(T,0) \approx \prod_{k=1}^{p} e^{-iu(k\Delta t)H_1\Delta t}e^{-iH_0\Delta t}$. Hence, we get a digitized formulation of the evolution in the form

$$|\psi_p\rangle = \prod_{k=1}^{p} e^{-iu(k\Delta t)H_1\Delta t}e^{-iH_0\Delta t} |\psi_0\rangle = \prod_{k=1}^{p} U_1(u_k)U_0 |\psi_0\rangle, \tag{3}$$

where $u_k = u(k\Delta t)$, $|\psi_k\rangle = |\psi(k\Delta t)\rangle$, $U_0 = e^{-iH_0\Delta t}$, $U_1(u_k) = e^{-iu(k\Delta t)H_1\Delta t}$.

For a drift Hamiltonian H_0 specified as a sum of Pauli strings as $H_0 = \sum_{q=1}^{N_0} c_q O_q$, where c_q's are real scalar coefficients, N_0 is given as a polynomial function of the number of qubits and $O_q = O_{q,1} \otimes O_{q,2} \otimes ... \otimes O_{q,n}$ with $O_{q,d} \in \{I, X, Y, Z\}$, the unitary U_0 can be efficiently implemented as a quantum circuit. Similarly, to be able to implement the operator U_1 as a quantum circuit efficiently, the Hamiltonian H_1 should be designed as $H_1 = \sum_{q=1}^{N_1} \hat{c}_q \hat{O}_q$.

The quantum circuit that implements $U(T,0)$ simulates the propagator $U(t)$, where choosing Δt sufficiently small, can guarantee that $\dot{V} \leq 0$ [12]. The following feedback law is adopted:

$$u_{k+1} = -Kf(\langle\psi_k| \, i[H_1, H_0] \, |\psi_k\rangle) = -Kf(\alpha_k), \tag{4}$$

where $\alpha_k = \langle\psi_k| \, i[H_1, H_0] \, |\psi_k\rangle$. This is a discrete version of the controller (2). In [12], the function $f(\cdot)$ is chosen to be the identity function i.e. $f(\alpha) = \alpha$, and the gain K is set to 1. This particular choice of the function $f(\cdot)$ is known as the standard Lyapunov control law. Hence, we obtain the following controller:

$$u_{k+1} = -\alpha_k \tag{5}$$

The implementation of this quantum algorithm follows the algorithmic steps outlined below. The initial step involves seeding the procedure with an initial value for u_1 and setting a value for the time step Δt. Subsequently, a group of qubits is initialized to an easy-to-prepare initial state $|\psi_0\rangle$, and a single circuit layer is applied to prepare the state $|\psi_1\rangle$. The controller for the next layer of the quantum circuit u_2 is estimated on the quantum computer. To estimate the controller, we expand α in terms of Pauli strings as follows:

$$\alpha_k = \langle\psi_k|\,\mathrm{i}[H_1, H_0]\,|\psi_k\rangle = \sum_{q=1}^{N} \bar{c}_q \langle P_q\rangle_k, \tag{6}$$

where we use the notation $\langle P_q\rangle_k = \langle\psi_k|\,P_q\,|\psi_k\rangle$, P_q is a Pauli string and N is the number of Pauli strings. Note that N depends on N_0 and N_1, and since N_0 and N_1 are given as polynomial functions of the number of qubits, then N is also a polynomial function of the number of qubits. A new layer is then added to the circuit, and this sequence is iteratively followed for a depth of p layers. The dynamically designed quantum circuit $\prod_{k=1}^{p} U_1(u_k)U_0$ along its parameters $\{u_k\}_{k=1}^{p}$ constitutes the output of the algorithm. This output can effectively approximate the ground state of the Hamiltonian H_0.

Algorithm 1. FALQON [12]

Input: H_0, H_1, Δt, p, $|\psi_0\rangle$
Output: circuit parameters $\{u_k\}_{k=1}^{p}$
1: Set $u_1 = 0$
2: **Repeat** at every step $k = 1, 2, 3, \ldots, p-1$
3: Prepare the initial state $|\psi_0\rangle$
4: Prepare the state $|\psi_k\rangle = \prod_{l=1}^{k} U_1(u_k)U_0\,|\psi_0\rangle$
5: Estimate α and calculate the controller u_{k+1} using (5)
6: **Until** $k = p$

---

3   Proposed Design of the Controller

In this section, we present our approach for modifying FALQON. We propose an alternative controller design using a switching control law.

In practice, for a finite number of samples $M < \infty$, we have a noisy estimate of the expectation $\tilde{\alpha}$. Hence the calculated controller becomes $\tilde{u}_{k+1} = -K\tilde{\alpha}_k$, where $\tilde{\alpha}_k$ is a noisy estimate of α_k. As the previous section shows, this controller will be directly fed to the next layer. Therefore, it will directly affect the performance of FALQON. To tackle this problem, we propose using a switching control law instead of the standard Lyapunov control law, where the control switches between a bang-bang controller and a standard Lyapunov control law. The bang-bang control switches between two states based on the sign of $\tilde{\alpha}$ and discards the noisy estimate of the expectation, thus mitigating its effect on the algorithm. In this work, we adopt the following switching control law. By defining $\epsilon > 0$ to be the additive error in the estimation of the expectation α_k, the noisy estimate of the controller ($M < \infty$) is given as, with probability $(1 - \delta)$:

$$\tilde{u}_{k+1} = -K \operatorname{sat}(\tilde{\alpha}_k) = \begin{cases} -W & \tilde{\alpha}_k - \epsilon > \phi \\ W & \tilde{\alpha}_k + \epsilon < -\phi \\ -K\tilde{\alpha}_k & \text{otherwise} \end{cases} \tag{7}$$

where $K > 0$ is the controller gain, $\phi > 0$ is a parameter to adjust the switching band, $W > 0$ and $0 < \delta < 1$. Note that, with probability $1 - \delta$, the estimated controller is totally equivalent to the noiseless controller in the region where the bang-bang controller is activated, i.e. in the region $|\tilde{\alpha}_k| > \phi + \epsilon$. Hence, we achieve sampling noise mitigation in this region by controller design.

We now quantify the error bound ϵ. From (6), to calculate $\alpha = \alpha_k$, we need to estimate N expectations of Pauli strings. For this case, performing a single-shot measurement of the qubits in a single circuit instance is the same as sampling an element from a distribution over $\{-1, 1\}$ with an expected value denoted as $\langle P_q \rangle_k$. To quantify the difference between the actual expectation and the estimated values, we employ Hoeffding's inequality. Hoeffding's inequality states that when provided with a sample of M independent and bounded random variables $\{X_i\}_{i=1}^M$ drawn from any distribution where $X_i \in [-\beta, \beta]$, the difference between the empirical expected value $\tilde{X}$ and the actual expected value satisfies the subsequent inequality:

$$\Pr\left(|\tilde{X} - \langle X \rangle| \geq \epsilon\right) \leq 2\exp\left(-\frac{M\epsilon^2}{2\beta}\right) =: \delta, \tag{8}$$

which implies that through a number of samples $M \geq 2\log(2/\delta)/\epsilon^2$, the expectation value $\langle P_q \rangle$ can be estimated within a precision of ϵ with probability $1 - \delta$ [2]. Since $\alpha = \sum_{q=1}^N \bar{c}_q \langle P_q \rangle$, we consider (8) substituting $\epsilon/(|\bar{c}_q|N)$ for ϵ and δ/N for δ. Namely, through $M_q \geq 2\log(2N/\delta)\bar{c}_q^2 N^2/\epsilon^2$ samples the expectation value $\langle P_q \rangle$ can be estimated within a precision of $\epsilon/(|\bar{c}_q|N)$ with probability $1 - \delta/N$, and therefore α can be estimated within a precision of ϵ with probability larger than or equal to $1 - \delta$. In fact,

$$P(|\tilde{\alpha} - \alpha| \geq \epsilon) = P(|\sum_q \bar{c}_q(\langle \tilde{P}_q \rangle - \langle P_q \rangle)| \geq \epsilon) \leq P(\sum_q |\bar{c}_q| \cdot |\langle \tilde{P}_q \rangle - \langle P_q \rangle| \geq \epsilon)$$

$$\leq P(\vee_q\{|\bar{c}_q| \cdot |\langle \tilde{P}_q \rangle - \langle P_q \rangle| \geq \frac{\epsilon}{N}\}) \leq \sum_q P(|\langle \tilde{P}_q \rangle - \langle P_q \rangle| \geq \frac{\epsilon}{|\bar{c}_q|N}) \leq \sum_q \frac{\delta}{N} = \delta.$$

Hence, by choosing δ and $M = \max_q M_q$, we can find the error bound on α as $\epsilon = \max_q |\bar{c}_q| \cdot N \sqrt{2\log(2N/\delta)/M}$. Based on this, we can calculate the switching control law defined by (7). From this inequality, we note that by increasing the number of samples, we can decrease the value of the error bound and, hence, increase the regions in which the bang-bang controller will be activated. In this way, we can adopt an adaptive number of samples, where we start with a small number of samples and check the condition $|\tilde{\alpha}_k| > \phi + \epsilon$ if it is not satisfied, we increase the number of samples to mM, where $m > 1$ and repeat till we satisfy the condition or reach a maximum number of samples $M_{\max}$. Simulation results suggest that this bound is conservative. To address this, we introduce a parameter K_e to adjust the bound, resulting in $\tilde{\alpha}_k \pm K_e\epsilon$ in the controller design. Note that better bounds could be achieved by adopting advanced estimation methods of the expectation α such as adaptive informational complete measurements [7] and classical shadows of quantum states [8]. We defer the analysis and comparison of these methods to future work. The implementation steps of our approach are outlined in Algorithm 2. This algorithm serves as a subroutine for FALQON, replacing step 5 in Algorithm 1.

Algorithm 2. Switching control law with adaptive number of samples

---

Input: $|\psi_k\rangle$, δ, M, m, $M_{\max}$, ϕ, K_e
Output: circuit parameter of the next layer u_{k+1}
1: Prepare the state $|\psi_k\rangle = \prod_{l=1}^{k} U_1(u_k)U_0 |\psi_0\rangle$
2: **while** $M \leq M_{\max}$
3:     Calculate ϵ using $\epsilon = \max_q |\bar{c}_q| \cdot N\sqrt{2\log(2N/\delta)/M}$
4:     Estimate $\tilde{\alpha}_k$ on the quantum computer using (6)
5:     **if** $|\tilde{\alpha}_k| > \phi + K_e\epsilon$
6:         Calculate the controller u_{k+1} using (7)
7:     **else**
8:         $M \leftarrow mM$
9: **end while**
10: Calculate the controller u_{k+1} using (7)

---

4   Application to the Maximum Clique Problem

We apply our proposed approach to MCP, known to be an NP-complete optimization problem [4]. For MCP, we are given a graph $G = (V, E)$ where V is the set of vertices and E is the set of edges. A clique refers to a group of vertices that form a fully connected subgraph, where every pair of vertices within this group is connected by an edge in G. The size of the clique corresponds to the number of vertices it contains. The objective of MCP is to identify a clique within G that consists of the greatest possible number of vertices. In [4], MCP is formulated as the following QUBO minimization problem:

$$\min_{x \in \{0,1\}^n} -A \sum_{i \in V(G)} x_i + B \sum_{(i,j) \in E(G^c)} x_i x_j, \tag{9}$$

where B/A is chosen to be sufficiently large. To apply FALQON to this problem, we first map it into the following Ising Hamiltonian [4]:

$$H_0 = \sum_{i \in V(G)} Z_i + 3 \sum_{(i,j) \in E(G^c)} (Z_i Z_j - Z_i - Z_j) \tag{10}$$

where we set $A = 1$, $B = 3$. For numerical simulation, we consider the instance of MCP with $V = \{0, 1, 2, 3, 4\}$ and $E = \{\{0, 1\}, \{0, 2\}, \{1, 4\}, \{1, 2\}, \{2, 3\}\}$. For more details on implementing FALQON for MCP, see [15]. To execute FALQON, we use the Qiskit Aer quantum simulator [1]. We design the cost Hamiltonian as $H_1 = \sum_{i=1}^{5} X_i$ and the initial state as the equal superposition state $|\psi_0\rangle = |+\rangle^{\otimes 5}$. We set the time step to be $\Delta t = 0.18$ and the number of samples to be $M = M_{\max} = 5$. We run the algorithm using the standard Lyapunov technique and our approach for circuit depth of 20 layers. The switching control law parameters are set to be $W = 3$, $K = 2$, $\phi = 0.02$, $K_e = 0.035$ and $\delta = 0.2$. The values of W and K are chosen to match the standard Lyapunov control while ϕ, K_e and δ are adjusted to increase the band where the bang-bang controller is activated. The results in Fig. 1 show that for a small number of samples $M = 5$, our proposed algorithm has better convergence to the ground state, which encodes the optimal solution to the problem $|00111\rangle$ with optimal value -4.75.

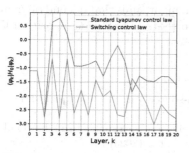 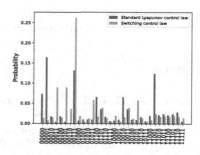

(a) Lyapunov function $\langle \psi_k | H_0 | \psi_k \rangle$ plotted versus the layer k.

(b) The measurements' results of the final state for both approaches.

Fig. 1. Simulation results of one run of FALQON applied to the MCP instance using the standard Lyapunov control law and the proposed switching control law for a number of samples $M = 5$.

In addition, we run the algorithm for 20 realizations and plot the mean and the corresponding standard deviation. The results are shown in Fig. 2. From Fig. 2, it is seen that for our proposed approach using the switching control law, the mean of the realizations decreases with the layer depth increase, while it fails for the standard approach. We note that for larger values of M, the Lyapunov function has better convergence for all realizations with increasing depth. However, both techniques have approximately similar performance.

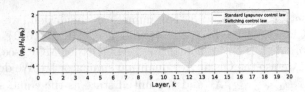

Fig. 2. Simulation results of applying FALQON to the MCP instance. FALQON is executed for 20 realizations and a number of samples $M = 5$, comparing the standard Lyapunov method with the proposed approach. The layer k is plotted versus the mean trajectory (solid line) and the corresponding standard deviation (shaded area) of the Lyapunov function $V_k = \langle \psi_k | H_0 | \psi_k \rangle$ for both approaches.

5   Conclusion and Future Work

In this work, we proposed an adaptive sampling noise mitigation technique. Simulation results show that our approach is robust against sampling noise in an example of MCP. Our future work will focus on adapting this adaptive sampling noise mitigation technique to other quantum algorithms that depend on quantum control theory, such as the quantum imaginary-time evolution algorithm [5].

References

1. Aleksandrowicz, G., et al.: Qiskit: an open-source framework for quantum computing (2019). Accessed 16 Mar 2019
2. Berg, V.D., et al.: Probabilistic error cancellation with sparse pauli–lindblad models on noisy quantum processors. Nat. Phys., 1–6 (2023)
3. Cerezo, M., et al.: Variational quantum algorithms. Nat. Rev. Phys. **3**(9), 625–644 (2021)
4. Chapuis, G., et al.: Finding maximum cliques on a quantum annealer. In: Proceedings of the Computing Frontiers Conference, pp. 63–70 (2017)
5. Chen, Y.C., et al.: Quantum imaginary-time control for accelerating the ground-state preparation. Phys. Rev. Res. **5**(2), 023087 (2023)
6. Cong, S., Meng, F.: A survey of quantum Lyapunov control methods. Sci. World J. (2013)
7. García-Pérez, G., et al.: Learning to measure: adaptive informationally complete generalized measurements for quantum algorithms. PRX quantum **2**(4), 040342 (2021)
8. Huang, H.Y., Kueng, R., Preskill, J.: Predicting many properties of a quantum system from very few measurements. Nat. Phys. **16**(10), 1050–1057 (2020)
9. Kuang, S., Dong, D., Petersen, I.R.: Rapid Lyapunov control of finite-dimensional quantum systems. Automatica **81**, 164–175 (2017)
10. Larsen, J.B., Grace, M.D., Baczewski, A.D., Magann, A.B.: Feedback-based quantum algorithm for ground state preparation of the fermi-hubbard model. arXiv preprint arXiv:2303.02917 (2023)
11. Magann, A.B., Rudinger, K.M., Grace, M.D., Sarovar, M.: Feedback-based quantum optimization. Phys. Rev. Lett. **129**(25), 250502 (2022)

12. Magann, A.B., Rudinger, K.M., Grace, M.D., Sarovar, M.: Lyapunov-control-inspired strategies for quantum combinatorial optimization. Phys. Rev. A **106**(6), 062414 (2022)
13. McArdle, S., Endo, S., Aspuru-Guzik, A., Benjamin, S.C., Yuan, X.: Quantum computational chemistry. Rev. Mod. Phys. **92**(1), 015003 (2020)
14. Simonetti, M., Perri, D., Gervasi, O.: Variational methods in optical quantum machine learning. IEEE Access (2023)
15. Wakeham, D., Ceroni, J.: Feedback-based quantum optimization (FALQON) (2021). https://pennylane.ai/qml/demos/tutorial_falqon/. Accessed 26 Feb 2024

Towards Federated Learning
on the Quantum Internet

Leo Sünkel[(✉)], Michael Kölle, Tobias Rohe, and Thomas Gabor

Institute for Informatics, LMU Munich, Munich, Germany
`leo.suenkel@ifi.lmu.de`

Abstract. While the majority of focus in quantum computing has so far been on monolithic quantum systems, quantum communication networks and the quantum internet in particular are increasingly receiving attention from researchers and industry alike. The quantum internet may allow a plethora of applications such as distributed or blind quantum computing, though research still is at an early stage, both for its physical implementation as well as algorithms; thus suitable applications are an open research question. We evaluate a potential application for the quantum internet, namely quantum federated learning. We run experiments under different settings in various scenarios (e.g. network constraints) using several datasets from different domains and show that (1) quantum federated learning is a valid alternative for regular training and (2) network topology and nature of training are crucial considerations as they may drastically influence the models performance. The results indicate that more comprehensive research is required to optimally deploy quantum federated learning on a potential quantum internet.

Keywords: Quantum Federated Learning · Quantum Internet ·
Quantum Machine Learning · Quantum Communication Networks

1 Introduction

Establishing a quantum communication network over large distances and thereby connecting a multitude of quantum devices with varying architectures and capabilities may allow the *quantum internet* to rise. However, what exactly such a potential network should look like remains an active research question [3, 21, 22, 27]. Given that the field is still largely in its infancy, it is vital to identify and examine the potential applications that such a large-scale network could enable. Distributed quantum computing [2], quantum key distribution, blind quantum computing, and quantum federated learning [6, 25] are all applications that have been discussed in recent years, and more are yet to be discovered. As quantum machine learning has sparked an interest in a wide range of disciplines in recent years, it is no surprise that, sooner or later, combining this area with the field of quantum communication will increasingly attract attention as this provides new opportunities and research avenues. Quantum federated learning

L. Franco et al. (Eds.): ICCS 2024, LNCS 14837, pp. 330–344, 2024.
https://doi.org/10.1007/978-3-031-63778-0_24

already is a step in this direction, and it is this field that is the main topic of discussion in this paper. The idea behind federated learning is to train a global model by a collection of clients that communicate over a network. The crucial point is that each client trains their model on their local dataset, i.e., they keep their data private. Rather than exchanging private data, clients only communicate their models weights; the global model is trained by aggregating weights from participating clients. This (classical) approach can easily be transferred to the quantum domain, resulting in quantum federated learning. However, this field is also still in its infancy and thus many open research questions remain. In this paper, we evaluate different approaches to quantum federated learning under various constraints that may be present in a quantum internet. More specifically, we run experiments using two different network topologies where the number of qubits of the quantum clients differ. Furthermore, we run experiments where clients are trained on subsets of the same dataset as well where each client is trained on a distinct one. We show that quantum federated learning is a compelling alternative to regular model training while it is crucial to take certain network constraints (e.g. each nodes qubit capacity) as well as the models training approach (e.g. nature of weight aggregation and exchange) into account. This paper is structured as follows. In Sect. 2 we cover the background of quantum communication networks, variational quantum circuits as well as quantum federated learning and discuss related work in Sect. 3. We illustrate our approach and architecture in Sect. 4 and discuss our experimental setup in Sect. 5. We present our results in Sect. 6 and conclude in Sect. 7.

2   Background

We begin this section with a brief recapitulation of the basic building blocks of quantum communication networks (QCNs) and what we understand as the quantum internet. We then discuss potential applications and what this novel form of communication means for quantum algorithms in general while also for quantum machine learning in particular. However, we will not be covering the basics of quantum computation and information (e.g. qubits, entanglement, superposition) and instead refer the reader to other resources [16,26] for an in depth introduction to these topics. Once we have reviewed the fundamental concepts of quantum networks, we will discuss federated quantum machine learning, the main topic of this paper.

2.1   Quantum Networks and the Quantum Internet

We define a QCN as a number of quantum computers that are connected via classical and quantum channels, i.e., they have the means to communicate classical messages as well as qubits or quantum states with each other. We will refer to these quantum computers as QPUs or nodes in the network, and we use these terms interchangeably. Furthermore, a QCN can be seen as a graph where the nodes are quantum devices (QPUs) and the edges are the communication channels. An example of such a simple network can be seen in Fig. 1. This example network consists of 9 nodes (QPUs) each with a different qubit capacity. Each edge in the network serves a dual purpose, facilitating both classical and quantum communications.

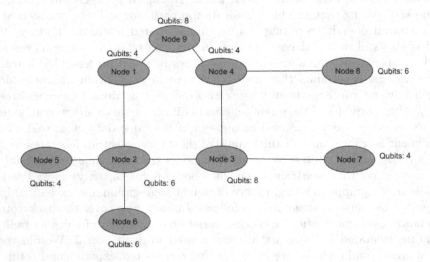

Fig. 1. Example of a network with random topology where nodes represent available QPUs. The QPUs in this network have varying capacity, i.e., number of qubits and the network is not fully connected. Both facts cause ramifications and constraints for applications and network design.

Unfortunately, quantum channels can be noisy, thus information loss is inevitable. Moreover, quantum information cannot be copied as this is prohibited by the *no cloning theorem*, and so classical protocols to overcome information loss as well as sending messages cannot be transferred into the quantum realm in a straightforward manner. In a QCN, qubits are *teleported* over quantum channels to other nodes in the network. However, qubits are not physically teleported, it is the state that is transferred to another qubit, however, the state of the original qubit is destroyed in this process, that is, the qubit cannot be copied. The teleportation protocol requires an *EPR-pair*, i.e., two qubits that are maximally entangled as well as a classical and a quantum channel. In order to teleport qubits over large distances, *quantum repeaters* can be deployed along the way to combat decoherence, i.e., the loss of information. Quantum repeaters perform *entanglement swapping* and possibly *entanglement distillation* that (1) transfer the quantum state and (2) increase the entanglement fidelity.

With these building blocks established, one can envision a large quantum network that is akin to a so-called quantum internet [9]. However, how concretely such a quantum internet exactly should look like is up to debate and a crucial research question in quantum communication and quantum computing alike [3,11,21,22,27]. A quantum internet may connect QPUs with vastly different architectures, i.e., consist of heterogeneous nodes, a fact that will become relevant later in this work.

In summary, a quantum network consists of multiple QPUs connected via classical and quantum channels that allow them to exchange classical as well as quantum information. However, due to various constraints imposed by the laws of quantum physics, many of the classical protocols cannot be transferred to the quantum setting. Moreover, decoherence and the loss of information are major challenges and while countermeasures have been proposed, more research is required to establish a working quantum internet.

This introduction is intended as a short recap of the fundamentals required to follow the work performed in this paper, for an in depth introduction we refer to [2,8,23].

2.2 Variational Quantum Circuits

Variational Quantum Circuits (VQCs) [4] stand out from classical circuits by leveraging quantum phenomena such as superposition and entanglement. These circuits consist of parameterized gates that are optimized through classical techniques, making them a hybrid approach suited for tasks like quantum machine learning. We summarize the essential components of VQCs in this section and describe how we apply them in the approach introduced in Sect. 4.

VQCs can, for example, be applied to classification problems [20] and are a promising hybrid approach to QML in the NISQ-era. As this is a classical-quantum hybrid approach it consists of essentially two parts, namely the quantum circuit and the classical optimization routine. The quantum part, i.e. the VQC, can loosely be divided into three distinct constituents: (i) feature encoding, (ii) a series (layers) of parameterized rotation gates followed by entangling gates and finally (iii) the measurement operations. An example circuit with two layers is depicted in Fig. 2. The parameters of the rotations correspond to the weights that are optimized by a classical optimizer while the measurement results are interpreted as predictions and are also used by the optimizer. This approach is iterative, i.e., it executes until a number of epochs or steps has been reached. For an in depth introduction to this topic we refer to [4,15,20].

2.3 Quantum Federated Learning

Before discussing QFL we will first give a short recap of the central idea behind federated learning in general, i.e., its origins in classical machine learning. We will then discuss quantum variants of this learning approach.

Federated learning (FL) [14] allows for the collective training of a global model by multiple clients, each contributing to the model without exposing their private data. This ensures that each client's data remains confidential and local to them. Rather than sharing data, clients transmit their model weights to a central server, where these weights are aggregated, updated, and then redistributed to all clients for further training iterations. Moreover, each client may be trained

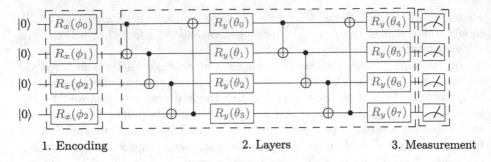

1. Encoding   2. Layers   3. Measurement

Fig. 2. Example of a variational quantum circuit (VQC): Features are embedded through rotations (depicted blue) in the first step. This is followed by two repeating layers of CNOT-gates and parameterized rotations (green), the θ values are the weights being optimized. In the final step, the qubits are measured resulting in a classical output (0 or 1) for each qubit. (Color figure online)

on different subsets or even entirely different datasets. Furthermore, many variants and approaches to FL have been proposed; the approach here is a basic and straightforward one, as we focus on QFL in this paper. For our purposes, it suffices to define FL as an decentralized learning approach in which a global model is trained by multiple clients without revealing their private data. For more comprehensive introduction to the topic we refer to [7,10,24,30]

QFL follows a similar line, however, many different approaches can also be employed here; though we will be focusing on simplicity. In QFL, the client models can be replaced by VQCs, and weights can be communicated through classical as well as quantum channels, though the former does not necessarily require a quantum network while the latter does. Analog to the classical approach, each client (e.g. VQC) is then trained on its own local dataset. We will discuss the peculiar details arising through the quantum internet in Sect. 4.

3   Related Work

Several different approaches and aspects of QFL have been explored by the research community. For example, privacy aspects are investigated in [12,19] while combining QFL with blind quantum computing is discussed by Li et al. in [13] and Zhang et al. propose a quantum method for parameter aggregation in [31]. Challenges of QFL in the context of quantum networks are discussed by Chehimi et al. in [5]. Wang et al. [25] apply QFL on a binary classification task using a ring topology, i.e., without a central model. Moreover, they use quantum weights in their approach. In [6] the authors evaluate a hybrid quantum-classical approach on a binary classification task using images.

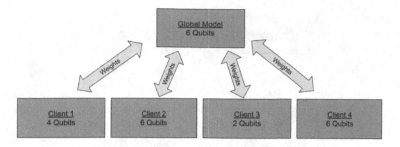

Fig. 3. Example of a QFL approach where a global model is collectively trained by 4 different clients with different qubit capacities. Clients send their weights to the global model that aggregates, updates and distributes the new weights for the next round of training.

4 Approach

In this section, we describe our approach to QFL. We discuss quantum clients and the overall architecture of our methods. This includes network constraints as well as how the models train and communicate.

4.1 Quantum Clients

In our proposed QFL approach, the client model is a VQC running on a quantum node in a (quantum) communication network. More specifically, we consider the quantum internet in which nodes are quantum computers each with potentially a different qubit capacity. It's important to note that the necessity for a quantum communication channel hinges on the specific methodology of weight exchange in the applied QFL approach. That is, weights can be exchanged classically and thus a simple classical channel connecting quantum nodes suffices. Furthermore, in our scenario each node executes the model of only a single client. Thus, the clients in our QFL approach execute VQCs with a different number of qubits and hence number of trainable parameters. Consider the example network in Fig. 1. Here, 9 nodes are connected to form a network where nodes have different qubit capacities.

4.2 Architecture

We evaluate two different approaches to QFL as part of this work that differ in the topology of the arrangement of clients and how weights are aggregated and exchanged. In the first approach, a global model is collectively trained by multiple clients who only exchange their weights with the global model. The global model is then responsible for aggregating, updating and distributing the weights among all participants in each training round. An example of this architecture is depicted in Fig. 3.

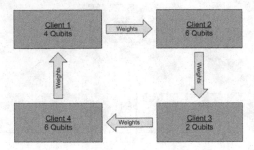

Fig. 4. Example of clients with varying qubit capacity arranged in a ring topology. In this scenario, when clients transfer their weights to the next, the number of weights is adjusted accordingly.

In the second approach, clients are arranged in a ring topology (cf. [25]) and no global model is trained. More specifically, clients train their model using their local dataset and send their weights to the succeeding client after completing a training round. This approach is shown in Fig. 4. As the clients capacity, i.e., number of qubits may vary from one to the next, the number of weights must be adjusted when sent to the following client. In our experiments we adjust the number of weights in the following way. If the next client contains less qubits (and thus weights), the client discards superfluous weights, i.e., it only uses the first n weights required. In the case where the next client contains more qubits, the client uses all weights from the previous one while filling up the missing weights with its own from the last round. Note that this is an ad-hoc approach and the appropriate aggregation of weights in the context of trainability is its own research topic and merits its own discussion, however, is not in the scope of this work.

4.3 Quantum Circuits

In our model, we apply two different VQCs that differ only in the embedding method applied and number of qubits. That is, we use angle embedding in one set of experiments and amplitude embedding in another.

5 Experimental Setup

This section outlines the models, datasets, and parameters used in our experiments.

5.1 Datasets

Our experiments utilize three datasets, detailed as follows. Note that we focus solely on binary classification in this work. In datasets that contain more than two classes, we divide the dataset into subsets such that each only contains images of two classes.

Moons: The moons dataset contains 2 features for a binary classification problem where each class is shaped as a half circle and is provided by scikit-learn [18]. We used a dataset with 3000 samples. The training set used is visualised in Fig. 5.

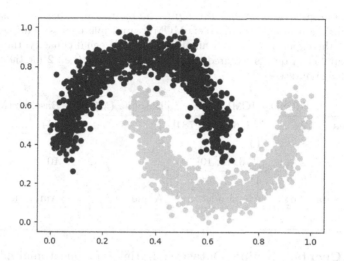

Fig. 5. Two half circles ("moons" dataset) with added noise factor of 0.1 created with sci-kit learn.

FashionMNIST: The FashionMNIST [28] dataset contains 70000 grayscale images for 10 different classes where each image has a size of 28 × 28. Note that there are 60.000 training and 10000 test images. We used this dataset provided by PyTorch [17].

PneumoniaMNIST: The PneumoniaMNIST dataset contains 5856 chest X-ray images with a size of 28 × 28 for binary classification. We used the dataset provided by [29]. Note that for all datasets, we used 1000 images per epoch in training.

Preprocessing: Images were resized for QPUs with lesser capacity than required to embed the entire image, the number of features used is shown in Table 1.

5.2   Training

In this section, we discuss the two training approaches we apply in our experiments. In the first, multiple clients are trained on a single dataset while in the second each client is trained on a different dataset.

Multiple Clients, Single Dataset: In this experiment, each client is trained on a single dataset (moons), however, the dataset is subdivided into distinct subsets prior, i.e., each client is trained on a different subset of the same dataset. Furthermore, in this experiment each client has an equal number of qubits (2) and uses a VQC with angle embedding as its model. The parameters used are listed in Table 1. This experiment was run using both topologies described in Sect. 4

Table 1. Parameters used in the QFL and quantum baseline experiments. (MCSD=multiple clients single dataset, MCMD=multiple clients multiple datasets). Note that in the experiments where all clients have an equal capacity, the capacity is set to the number of qubits required to embed all features (i.e., 2 for the moons and 10 for the image datasets).

	MCSD	MCMD	Baseline (MCSD)	Baseline (MCMD)
Features	2	[16, 64, 784, 784]	2	784
Depth	8	10	8	10
Qubits	2	[4, 6, 10, 10]	2	10
Clients	3	4	–	–
Embedding	Angle	Amplitude	Angle	Amplitude

Multiple Clients, Multiple Datasets: In this experiment, multiple datasets are used in training, that is, each client is trained on a different trainingset or subset. More specifically, we used three distinct subsets of FashionMNIST ("Trouser vs. Pullover", "Sandal vs. Sneaker" and "Dress vs. Coat") and the pneumonia dataset, where each client was trained on one of these sets. For example, client 1 is trained on images depicting trousers and pullovers, client 2 on sandals and sneakers while the third client is trained on chest X-ray images and the fourth on images of dresses and coats, each in a binary classification task. Moreover, each client may have a different capacity (i.e. qubits). This means that each client also embed, and thus are trained, using a different number of features of the input image. Parameters of this approach are depicted in Table 1. Like the first experiment, this one was also run using both topologies described in Sect. 4.

5.3   Simulation

All circuits were implemented and simulated using PennyLane [1] while training was performed using Pytorch [17]. No noise was used in the simulations, and we assume ideal communication between nodes. Running experiments under noise and integrating quantum communication for weight exchange are potential avenues for future work.

6   Results

We discuss the results of all experiments conducted as part of this work in this section. We first examine the experiments with multiple clients, single dataset using both QFL approaches described earlier. We then move on to the experiments with multiple clients, multiple datasts. All experiments were run with 5 different seeds and plots depict the mean of the aggregated results. Note that for baselines, we aggregated the results of all individual binary classification experiments, which is a crucial fact to keep in mind when comparing all approaches. Moreover, in the test results of the experiments using the ring topology, each client uses the final parameters after training for its own model on its respective test dataset and the results depicted below are aggregated over all clients.

6.1   Multiple Clients, Single Dataset

Recall that in this setting, we train multiple clients on a single dataset (we abbreviate this approach as MCSD), however, each client receives its own distinct subset. Furthermore, we evaluated this approach on two different QFL architectures, i.e., topologies, namely star and ring. Training and test results for this experiment with the star topology are shown in Fig. 6a and Fig. 6b respectively. The classical baseline delivers the best training and test accuracy, though this was to be expected as this approach consists of vastly more parameters and a problem of this scale generally is not problematic for classical neural networks. The QFL approach achieves similar performance as the quantum baseline; the difference being only minuscule.

The train and test results using a ring topology with the moons data is shown in Fig. 7a and 7b respectively. While the quantum baseline performs slightly better earlier in training, both approaches perform almost identical and converge relatively fast. Testing results are in a similar range for both quantum methods.

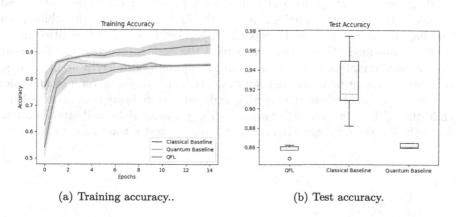

(a) Training accuracy..   (b) Test accuracy.

Fig. 6. Results for the MCSD experiments using the star topology.

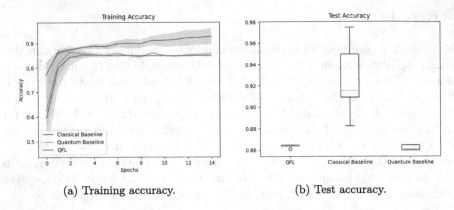

(a) Training accuracy.   (b) Test accuracy.

Fig. 7. Training and test results on MCSD experiments with the ring topology.

6.2 Multiple Clients, Multiple Datasets

We discuss the experiments using multiple clients, multiple datasets (MCMD) next. Recall that in these experiments, each client is trained on a different dataset, that is, the first client is trained using images of two distinct classes from FashionMNIST, the second uses two different classes from the same dataset while the third client is trained using the PneumoniaMNIST dataset and a fourth client is trained from two different classes from FashionMNIST. Note that the baselines were also trained on these datasates individually, the results depicted in the plots are aggregated over the individual experiments. Moreover, we ran the QFL experiments in two different settings in each topology; results of both are shown in each plot. In one setting, the QPU capacity (i.e., number of qubits) varies for each client. As QPUs therefore run slightly different circuits, the number of features embedded also vary. To accommodate this, the images were resized for each client individually such that it fits the capacity of the respective client. Details on number of qubits and embedded features for each client are listed in Table 1. In the second setting, we ran the QFL experiments in which each client has the same capacity; the capacity was chosen such that the entire image could be embedded. The motivation behind the first approach is the fact that in a potential quantum internet, quantum computers of different architectures, capacity, etc. might be connected; we evaluated such a setting on a small scale.

Figures 8a and 8b show the training and test results using the star topology. While the QFL approach with varying QPU capacity delivers better training results than with equal capacity, its results on the test set are less stable.

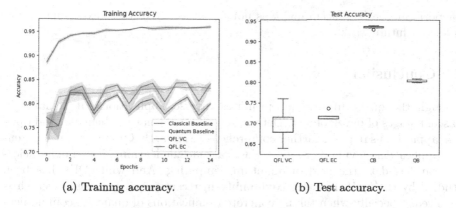

(a) Training accuracy.   (b) Test accuracy.

Fig. 8. MCMD experiment results using the star topology.

The results using the ring topology are shown in Fig. 9a and 9b. In this approach, the quantum baseline delivers the best quantum results in both training and test accuracy. All approaches, including the classical baseline, converge relatively early though.

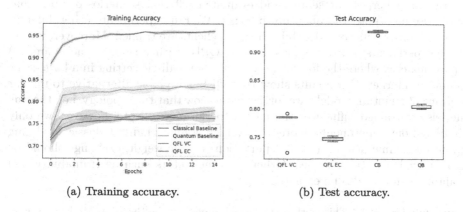

(a) Training accuracy.   (b) Test accuracy.

Fig. 9. MCMD results using the ring topology.

Overall, the baselines deliver the best results in our experiments on image data, however, the QFL approaches produce acceptable results all the same. Though the differences in accuracy between both training approaches in QFL, i.e., ring and star, require further investigation. While the star approach seems less stable during training, the training accuracy nonetheless performs much better than the ring approach. However, on the test set the ring approach performs slightly better. Adjusting hyperparamters and the nature of weight exchange and aggregation are likely to affect the models performance, though optimizing for performance (i.e., accuracy) was not the objective of these experiments and is left for future work.

7 Conclusion

Though the quantum internet promises diverse applications, it remains in nascent stages of development. Determining its exact nature, benefits, and optimal applications requires further comprehensive research. QFL is one such potential application, having its roots in classical machine learning, it can nonetheless be transferred to the field of quantum computing. And while QFL has been studied by the QML research community in recent years, far more research is required, especially when taking concrete ramifications of quantum communication into account as new challenges will inevitably arise through the incorporation of this novel communication medium. QFL allows the collective training of a global model while participating clients are not required to reveal their data. By allowing clients to keep their data local, QFL can enhance privacy, however, incorporating quantum communication may strengthen security and privacy even further. Though the main advantage is privacy, enhancing the models performance in comparison to non-distributed training requires further investigation.

In this paper, we presented and evaluated different scenarios of QFL that may arise on a potential quantum internet. We ran experiments using different network constraints that also differ how the clients are trained. Moreover, we ran these experiments in two settings, i.e., (1) QPUs with varying capacity and (2) equal capacity, where the former may be a more realistic setting in a large-scale quantum internet. Our results show that QFL is a viable alternative to regular training of quantum models, we furthermore show that the topology, i.e., the way models are trained influences the models performance. Note, however, we only simulated our experiments, moreover, weights were exchanged classically. Using quantum communication (e.g. teleportation) for the weight exchange should be investigated in future studies. Other relevant factors such as trainability and scalability should also be explored.

Acknowledgments. This work is sponsored in part by the Bavarian Ministry of Economic Affairs, Regional Development and Energy as part of the 6GQT project (https:// 6gqt.de)

Disclosure of Interests. The authors have no competing interests to declare.

References

1. Bergholm, V., et al.: Pennylane: Automatic differentiation of hybrid quantum-classical computations. arXiv preprint arXiv:1811.04968 (2018)
2. Caleffi, M., et al.: Distributed quantum computing: a survey. arXiv preprint arXiv:2212.10609 (2022)

3. Caleffi, M., Cacciapuoti, A.S., Bianchi, G.: Quantum internet: From communication to distributed computing! In: Proceedings of the 5th ACM International Conference on Nanoscale Computing and Communication, pp. 1–4 (2018)
4. Cerezo, M., et al.: Variational quantum algorithms. Nature Rev. Phys. **3**(9), 625–644 (2021)
5. Chehimi, M., Chen, S.Y.C., Saad, W., Towsley, D., Debbah, M.: Foundations of Quantum Federated Learning Over Classical and Quantum Networks (Oct 2023). arXiv:2310.14516 [quant-ph]
6. Chen, S.Y.C., Yoo, S.: Federated quantum machine learning. Entropy **23**(4), 460 (2021)
7. Geyer, R.C., Klein, T., Nabi, M.: Differentially private federated learning: A client level perspective. arXiv preprint arXiv:1712.07557 (2017)
8. Illiano, J., Caleffi, M., Manzalini, A., Cacciapuoti, A.S.: Quantum internet protocol stack: a comprehensive survey. Comput. Netw. **213**, 109092 (2022)
9. Kimble, H.J.: The quantum internet. Nature **453**(7198), 1023–1030 (2008)
10. Konečnỳ, J., McMahan, H.B., Yu, F.X., Richtárik, P., Suresh, A.T., Bacon, D.: Federated learning: Strategies for improving communication efficiency. arXiv preprint arXiv:1610.05492 (2016)
11. Kozlowski, W., Wehner, S.: Towards large-scale quantum networks. In: Proceedings of the Sixth Annual ACM International Conference on Nanoscale Computing and Communication, pp. 1–7 (2019)
12. Li, C., Kumar, N., Song, Z., Chakrabarti, S., Pistoia, M.: Privacy-preserving quantum federated learning via gradient hiding (2023)
13. Li, W., Lu, S., Deng, D.L.: Quantum federated learning through blind quantum computing. Sci. China Phys., Mech. Astronomy **64**(10), 100312 (2021). https://doi.org/10.1007/s11433-021-1753-3, arXiv:2103.08403 [quant-ph]
14. McMahan, B., Moore, E., Ramage, D., Hampson, S., y Arcas, B.A.: Communication-efficient learning of deep networks from decentralized data. In: Artificial Intelligence and Statistics, pp. 1273–1282. PMLR (2017)
15. Mitarai, K., Negoro, M., Kitagawa, M., Fujii, K.: Quantum circuit learning. Phys. Rev. A **98**(3), 032309 (2018)
16. Nielsen, M.A., Chuang, I.L.: Quantum computation and quantum information. Cambridge university press (2010)
17. Paszke, A., et al.: Pytorch: An imperative style, high-performance deep learning library. In: Advances in Neural Information Processing Systems, vol. 32 (2019)
18. Pedregosa, F., et al.: Scikit-learn: machine learning in Python. J. Mach. Learn. Res. **12**, 2825–2830 (2011)
19. Rofougaran, R., Yoo, S., Tseng, H.H., Chen, S.Y.C.: Federated Quantum Machine Learning with Differential Privacy (2023)
20. Schuld, M., Bocharov, A., Svore, K.M., Wiebe, N.: Circuit-centric quantum classifiers. Phys. Rev. A **101**(3), 032308 (2020)
21. Simon, C.: Towards a global quantum network. Nat. Photonics **11**(11), 678–680 (2017)
22. Van Meter, R., et al.: A quantum internet architecture. In: 2022 IEEE International Conference on Quantum Computing and Engineering (QCE), pp. 341–352. IEEE (2022)
23. Van Meter, R., Touch, J.: Designing quantum repeater networks. IEEE Commun. Mag. **51**(8), 64–71 (2013)
24. Wang, H., Yurochkin, M., Sun, Y., Papailiopoulos, D., Khazaeni, Y.: Federated learning with matched averaging. arXiv preprint arXiv:2002.06440 (2020)

344 L. Sünkel et al.

25. Wang, T., Tseng, H.H., Yoo, S.: Quantum federated learning with quantum networks. arXiv preprint arXiv:2310.15084 (2023)
26. Watrous, J.: The theory of quantum information. Cambridge university press (2018)
27. Wehner, S., Elkouss, D., Hanson, R.: Quantum internet: A vision for the road ahead. Science **362**(6412), eaam9288 (2018)
28. Xiao, H., Rasul, K., Vollgraf, R.: Fashion-mnist: a novel image dataset for benchmarking machine learning algorithms (2017)
29. Yang, J., et al.: Medmnist v2-a large-scale lightweight benchmark for 2D and 3D biomedical image classification. Sci. Data **10**(1), 41 (2023)
30. Yang, Q., Liu, Y., Chen, T., Tong, Y.: Federated machine learning: concept and applications. ACM Trans. Intell. Syst. Technol. (TIST) **10**(2), 1–19 (2019)
31. Zhang, Y., Zhang, C., Zhang, C., Fan, L., Zeng, B., Yang, Q.: Federated Learning with Quantum Secure Aggregation (2023)

Statistical Model Checking for Entanglement Swapping in Quantum Networks

Anubhav Srivastava and M. V. Panduranga Rao[✉]

Indian Institute of Technology Hyderabad India, Sangareddy, India
{cs21mtech02001,mvp}@iith.ac.in

Abstract. Given the fragile, stochastic and time critical nature of quantum communication systems, it is useful to analyse them with the rigour of formal methods. However, computationally expensive methods like exact probabilistic model checking do not scale with the size of the quantum network. In this work, we analyse entanglement swapping, an important primitive in quantum networks, using statistical model checking. We investigate the robustness of entanglement swapping against important parameters like longevity of quantum memory, success probability of entanglement generation and Bell State Measurements, and heterogeneity of the quantum network nodes. We demonstrate the usefulness of the approach using the MultiVeStA statistical model checker and the SeQUeNCe quantum network simulator.

Keywords: Quantum Networks · Statistical Model Checking · Discrete Event Simulators for Quantum Networks

1 Introduction

While quantum communications promise to be (unconditionally) secure, they involve transfer of fragile qubits across long distances. While transmission of classical bits over large distances is relatively easy, some peculiarities of quantum mechanics like the *no-cloning* theorem precludes the use of conventional classical repeater and signal boosting approaches.

Fortunately, it has been shown that quantum repeaters are possible to envisage, building upon the so-called *entanglement swapping* protocol [7,14,32]. Indeed, this forms the basis of future quantum *networks* [30]. This protocol allows (geographically distant) distant nodes to share maximally entangled pairs (also called EPR pairs: see Sect. 2 for a brief introduction) via intermediate quantum repeaters. Once this is achieved, several protocols like teleportation of quantum information are possible between the quantum nodes [8].

At an abstract level, a quantum network can be represented by an undirected graph with the nodes called quantum nodes–we do not distinguish between "end nodes" and intervening repeaters. The edges represent a quantum channel, say a fibre optic channel, between two quantum nodes. A physical qubit can be transmitted from one node to another directly if and only if they are adjacent quantum nodes.

Let us consider the simple case of three nodes a, b and c where a is adjacent to b and b is adjacent to c, but not a. To distribute an EPR pair between a and c, the following two steps need to be performed. First, an EPR pair each is generated between a and b, and b and c through the physical quantum channel. Then, a *Bell State Measurement* (BSM) performed at b, to convert the $a-b$ and $b-c$ EPR pairs to an $a-c$ EPR pair. This sequence of operations can be extended in principle to arbitrary path lengths. Once a route (cf [9,10,21,25]), that is, a path from source to destination quantum nodes, is determined on the quantum network, the main challenge is to establish end-to-end entanglement through entanglement swapping. However, as we will see, entanglement swapping over long paths itself presents several design and implementation challenges. Therefore, we focus on issues pertaining to entanglement swaps along line graphs in this work.

Components of quantum communication systems are difficult to build as of today. For example, for the system to be of use practically, quantum memory (that is, registers of qubits) of reasonable longevity is needed–at least of the order of a few minutes. Secondly, quantum operations like unitary gates and measurements need to be performed on the qubits. These operations are currently very error-prone and in general stochastic in nature. Finally, these protocols involve classical communications as well. This also brings into play synchronization issues. As such, design and implementation of these systems is not straightforward. Specifically, it is not easy to decide whether a given configuration and figure of merit of the system can yield desired performance, in terms of parameters of interest. Examples of such parameters include success probability of sharing an entangled pair between two nodes of the quantum communication network within a stipulated time. More importantly, synchronization properties and scheduling in the context of heterogeneous nodes need to be investigated.

Formal methods, in particular (probabilistic) model checking techniques, offer an approach for analyzing such (stochastic) systems [5]. Indeed, the use of formal methods to study correctness of quantum programs is catching on [17,18,20]. Even quantum cryptography protocols have been subjected to analysis through formal methods [6,15]. Desu et al. showed an approach to study quantum communications for timeliness through Probabilistic Timed Automata (PTA) and Probabilistic Timed Computational Tree Logic (PTCTL) [12]. Unfortunately, state space exploration based exact model checking approaches are computationally expensive and do not scale well [16,24]. Moreover, the modeling effort is non-trivial.

Statistical Model Checking (SMC) offers an inexpensive alternative, with the facility of using a (discrete event) simulator for a model of the system being analyzed [2,23,27,28]. Assuming that the simulator is faithful to the actual stochastic system in the relevant parameters, and a probability measure is well defined on its runs, it is a good substitute for the actual system. In these circumstances, it can be subjected to analysis through Statistical Model Checking and the results can be thought of holding for the actual system.

Fortunately, while development of quantum communication hardware has been somewhat slow, there has been significant progress in the design and implementation of sophisticated quantum networks simulation software [11,13,31].

In this work, we explore SMC for studying quantum communication systems. We use the MultiVeStA model checker [26] in conjunction with the quantum network simulator SeQUeNCe [31]. Our contributions are two-fold:

- We integrate MultiVeStA with SeQUeNCe to perform statistical model checking on quantum network protocols—we discuss some important aspects of this integration. We make the integration software available at [1] for further use by Quantum Networks researchers and developers.
- We perform an extensive study of some queries that we think are important for understanding entanglement swap protocols in quantum networks. The study yields interesting insights. The queries that we discuss have the following (not mutually exclusive) objectives:
  - comparison with Probabilistic Timed Automata model based results of Desu et al [12].
  - some additional queries on swap scheduling that are of interest.
  - a query on entanglement swap synchronization that is not only of extreme importance but also shows the usefulness of the tool-chain in performing complex analysis through nested queries.

We hope that this work and the tool-chain reported will kick-start a greater effort in analyzing quantum communication protocols and hardware performance through model checking.

The rest of the paper is arranged as follows. We discuss briefly the necessary preliminaries in the next section. In Sect. 3, we describe the integration of the statistical model checker and the quantum network simulator. Section 4 discusses the results in detail and the insights that they provide. We conclude the paper in Sect. 5 with a discussion of some future directions.

2   Relevant Preliminaries and Previous Work

We assume a basic familiarity with quantum mechanics, like the state, evolution and measurement postulates [22]. In this work, we will be particularly interested in the "maximally entangled" EPR pair, or simply EPR pair defined by the quantum state vector $\frac{|00\rangle + |11\rangle}{\sqrt{2}}$.

As mentioned in the previous section, the central objective in quantum networks is to distribute an EPR pair between two quantum nodes. While this is straightforward for adjacent nodes, *entanglement swapping* helps in achieving this objective between non-adjacent nodes. The protocol proceeds as follows. Consider a line graph quantum network with three nodes a, b and c. The nodes a and b are adjacent to each other in the sense that they share a link over which qubits can be transported without any loss. They generate and share the (maximally entangled) EPR pair $\frac{|0_a 0_b\rangle + |1_a 1_b\rangle}{\sqrt{2}}$. Similarly, b and c are adjacent to each other and share $\frac{|0_b 0_c\rangle + |1_b 1_c\rangle}{\sqrt{2}}$. Thus, the node b has two qubits, one from the EPR shared with a and one with b. As per the entanglement swapping protocol b performs a Bell State Measurement [7] on these two qubits, yielding $\frac{|0_a 0_c\rangle + |1_a 1_c\rangle}{\sqrt{2}}$ an EPR pair shared between a and c.

If the quantum memory of even one of two nodes that share an EPR pair is not long-lived, the EPR pair itself is not long-lived. Thus, we can associate the notion of *quality* with such an "EPR edge"–the ability to sustain an EPR pair for long duration. The entanglement swap protocol is error-prone due to several reasons. For the simplest case discussed above, it depends on the quality of $a - b$ and $b - c$ EPR edges. Similarly, it involves unitary gates like the single qubit Hadamard gate and the two qubit CNOT gate, and a measurement operation, the implementation of which is also error-prone.

Considering this, a rigorous analysis of a system model has the potential to yield deeper insights into design and implementation, and fine tuning of system parameters. Example analyses possible for entanglement swapping are as follows:

- External behaviour: The most important metric for users and applications on quantum networks is the time-limit μ within which the EPR pairs are distributed: Does the hardware and software configurations have the capability of sharing an EPR pair between two nodes of a quantum network within μ time-steps?
- Internal behaviour: (a) How does the success probability improve with quantum memories of increasing longevity? (b) A network designers' primary task is to find optimal hardware and software parameter values for components according to their use case. For example, how many EPR pair generation attempts have to be made between nodes of low coherence (informally, fidelity or closeness to the original quantum state) time quantum memories, when compared to nodes of high coherence time quantum memories, to obtain a similar end-to-end EPR distribution probability? This provides network designers with insights about the trade off between the *lifetime* of quantum memories and *retrial_attempts* and therefore time-limit to be allocated to different parts of the network. (c) Entanglement swap along a path entails scheduling the swaps at different nodes. What sequence of swaps gives the best success probability of sharing end-to-end entanglement? Given that all quantum nodes may not be equal in their properties like availability at a particular instant and memory lifetimes, is it possible to perturb the schedules with reducing success probabilities by too much? (d) Given the time criticality of the quantum operations, parallel sub-operations can cause synchronization errors. For example, for a Bell State Measurement to be performed at quantum node b, both $a - b$ and $b - c$ EPR pairs need to have been created within a short time-gap of each other. What is the probability that this indeed happens?

2.1   (Statistical) Model Checking

Model checking offers a powerful tool for performing this kind of analysis. Given a mathematically precise model of the system, the problem is to automatically check if it satisfies a property that is also specified formally as a statement in an appropriate logic [5]. Indeed, Desu et al [12] employ such a technique to study quantum networks. Modeling the quantum network as a network of Probabilistic Timed Automata and specifying properties in Probabilistic Timed Computational Tree Logic, they report a simple timing analysis of the system using PRISM [19] and validate it using the quantum network simulator SeQUeNCe [31]. However, in general, exact model checking for Probabilistic Timed Automata is notorious for being computationally expensive –EXPTIME-complete even for two clocks [16]. Moreover, as the system grows larger and more complex involving more features, modeling it as Probabilistic Timed Automata also becomes difficult.

Statistical model checking provides an alternative. Given an executable model of a stochastic system, Statistical Model Checking involves generating a Monte-Carlo sample of the runs of the model and evaluating a property of interest stated in an appropriate system of logic on each run. The number of runs is decided so as to allow the size of the $(1 - a) * 100\%$ confidence interval to be within $d/2$ distance of the estimated mean, for some chosen a and d [2,27,28].

MultiVeStA is a statistical model checking tool, which builds on a series of such tools. VeSta [29] supported a variety of modeling formalisms like (discrete and continuous time) Markov chains and the executable specification language PMaude for probabilistic read-write theories [3]. Further, not only were specification languages like PCTL supported, but also the QUAntitative Temporal EXpressions language (QuaTEx). The only requirements are that discrete event simulations can be performed on the models and the probability measures are well defined on the paths of the model. PVesTa provides supports for parallelism [4]. MultiVeStA provides for direct integration with discrete event simulators and supports counter-factual analysis [26]. The specification language MultiQuaTEx is a minor variant of QuaTEx.

2.2   SeQUeNCe

SeQUeNCe [31] is an open source discrete event quantum network simulator that aims at realistic simulation of quantum hardware and provides a suite of pre-built quantum networking protocols. It follows a layered architecture where each layer is implemented as a separate module.

It has a dedicated hardware layer that provides simulation components such as light sources, quantum memory and photon detectors. On top of the hardware layer, it provides an entanglement layer which includes protocols for entanglement generation, purification and swapping. SeQUeNCe also provides a network management module that performs *routing* to obtain an entanglement swap path between a source and destination and *reservation* to reserve the resources along the path for a request. SeQUeNCe provides other modules for quantum state management, circuit execution, and event creation and execution during the simulation.

3   Integrating MultiVeStA with SeQUeNCe

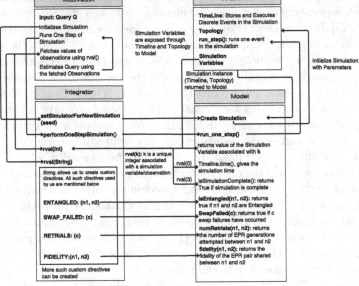

Fig. 1. Integration of MultiVeStA with SeQUeNCe

Figure 1 shows a high-level block diagram of the integration of MultiVeStA with SeQUeNCe. Briefly, we need to write two modules: the *Integrator* and the *Model*. MultiVeStA accesses SeQUeNCe through the *Integrator* to initialize, run the simulations and access the values of simulation variables. The *Integrator* implements a MultiVeStA specified function called *rval()* that fetches the values of the simulation variables that are required to evaluate the queries, through the interface called *Model*. We also introduce custom *Directives* of the form of "*ACTION* : $(arg1, arg2, ..., argN)$" through the *String* arguments of *rval()*. These are extensible and new directives can be added depending on the requirements of the user. Please see the github repository for details [1].

4   Queries and Results

4.1   Experimental Setup

To demonstrate and exposit on different queries, we report experiments on linear topologies of three and five nodes and demonstrate how MultiQuaTEx queries can be used to obtain state information about the quantum network at any given time. To illustrate scalability, we use longer line graphs of up to hundred nodes.

The metric of primary importance is the success probability of end-to-end entanglement. We vary various parameters and study their impact on this overall probability. We choose the same quantum channel attributes as in [12]:

```
quantum channel between adjacent nodes: 50km
attenuation of quantum channel: 0.2 dB/km
classical channel delay for adjacent nodes: 1ms
classical channel delay for path length 2: 2ms
quantum channel frequency =  100 GHz.
```

We focus on the following parameters in our *model*:

- τ is the lifetime of quantum memory
- μ is the time-limit within which an EPR pair should be shared between the end nodes
- p_{gen} is the success probability of EPR pair generation between neighbouring nodes
- p_{bsm} is the success probability of bell state measurement
- *retrial_limit* is the maximum number of times an operation can be attempted

In some experiments described in the next section we compare the results of MultiVeStA queries with earlier results of [12]. In such cases, we use the time t_{gen} taken for EPR generation between neighboring nodes to translate the time units between the MultiVeStA-SeQUeNCe implementation and the PTA model. In our MultiVeStA-SeQUeNCe experiments t_{gen} is 0.002 s and in the PTA model it is 5 time units. Therefore 0.002 s of the MultiVeStA-SeQUeNCe experiments are equivalent to 5 time units of the PTA model. We run MultiVeStA with the following parameters for all the experiments:

```
a (of the confidence interval) = 0.05
d (of the confidence interval)= 0.1
batch size = 100
```

4.2 Impact of Time-Limit (μ) on the Success Probability

Given a quantum network, a user would wish to ascertain if the hardware, software and the configurations are such that it is possible to distribute an EPR pair within a stipulated time. The MultiQuaTEx query that corresponds to this, is shown in the listing below. We consider a simple network of three nodes a, b and c and seek to distribute an EPR pair between nodes a and c. While this can be easily extended to longer paths, we stick to three nodes because of ease of explanation and comparison with a previous work that report similar experiments [12]. It shows a *parametric* query that estimates the probability of distributing an EPR pair between the nodes a and c by varying the time-limit μ. We fix τ to 0.006 and vary μ from 1.003 to 1.040 s. We choose these values for the following reason: in the PTA Model of [12], τ was set to 15 time-units which is 3×5 time-units and using the conversion factor that gives $\tau = 3 \times 0.002 = 0.006$ seconds for the MultiVeStA implementation. Similarly μ ranges from 10 to 100 for the PTA Model, which is equivalent to μ ranging from $1 + 0.002 \times 2$ to $1 + 0.002 \times 20 = 1.004$ to 1.040 seconds for the MultiVeStA implementation. From Fig. 2a we observe the success probability increases when μ increases. This is because when time-limit μ increases, more EPR generations can be attempted and one of them finally leads to a successful entanglement. This leads to EPR pairs becoming available for the Bell State Measurement to be attempted. The expected success probability however only reaches 0.5 because p_{gen} and p_{bsm} are set to 0.5. This setting limits the overall success probability. With better hardware, the probabilities of successful generation of an EPR pair and Bell State Measurement increases, and will result in higher probability of successfully sharing an EPR pair between a and c.

```
P_ac(mu) = if (s.rval(0) > mu*0.001
               || s.rval("Entangled: (a, c)") == 1)
            then s.rval("Entangled: (a, c)")
        else if (s.rval("SWAP_FAILED: 1") == 1)
            then 0
            else
            #P_ac(mu)
        fi
    fi;
    eval parametric(E[ P_ac(mu) ], mu, 1004, 4, 1041);
```

4.3   Impact of Quantum Memory Lifetime (τ) on the Success Probability

The query that we discuss in this subsection aims to explore the impact of increasing the lifetime of quantum memory τ on the success probability of distributing an EPR pair.

To perform this experiment, we set p_{gen} and p_{bsm} to 0.5 and the time-limit $\mu = 1.020$. These values are chosen to allow comparison with previous work [12]. The probabilities p_{gen} and p_{bsm} are kept the same as the PTA model whereas the time-limit μ has to translated in accordance with the MultiVeStA implementation. In the PTA model, μ was chosen as $10 \times \tau_{init} = 10 \times 5 = 50$ time-units. Following the same logic, we get $\mu = 1 + 10 \times \tau_{init} = 1 + 10 \times 0.002 = 1.020$ seconds. We add an offset of 1 s because the EPR distribution process starts at 1 s in current implementation. The query estimates the probability of an entanglement being distributed between a and c within the time-limit μ. If the time is less than 1.020 and a and c do not share an EPR pair, MultiVeStA moves the simulation to the next state using the next operator #. From Fig. 2b, we observe that the success probability increases with an increase in τ. This is because, by increasing τ, we allow EPR pairs to maintain fidelity for a longer time, thereby ensuring BSM is attempted. If successful, it yields an EPR pair between a and c. Similar to the last query, the success probability is limited to 0.5 because p_{gen} and p_{bsm} are 0.5. We see that the results from current implementation follow the trends previously discovered through the PTA Model.

```
P_ac()= if (s.rval(0) > 1.02 || s.rval("Entangled: (a, c)") == 1)
            then s.rval("Entangled: (a, c)")
        else if (s.rval("SWAP_FAILED: 1") == 1)
                then 0
            else
                #P_ac()
            fi
    fi;
eval E[ P_ac() ];
```

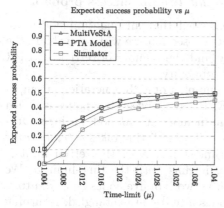

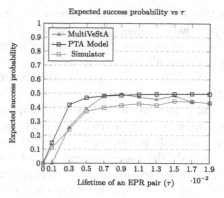

(a) $\tau = 0.006s$, $p_{gen} = 0.5$, $p_{bsm} = 0.5$.

(b) $\mu = 1.02s$, $p_{gen} = 0.5$, $p_{bsm} = 0.5$.

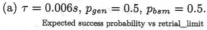

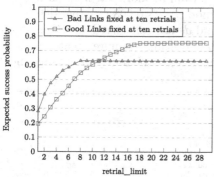

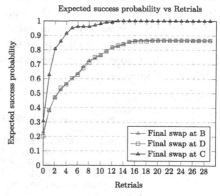

(c) Impact of retrial_limit and quality of links on overall success probability. $p_{gen} = 0.8$, $p_{bsm} = 1$, and $\mu = 1.1s$.

(d) Impact of schedule on final success probability

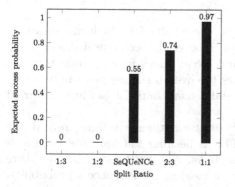

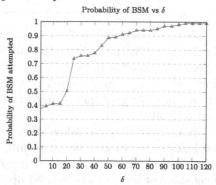

(e) Impact of Swap Schedule on Success Probability for longer paths

(f) Impact of δ on probability of swap being attempted

Fig. 2. Results of Queries

4.4   Impact of *retrial_ limit* and *link_ quality* on success probability

While in the previous experiments, we assumed that all quantum nodes are equal, we now drop that assumption. In this subsection, we wish to evaluate the impact of the quality of quantum memory and the number of retrials of EPR generation allowed on the overall success probability.

We define *retrial_ limit* as the maximum number of EPR generation attempts allowed between two *adjacent* nodes. An EPR pair between two nodes is called *Good* if and only if the longevity of quantum memory at *both* nodes is high. Otherwise, it is called as *Bad*. The query in the following listing takes *retrial_ limit* of each link as an argument. If at any time-step, any of the retrial limit is exceeded, the query returns 0. Otherwise, it runs until the time-limit μ. We report this experiment on a linear quantum network of five nodes labeled a to e. At every time-step, the query checks if nodes a and e are entangled, returning 1 if they are. At the end of the time-limit, the query returns 1 if a and e are entangled or 0 otherwise. The lifetime of quantum memories at nodes a and b is 0.010 s, at nodes d and e it is 0.006 s and at node c it is 0.008 s. As per definition, links ab and bc are *good*, whereas links cd and de are considered *bad*. Our objective through the query below is to show that *bad* links require greater number of retrials than *good* links to provide same success probability of entanglement distribution between nodes a and e.

```
P_ae(ab, bc, cd, de) =
    if (s.rval(0) > 1.1
    || s.rval("Entangled: (a, e)") == 1)
    then s.rval("Entangled: (a, e)")
    else if (s.rval("RETRIALS: (a, b)") <= ab &&
             s.rval("RETRIALS: (b, c)") <= bc &&
             s.rval("RETRIALS: (c, d)") <= cd &&
             s.rval("RETRIALS: (d, e)") <= de)
        then #P_ae(ab, bc, cd, de)
        else
            0
        fi
    fi;
```

We run this query with two scenarios, the results for which are shown in Fig. 2c. The blue plot shows the variation in overall success probability when *retrial_ limit* for *bad* links is fixed at 10 and *retrial_ limit* for *good* links is varied from 1 to 30. Similarly the red plot shows the overall success probability when *retrial_ limit* for *good* links is fixed at 10 and *retrial_ limit* for *bad* links is varied from 1 to 30.

We observe that for values of *retrial_ limit* less than ten, fixing retrials for bad links seems to give better results. This is because *bad* links are fixed at a higher *retrial_ limit* (=10) than the good links. Any EPR pairs that decohere along the *bad* links can be retried up to 10, yielding higher success probability. However, fixing *retrial_ limit* for *good* links does not help much, as can be seen in the red plot. The *retrial_ limit* for *bad* links is less than ten, and as such the decohering EPR pairs do not get many retrial opportunities for regeneration. If we look at *retrial_ limit* greater than 10, we observe that red plot increases steadily with an increase in *retrial_ limit* whereas blue plot plateaus out.

The reason for the plateauing is as follows. For the blue plot, any further increase in *retrial_limit* for *good* links does not help because *good* links already hold entanglements for a long time and the failures are occurring due to entanglements decohering on *bad* links that do not have more attempts available.

On the other hand, the red plot increases steadily with an increase in *retrial_limit* as *bad* links are getting more attempts to generate the expired entanglements. The *good* links, fixed at 10 retrials, already have sufficient time to hold the entanglement until all the required entanglements become available for a Bell State Measurement. The two plots intersect at *retrial_limit*=10 where the *retrial_limit* of both the *good* and the *bad* links is 10.

4.5   Impact of *schedule* on success probability

Schedule refers to the order in which nodes attempt BSM operations along the path to distribute EPR pairs between the end nodes. Through this experiment we want to study variations in success probability for different values of retrials for two schedules.

For this experiment, we select a linear topology $a - b - c - d - e$ of five good nodes, all having a memory lifetime of 0.010 s. We set the p_{gen} to 0.8 as we want to study the impact of EPR generation failures and therefore retrials on the final success probability. We set the p_{bsm} to 1 because swap failures are independent of retrials along the edges or the coherence time of quantum memories. From Fig. 2d, we can see the success probability is higher when the final swap happens at c compared to when final swap happens at b or d.

This is because when final swap happens at c, the a-c and c-e entanglement operations can be attempted simultaneously in parallel. This allows a-c and c-e EPR pairs to be available for a BSM to be attempted to provide a-e. In case of a final swap at b, the first swap happens at d to give c-e and second swap happens at c to give b-e. This takes significant time within which the entanglement generated between a-b decoheres. This makes it difficult for EPR pairs a-b and b-e to be available at the same time for the final Bell State Measurement to be attempted at b. Symmetrically, the schedule with the final swap at d also faces the same problem.

Swap Schedules for Longer Paths. It turns out that the choice of schedule plays a critical role in general in the end-to-end entanglement distribution probability. We scale to line graphs of fifty nodes where each node has a memory lifetime of $0.1s$ and we set p_{gen} and p_{bsm} to 1. We compare five different schedules; the results of this comparison are shown in Fig. 2e. We refer to the node at which a Bell State Measurement takes place as the *swap-node*. The split is obtained recursively around this node. We define *split-ratio*, as the ratio of number of nodes to the left and right of the swap node.

We begin with a schedule called split-in-the-middle where the swap-node has equal number of nodes to the left and right (thus split ratio of 1:1). We see in Fig. 2e that this performs best (success probability of 0.97), this is because independent swap operations can be attempted in parallel to the highest degree in this split.

We skew the split by decreasing the *split-ratio* to 2:3. In this case we see that the success probability decreases to 0.74. This is expected, since we have moved onto an asymmetric scenario where there is a possibility that one of the sides of the *swap-node* is not available for the swap to be performed. As we keep on skewing the split (by decreasing the split-ratio to 1:2 and then 1:3), we see that the probability drops to 0. This is because the entanglements on the final swap-node are not becoming available simultaneously due to different number of nodes on both sides. The default schedule created by SeQUeNCe results in the success probability of 0.55. These results show that a trade-off between a schedule perturbed away from 1:1 and successful probability can be struck, taking into account potential temporary unavailability of the nodes involved.

4.6   Impact of δ on Probability of BSM Being Attempted

A related challenge while scheduling swaps, is to understand the impact of the time interval between parallel operations. Consider a three node quantum network as before. Although $a - b$ and $b - c$ EPR pairs can be generated in parallel, a significant lag in their completion times could lead to problems. Without loss of generality, let us say $a - b$ is completed, and waiting for $b - c$. It is possible that by the time $b - c$ is completed, $a - b$ decoheres and the subsequent Bell State Measurement (BSM) at b cannot be performed.

Therefore, we wish that both EPR formations happen within δ time of each other, for small δ. Therefore δ is an important parameter because EPR pairs are short-lived and we want the BSM to proceed as soon as they are generated. For example in the three node linear topology, if $a - b$ is generated first at time-step t, we would want $b - c$ to be generated within $t + \delta$. The query in the following listing returns 1 if both EPR pairs are available within $t + \delta$ time of each other and 0 otherwise. The query does this for different values of δ from 0 to 0.012 s, where 0 enforces both the entanglements should be available at the same time-step.

```
P_swap(T, delta) =
            if(s.rval(0)>T)
                then 0
            else if((s.rval("Entangled: (a, b)") == 1)
                    && (s.rval("Entangled: (b, c)") == 1))
                    then 1
                else if(((s.rval("Entangled: (a, b)") == 1)
                    || (s.rval("Entangled: (b, c)") == 1))
                    && (T == 1000))
                    then
                        #P_swap(s.rval(0)+delta*0.0001, delta)
                    else
                        #P_swap(T, delta)
                    fi
                fi
            fi;
eval parametric(E[ P_swap(1000, delta) ], delta, 0, 5, 120);
```

We can see from Fig. 2f that the probability of swap being attempted increases steadily with increase in δ. This is because an increase in δ increases the time window within which both EPR pairs have to become available for the Bell State Measurement to be attempted. We also observe that at $\delta = 0$, 20% of the time the EPR pairs a-b and b-c are available at the same time-step. When δ is 0.012, the probability of swap being attempted is almost 1, that is the EPR pairs $a - b$ and $b - c$ are almost always available for the Bell State Measurement to proceed.

4.7   Running Time

Figure 3 shows the running time of a simple query regarding success probability of end-to-end EPR pair establishment with increasing path length, in increments of 10. We discuss the results in three regimes. The first two regimes are from path length 10 to 30, and 70 to 100. In these regimes, the probability of success is 1 and 0 respectively. Further, the CI conditions are satisfied with a small number of simulations. Therefore, the rise in running time is linear in path length in both the regimes. The second regime is from path length 40 to 60, where the probability of success lies strictly more than 0 and strictly less than 1. For these cases, to satisfy the CI conditions, more simulations need to be sampled. Therefore, there is a rise in running time in this regime. Nevertheless, an exact model checking approach would show a steep increase due to state space explosion for a model complex enough to capture such a system. These simulations were run on an AWS instance with 16 GB of RAM and four virtual CPUs each having a clock speed of 2.3 GHz.

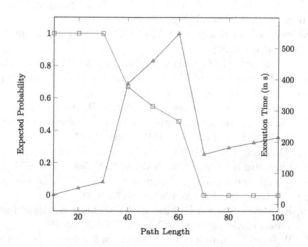

Fig. 3. Execution Time vs Path Length (Number of nodes). The blue plot shows the execution time (in seconds) and the red plot shows the corresponding probabilities of successfully establishing entangled pairs between the end nodes. (Color figure online)

5   Conclusions and Future Work

In this work, through an integration of a statistical model checker and quantum network simulator, we show how formal methods can be used for timing and performance analysis of quantum networks. A more detailed analysis need to be carried out that helps in identifying which hardware parameters make the most impact on price and performance. We believe this will help in planning for trade-offs in design of future quantum networks. An investigation through more expressive logics and more powerful model checkers would also yield rich dividends. Finally, can we borrow synthesis techniques from formal methods for developing high throughput, secure quantum networks?

Acknowledgments. The authors thank QCAL MeitY, Govt of India, for providing AWS quantum computing credits.

References

1. https://github.com/anubhavsrivastavaa/multivesta-qnetworks
2. Agha, G., Palmskog, K.: A survey of statistical model checking. ACM Trans. Modeling Comp. Simul. (TOMACS) **28**(1), 6 (2018)
3. Agha, G.A., Meseguer, J., Sen, K.: PMaude: rewrite-based specification language for probabilistic object systems. Electron. Notes Theor. Comput. Sci. **153**(2), 213–239 (2006)
4. AlTurki, Musab, Meseguer, José: PVESTA: a parallel statistical model checking and quantitative analysis tool. In: Corradini, Andrea, Klin, Bartek, Cîrstea, Corina (eds.) CALCO 2011. LNCS, vol. 6859, pp. 386–392. Springer, Heidelberg (2011). https://doi.org/10.1007/978-3-642-22944-2_28
5. Baier, C., Katoen, J.P.: Principles of Model Checking (Representation and Mind Series). The MIT Press (2008)
6. Bennett, C.H., Brassard, G.: Quantum cryptography: Public key distribution and coin tossing. In: Proceedings of IEEE International Conference on Computers, Systems, and Signal Processing, p. 175. India (1984)
7. Bennett, C.H., Brassard, G., Crépeau, C., Jozsa, R., Peres, A., Wootters, W.K.: Teleporting an unknown quantum state via dual classical and Einstein-Podolsky-Rosen channels. Phys. Rev. Lett. **70**, 1895–1899 (1993)
8. Bouwmeester, D., Pan, J.W., Mattle, K., Eibl, M., Weinfurter, H., Zeilinger, A.: Experimental quantum teleportation. Nature **390**(6660), 575–579 (1997)
9. Caleffi, M.: Optimal routing for quantum networks. IEEE Access **5**, 22299–22312 (2017). https://doi.org/10.1109/ACCESS.2017.2763325
10. Chakraborty, K., Rozpedek, F., Dahlberg, A., Wehner, S.: Distributed routing in a quantum internet. arXiv preprint arXiv:1907.11630 (2019)
11. Dahlberg, A., Wehner, S.: SimulaQron-a simulator for developing quantum internet software. Quantum Sci. Tech. **4**(1), 015001 (2018)
12. Desu, S.S.T., Srivastava, A., Rao, M.V.P.: Model checking for entanglement swapping. In: Bogomolov, S., Parker, D. (eds.) Formal Modeling and Analysis of Timed Systems - 20th International Conference, FORMATS 2022. LNCS, vol. 13465, pp. 98–114. Springer (2022). https://doi.org/10.1007/978-3-031-15839-1_6

13. Diadamo, S., Nötzel, J., Zanger, B., Beşe, M.M.: QuNetSim: a software framework for quantum networks. IEEE Trans. Quantum Eng. **2**, 1–12 (2021)
14. Goebel, A.M., et al.: Multistage entanglement swapping. Phys. Rev. Lett. **101**, 080403 (2008)
15. Huang, B., Huang, Y., Kong, J., Huang, X.: Model checking quantum key distribution protocols. In: 2016 8th Intl Conference on Information Technology in Medicine and Education (ITME), pp. 611–615 (2016)
16. Jurdziński, Marcin, Laroussinie, François, Sproston, Jeremy: Model checking probabilistic timed automata with one or two clocks. In: Grumberg, Orna, Huth, Michael (eds.) TACAS 2007. LNCS, vol. 4424, pp. 170–184. Springer, Heidelberg (2007). https://doi.org/10.1007/978-3-540-71209-1_15
17. Kakutani, Y.: A logic for formal verification of quantum programs. In: Datta, A. (ed.) ASIAN 2009. LNCS, vol. 5913, pp. 79–93. Springer, Heidelberg (2009). https://doi.org/10.1007/978-3-642-10622-4_7
18. Khatri, S.: On the design and analysis of near-term quantum network protocols using Markov decision processes (2022). https://arxiv.org/abs/2207.03403
19. Kwiatkowska, M., Norman, G., Parker, D.: PRISM 4.0: verification of probabilistic real-time systems. In: Gopalakrishnan, G., Qadeer, S. (eds.) CAV 2011. LNCS, vol. 6806, pp. 585–591. Springer, Heidelberg (2011). https://doi.org/10.1007/978-3-642-22110-1_47
20. Liu, J., et al.: Formal verification of quantum algorithms using quantum hoare logic. In: Dillig, I., Tasiran, S. (eds.) CAV 2019. LNCS, vol. 11562, pp. 187–207. Springer, Cham (2019). https://doi.org/10.1007/978-3-030-25543-5_12
21. Meter, R.V.: Quantum Networking. Wiley, London (2014)
22. Nielsen, M.A., Chuang, I.L.: Quantum Computation and Quantum Information. Cambridge University Press (2000)
23. Nimal, V.: Statistical approaches for probabilistic model checking. Ph.D. thesis, University of Oxford (2010)
24. Norman, G., Parker, D., Sproston, J.: Model checking for probabilistic timed automata. Formal Methods Syst. Des. **43**(2), 164–190 (2013)
25. Schoute, E., Mancinska, L., Islam, T., Kerenidis, I., Wehner, S.: Shortcuts to quantum network routing. arXiv preprint arXiv:1610.05238 (2016)
26. Sebastio, S., Vandin, A.: MultiVeStA: statistical model checking for discrete event simulators. In: 7th Intl Conference on Performance Evaluation Methodologies and Tools, ValueTools 2013, pp. 310–315. ICST/ACM (2013)
27. Sen, K., Viswanathan, M., Agha, G.: Statistical model checking of black-box probabilistic systems. In: Alur, R., Peled, D.A. (eds.) CAV 2004. LNCS, vol. 3114, pp. 202–215. Springer, Heidelberg (2004). https://doi.org/10.1007/978-3-540-27813-9_16
28. Sen, K., Viswanathan, M., Agha, G.: On statistical model checking of stochastic systems. In: Etessami, K., Rajamani, S.K. (eds.) CAV 2005. LNCS, vol. 3576, pp. 266–280. Springer, Heidelberg (2005). https://doi.org/10.1007/11513988_26
29. Sen, K., Viswanathan, M., Agha, G.A.: VESTA: a statistical model-checker and analyzer for probabilistic systems. In: Second Intl Conference on the Quantitative Evaluation of Systems (QEST 2005), pp. 251–252. IEEE (2005)
30. Wehner, S., Elkouss, D., Hanson, R.: Quantum internet: a vision for the road ahead. Science **362**(6412), eaam9288 (2018)
31. Wu, X., et al.: SeQUeNCe: a customizable discrete-event simulator of quantum networks (2020)
32. Żukowski, M., Zeilinger, A., Horne, M.A., Ekert, A.K.: "Event-ready-detector" bell experiment via entanglement swapping. Phys. Rev. Lett. **71**, 4287–4290 (1993)

Noise Robustness of a Multiparty
Quantum Summation Protocol

Antón Rodríguez-Otero, Niels M. P. Neumann[✉], Ward van der Schoot,
and Robert Wezeman

The Netherlands Organisation for Applied Scientific Research,
The Hague, The Netherlands
antonr.o.98@gmail.com, niels.neumann@tno.nl

Abstract. Connecting quantum computers to a quantum network opens a wide array of new applications, such as securely performing computations on distributed data sets. Near-term quantum networks are noisy, however, and hence correctness and security of protocols are not guaranteed. To study the impact of noise, we consider a multiparty summation protocol with imperfect shared entangled states. We study analytically the impact of both depolarising and dephasing noise on this protocol and the noise patterns arising in the probability distributions. We conclude by eliminating the need for a trusted third party in the protocol using Shamir's secret sharing.

Keywords: Distributed Quantum Computing · Noisy Quantum Communication · Multi-Party Computation · Shamir Secret Sharing

1 Introduction

Quantum computing is an emerging field where advances are made on the hardware-side, software-side, as well as applications. Many companies and universities are working on building better quantum hardware with more resources of better quality. At the same time, new algorithms are being discovered, and these new quantum algorithms are applied in various new settings.

The theoretical speedup quantum computers offer for various problems discerns them from classical alternatives. Amongst these are some of the most complicated problems encountered in every-day life. Examples where quantum computers outperform classical alternatives include breaking certain asymmetric encryption protocols [21], developing new materials and personalised medicines [10], and solving complex systems of linear Eq. [12].

Another aspect at which quantum computers distinguish themselves from classical alternatives, is the security of a quantum state: Opposed to classical information, in general, quantum information cannot be read out or copied faithfully. Reading out a quantum state destroys the state irrevocably and loses information, whereas trying to copy a quantum state leaves the state and its copy entangled, and operations performed on an entangled copy differ from those

L. Franco et al. (Eds.): ICCS 2024, LNCS 14837, pp. 360–374, 2024.
https://doi.org/10.1007/978-3-031-63778-0_26

applied to the original unentangled state. Because of this, sharing information via quantum states is secure. This idea underlies the field of quantum communication and its subfield quantum key distribution.

Combining quantum computing with quantum communication joins the best of both worlds: by using quantum communication between different quantum computers, these devices can collaboratively solve larger problems, while the information shared between the devices remains secure. This field is called *Distributed Quantum Computing* (DQC).

Distributed quantum computing entails the collaborative execution of quantum algorithms using multiple quantum devices. Distributed computations can occur at various levels: for example, the devices may independently run their own quantum circuits, after which the outputs are combined to obtain the final results. Alternatively, the devices may cooperate intricately through quantum communication to execute a single overall circuit. This study concentrates on the latter scenario, specifically exploring the execution of a distributed quantum addition circuit.

The key challenge in this form of distributed quantum computing, is the application of non-local multi-qubit gates. As any multi-qubit gate can be decomposed into CNOT gates with additional local one-qubit gates [3], it suffices to implement the CNOT-gates in a non-local fashion. Eisert et al. gave the first description of how to perform operations between different quantum devices through the use of local operations and classical communications (LOCC) and shared entanglement between the different devices [9]. Later, this work was extended and a distributed version of Shor's algorithm was theorised [24,25].

Distributed quantum computing works by transforming traditional quantum algorithms to their distributed version. In these distributed versions, operations performed between qubits located on different devices are called non-local and are replaced by a non-local quantum gate established using shared entangled states. In comparison, operations between qubits on the same device are called local, and are unchanged. These three works consider all operations, both local and non-local, to be perfect. Beals et al. later proved that distributing an algorithm over different resources incurs only a small overhead in the cost [4]. Hence, when programming quantum algorithms on a higher level, the underlying structure of the hardware, local or distributed, has only a marginal effect.

Follow-up work mainly focused on applications run using a distributed quantum network [7], or on how to best implement a distributed quantum computer network [5,11]. One aspect to take into account in these distributed networks is the robustness against noise, as current hardware is noisy and will remain so for the foreseeable future. It is therefore interesting to consider the effect of imperfect operations in such distributed settings. A first work on this topic computed the fidelity of a distributed and imperfect quantum phase estimation algorithm, when distributed over a varying number of devices [16]. Another example is the work by Khabiboulline et al. where a secure quantum voting protocol is presented [14].

In this work, we extend this line of research by considering imperfect non-local operations as well, but applied to the distributed quantum summation protocol [17], which extends the algorithm proposed by Draper [8] and later improved by Ruiz-Perez and Garcia-Escartin [19]. The quantum summation algorithm uses the Quantum Fourier Transform to map the states to their phase state representation. In the phase space, addition corresponds to specific controlled phase gates.

In this protocol, we consider different parties which aim to compute the sum of their inputs, without revealing these inputs. Each party has access to a local quantum computer, which can generate shared entangled states with other devices. In practice, quantum hardware remains noisy and it is necessary to consider decoherence effects when developing applications. Currently, the fidelity of state teleportation between non neighbouring nodes is around 0.7 [13] while the fidelity of quantum operations on quantum devices is around 0.95 to 0.99 [15]. For this reason, we omit in this work the effect of imperfect quantum operations, and focus on the impact of an imperfect quantum network links. Concretely, we consider how dephasing and depolarising noise on the shared entangled states affects the output fidelity of this distributed summation protocol.

We also extend this line of research by combining it with a primitive from cryptography called *Shamir Secret Sharing*. In earlier works, multiparty protocols are considered with the use of a central server party which is trusted by everyone. This is not a realistic assumption in practical use cases. In this work, we show how the considered multiparty summation protocol can be extended to a setting without the requirement of a trusted server party. We show that the protocol yields the same output as the original protocol, while none of the parties learns inputs from other parties.

Section 2 explains the multiparty summation protocol and the two considered noise models. Next, Sect. 3 presents the results of simulations for both noise models. Section 4 contains an analytical study of the noise patterns and the periodicity therein. Afterwards, Sect. 5 details the extension of the protocol to a version without the need of a trusted server party. Section 6 concludes with a summary and an outlook to future distributed quantum computing work.

2 Preliminaries

2.1 Distributed Quantum Computing

We start by describing the non-distributed version of the quantum summation protocol, after which we explain the distributed version from [17]. Suppose we have two integers $a, b < 2^n$ that we wish to add and that we have their corresponding quantum states $|a\rangle$ and $|b\rangle$ that are their binary representation using n qubits. The protocol first applies the quantum Fourier transform of size n, denoted by QFT_n to $|b\rangle$. This yields the phase state representation of b, given by $|\phi(b)\rangle$. Then, applying phase gates to the qubits of $|\phi(b)\rangle$ controlled by the qubits of $|a\rangle$ gives the quantum state $|\phi(a+b)\rangle$. After applying an inverse quantum Fourier transform, the state $|a+b\rangle$, describing the binary representation of $a+b$, is obtained.

This summation protocol can be easily extended to allow a server party to do the addition of k different numbers held by k different computing parties. Figure 1 showcases the extended protocol for two computing parties with an additional server party. The server party holds the result at the end of the protocol. Note that the phase gates applied by different parties commute and hence, every party can apply their local phase gates simultaneously.

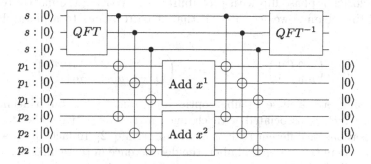

Fig. 1. Example of **DQA** [17]. A server party adds integers from two computing parties. The blocks Add x^k denote the phase-gates needed to add integer x^k as described above. The final quantum state in the first register is $|x^1 + x^2\rangle$.

The above multiparty protocol translates to a non-local protocol by replacing every CNOT gate by a non-local CNOT gate. The resulting protocol is called the *DISTRIBUTED-QFT-ADDER* (**DQA**). Multiple implementations for the CNOT-gates exist, some of which even allow simultaneous implementation of the phase gates by all parties [17].

2.2   Noise Models

The quantum network distributes entangled GHZ states between the server node and the different party nodes. The presence of noise translates to imperfect entanglement between the nodes.

Current state-of-the-art protocols for entanglement generation between nodes of a quantum network are heralded, which allows to deterministically know whether the entanglement distribution process succeeded. Typically, heralding work in experiments via a photon measurement such that entanglement is established if and only if a photon has been detected. We therefore disregard lossy quantum channels by assuming that the distribution is done in a heralded way.

In this study, we consider two types of noise in the quantum links, namely dephasing and depolarising noise. These types of noise arise often in physical implementations and current software packages allow for easy simulation of these noise type, whilst the analytic study remains feasible at the same time. The respective noise channels are applied to the quantum links by applying them to all qubits in the GHZ state independently.

Noise Model A: Dephasing Channel. A dephasing channel is a *completely positive and trace-preserving* (CPTP) map that represents the decay of the quantum phase of a system, that is, the off-diagonal elements of the density matrix. A one-qubit dephasing channel ε_{depha} is usually represented by the map

$$\varepsilon_{depha} : \rho \mapsto \left(1 - \frac{p}{2}\right)\rho + \frac{p}{2}\mathbf{Z}\rho\mathbf{Z} \tag{1}$$

which performs a phase flip with probability $p/2$. Writing the matrix representation of this channel we see, indeed, that it corresponds to a phase damping process:

$$\rho = \begin{pmatrix} \rho_{00} & \rho_{01} \\ \rho_{10} & \rho_{11} \end{pmatrix} \rightarrow \varepsilon_{depha}(\rho) = \begin{pmatrix} \rho_{00} & (1-p)\rho_{01} \\ (1-p)\rho_{10} & \rho_{11} \end{pmatrix}. \tag{2}$$

Dephasing errors arise in fiber optic links due to the birefringence phenomenon, which is associated with changes in the refractive index for different polarisations or different regions of the material [22]. In fact, using a special polarisation maintaining optical fiber, the decoherence in these links can be completely described via dephasing processes [23]. Assuming heralded entanglement distribution, we can ignore the high attenuation arising in these links. Our focus is the impact of noise on the fidelity of the protocol, so we omit the entanglement generation rate and the lower transmission rates.

Noise Model B: Depolarising Channel. The depolarising channel is usually seen as the quantum equivalent of white noise. Depolarising channels model processes that completely scramble the starting state with some probability. As a result, both quantum and classical information is lost. Given a valid n-qubit quantum state ρ, an n-qubit depolarising channel ε_{depol} can be written as

$$\varepsilon_{depol} : \rho \mapsto (1-p)\rho + \frac{p}{d}\mathbf{I}, \tag{3}$$

where $d = 2^n$ is the dimension of the Hilbert space ρ lives in.

Depolarising channels can, amongst other things, model the misalignment of reference frames between the nodes in a quantum network [26]. Moreover, depolarising channels can also model what happens if heralding fails, for instance when the detector wrongfully measures a photon. Such an event is called a dark count and leads to reading out an empty quantum memory [6].

3   Simulations with Noise

We simulated the quantum summation protocol **DQA** [17] and included the noise models discussed above to see how well the protocols perform in noisy settings. By adding dephasing or depolarising noise to the protocol, we expect incorrect outcomes found by the server party at the end of the protocol. Therefore, we report the results as probability distributions in histograms or polar plots.

We implemented both noise models using Qiskit [1] where we applied the noise models only to the quantum links, the part where the GHZ-states are generated. We implemented these noise models by inserting local noisy identity gates at the end of the GHZ generation block. We ran experiments for varying number of parties and inputs, as well as noise levels. Each simulation consists of 9,000 independent runs of the circuit[1].

3.1 Dephasing Noise

We first consider the impact of the dephasing noise by analysing four parties, each with input 1 and a dephasing noise of $p/2 = 0.07$ for each of the four quantum links.

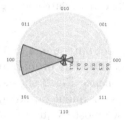

Fig. 2. Polar representation of the distribution .

Two parties, dephasing noise

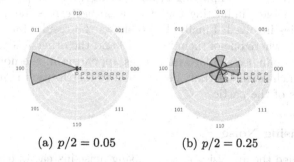

(a) $p/2 = 0.05$ (b) $p/2 = 0.25$

Fig. 3. Probability distribution for the case of two parties inputting 2, with identical dephasing noise applied on each entangled qubit pair for varying noise levels, $p/2$, as indicated. The correct outcome is given by 100 in binary.

---

[1] The implementations are available upon reasonable request to the authors.

Two parties, dephasing noise

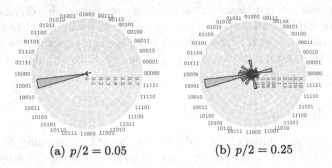

(a) $p/2 = 0.05$ (b) $p/2 = 0.25$

Fig. 4. Probability distribution for the case of two parties with inputs 10 and 7, respectively, with identical dephasing noise applied on each entangled qubit pair for varying noise levels, $p/2$, as indicated. The correct outcome is given by 10001 in binary.

In the noisy setting, the correct outcome, 4 or 100 in binary, has the highest probability, followed by outcome 000. The probability distribution seems to have some symmetry (cf. Section 4), hence Fig. 2 shows the probability distribution in a polar plot. The second most frequent outcome is diametrically opposed to the correct one; and the next two most frequent outcomes are $\pi/2$ radians away from the highest probability and diametrically opposed to each other.

Interestingly, this symmetry emerges also for a different number of parties and for varying inputs. First, Fig. 3 shows the results for a protocol run with two parties, both of them inputting 2, for varying values of p. Again, a similar noise pattern emerges, indicating that the noise pattern is independent of the number of parties involved. Finally, Fig. 4 shows the results for a protocol run with two parties, where one of them inputs 10 and the other 7, again for varying values of p. We again see similar symmetries appearing in the noise probability distribution, which indicates that the probability distribution is independent of the input values of the parties.

3.2 Depolarising Noise

We also performed the analysis for depolarising noise instead of dephasing noise. Interestingly, the same probability distributions were found as for dephasing noise, with the same symmetry patterns emerging. We hence omitted the figures, as they give no additional information compared to the figures in the previous circuit. In the next section, we proof that indeed the probability distributions follow a specific pattern with symmetries, independent of the type of noise.

4   Analytical Study

The probability distributions shown in the previous sections show some symmetry. In this section we analyse this symmetry effect and show that, for fixed error probability, indeed the weighted Hamming distance with the correct output string determines the probability of being measured. We derive an expression for the probability distribution that applies to both dephasing and depolarising noise, and then discuss the intuition on the relation between the analytical expression and the observed pattern. Proofs of the results presented in this section can be found in Appendix A and in the full version of this paper [18].

4.1   Proof of Probability Distribution

The probability distribution of the **DQA** under dephasing or depolarising noise follows a specific probability distribution.

Lemma 1. *Under depolarising or dephasing noise, the server party state right before the application of the Inverse Quantum Fourier Tranform on a **DQA** for n parties can be written as*

$$\rho = \bigotimes_{s=0}^{n-1} \frac{1}{2}(|0\rangle\langle 0| + |1\rangle\langle 1| + ae^{-i\theta_s}|0\rangle\langle 1| + ae^{i\theta_s}|1\rangle\langle 0|), \tag{4}$$

where $a = \prod_{i=0}^{n-1} a_i$, with $a_i = (1-p_i)^2$ for dephasing noise and $a_i = 1-(1-p_i)^2$ for depolarising noise, with p_i the noise parameter for party i.

Now, to characterise the output probability distribution of the distributed adder protocol, we need to look at the product of the a_i factors. As we consider the same error rates for every qubit, we define the fidelity parameter a as $a = \prod_{i=0}^{n-1} a_i$. In particular, $a = 1$ corresponds to a noiseless GHZ-state, whereas $a = 0$ corresponds to a completely dephased or depolarised GHZ state.

Theorem 1. *Let $\{t^{(i)}\}_{i=1}^{m}$ be the inputs of m different parties and let for each $s \in \{0, 1, \ldots n-1\}$*

$$\theta_s = \frac{\pi}{2^{n-1-s}} \sum_{i=1}^{m} t^{(i)} = \frac{2\pi}{2^{n-s}} \sum_{i=1}^{m} t^{(i)}. \tag{5}$$

*Then, the m-player **DQA** protocol produces the output probability distribution such that for each potential output x*

$$P(x) = \frac{1}{2^n} \prod_{s=0}^{n-1} [1 + a\cos(\theta_s - 2\pi x/2^{n-s})], \tag{6}$$

with fidelity parameter $a \in [0,1]$ related to the depolarising or dephasing noise level.

4.2   Understanding the Noisy Distribution

This section provides intuition for what the proven theoretical distribution in Eq. (6) actually looks like and how it translates to the distribution observed in the simulations. From the equation, it follows that the probability is maximised if all cosines evaluate to 1, which happens precisely if the argument of the cosines is an integer multiple of 2π. Setting $x = \tilde{x}$, with $\tilde{x}$ the correct outcome of the summation, this indeed maximises the probability. The resulting probability then equals

$$P(\tilde{x}) = \left(\frac{1+a}{2}\right)^n.$$

A noiseless setting where $a = 1$, indeed gives a probability of 1.

Now, if $y \in \mathbb{Z}_{2^n}$ is a different outcome, y can be written as $y = \tilde{x} + z$ for some error $z \in \mathbb{Z}_{2^n}\backslash\{0\}$. The probability to observe y then equals

$$P(y) = \frac{1}{2^n}\prod_{s=1}^{n}[1 + a\cos(2\pi z/2^s)] \tag{7}$$

Note, we relabeled the counter with respect to Eq. (17). For every s, we see a periodic behavior in z resulting from the cosines. Combined, we get a complex periodic behavior in the probability distribution of the possible outcomes.

We can now prove that the probability distribution is symmetric around x:

Lemma 2. *Let $z \leq 2^{n-1}$, then $P(x + z) = P(x + (2^n - z))$.*

By the previous lemma, information on noise strings $z \leq 2^{n-1}$ gives sufficient information on all possible noise strings.

In addition, this allows us to show the behaviour observed in Sect. 3 regarding the second and third most frequent outcomes:

Lemma 3. *For any integer $k \in \{1,\ldots,n-1\}$ and any error string $z \in \{1,2,\ldots 2^k\}$, we have that*

$$P(x + 2^k) \geq P(x + z)$$

This lemma yields indeed that the state diametrically opposite to the correct outcome has the second largest probability, the states $\pi/2$ radians away from the correct outcome have the third largest outcome, and so forth.

In addition, we see that the probabilities closer to the correct value are larger. To be more precise:

Lemma 4. *For any integer $k \in \{1,2,\ldots,n-1\}$ and error string $z \in \{1,2,\ldots,2^k\}$, we have*

$$P(x + z) \geq P(x + (2^{k+1} - z))$$

Running the circuit multiple times gives samples from the probability distribution. It would be natural to try and use multiple circuit runs to increase the probability of obtaining the correct answer, for example by comparing the obtained distribution with the theoretical distribution derived above. However, as for fixed fidelity parameter a the probabilities are exponentially small in n, standard techniques using Chernoff bounds require an exponential number of samples to lower bound the success probability of retrieving x.

5   Protocol Without Trusted Server

Like most multiparty protocols, the **DQA** protocol requires a trusted third party. Although this is a common assumption in some classical multiparty computation protocols, it is unrealistic in practice. This section therefore introduces a modification to the protocol which eliminates the necessity of this reliable authority.

For this modification to work, a certain primitive from cryptography is required, called *(Treshold) Shamir's Secret Sharing* [20] (SSS), a well-known classical protocol originally intended to distribute secrets between several entities, with the ability to reconstruct them.

Suppose an agent desires to distribute a secret X among k parties. Then the agent would choose a polynomial of order $t < k$ over a finite field $GF(q)$

$$g(x) = X + a_1 x + \ldots + a_{t-1} x^{t-1} \tag{8}$$

with a prime number q such that $X < q$. Next, the secret sharer would choose a set of k different points $\{x_1, \ldots, x_k\}$ to evaluate the polynomial on and would send one polynomial evaluation $g(x_j)$ to each party. Now, any subset of t parties can reconstruct the secret by simply performing polynomial interpolation using the polynomial evaluations that they received and then determining $g(0) = X$. Interpolation requires t parties, as the polynomial has degree $t - 1$. Knowing $t - 1$ shares $\{(x_i, g(x_i))\}$ does not provide any information about the secret.

SSS can be utilized to perform multiparty summation in a secure manner by combining it with repeated usage of **DQA** in multiple rounds. In each round, a different party acts as server of **DQA**, the other parties input one of their shares for the quantum protocol. The party acting as server then receives the sum of the shares at the end of the round. As shares are only used as input in the quantum protocol, no party learns the shares of other parties By combining the shares of any subset of at least t parties, the parties can reconstruct the summation result. This results in the following protocol:

NO-TP-ADDER (NTPA)

Consider m parties, each holding a number X_i.

- **Step 1:** the parties agree on a sufficiently large prime q and make q public;
- **Step 2:** each party i chooses a degree-t polynomial over the finite field $GF(q)$

$$g_i(x) = X_i + a_1^i x + \ldots + a_{t-1}^i x^{t-1} \tag{9}$$

where the coefficients are chosen at random but non-zero, except for a_0^i, which corresponds to the real input from the party ($a_0^i = X_i$);
- **Step 3:** In each round, a different party will act as the server party, which requires agreement on the order for the parties to act as server;
- **Step 4:** m rounds of **DQA** are performed. For each $r \in \{1, \ldots, m\}$:
  - Party r acts as server;
  - Party i inputs share $g_i(r)$;
  - Party r receives the partial summation $\sum_i g_i(r)$

At the end of all rounds, each party r has received the partial summation $\sum_i g_i(r)$. It hence knows $G(r)$ for $G(x) = \sum_{i=1}^{m} g_i(x)$

- **Step 5:** The summation result can be restored by having t parties cooperate to share their intermediate results $G(r)$. As $G(x)$ has degree at most $t - 1$, these t evaluation points of $G(x)$ are hence sufficient to reconstruct $G(x)$, from which the summation result can be computed by evaluating $G(0) = \sum_{i=1}^{m} g_i(0) = \sum_{i=1}^{m} X_i$. Note that just like in the original SSS protocol, at most $t - 1$ shares are not sufficient to conclude anything about $\sum_{i=1}^{m} X_i$.

As each party acts as a server once, no party holds more power or knowledge than any other, eliminating the need for a trusted third party.

Note that as the parties only know the intermediate shares $G(r)$, they still need t shares to recover the result of the summation.

6   Conclusions and Outlook

As current quantum hardware is noisy, and is expected to remain so for a while, it is important to study the impact that imperfect operations have on the fidelity of quantum protocols. Multiple works on noisy quantum algorithms are available in literature. Similarly, a few works on distributed quantum computing are available. This work focuses on the combination of distributed noisy operations and analyses the effect of imperfect operations on the outcome. We restricted ourselves to imperfect shared entangled states with perfect operations on the individual devices, as the errors within local devices are generally smaller than the ones seen on quantum communication.

We considered a practical implementation of the distributed multiparty quantum summation protocol [17]. We apply depolarising or dephasing noise on the shared entangled states and analytically study the behaviour of the protocol. The probability distributions corresponding to the final state of a noisy summation protocol given these noise models show a clear symmetric pattern, proved in the analytic study. The probability to find an erroneous state depends on the amount of noise affecting the execution of the protocol and the weighted Hamming distance between the erroneous string and the correct outcome.

The protocol initially uses a trusted third party. Building upon the classical Shamir's Secret Sharing protocol, we could remove the need for a trusted party. As an added benefit, all parties automatically learn the outcome of the protocol.

Future work should address the effects of other sources of noise that are present in the protocol, such as imperfect local operations. More importantly, a proper study of the effects that noise has on the security of the protocol needs to be done. In a perfect noiseless setting, the quantum no-cloning theorem ensures that no information is leaked to the outside world without corrupting the states that the parties share; thus, the parties could detect the presence of an adversary and abort the protocol before inputting their secrets. However, the signature on the shared entangled states of the action of an eavesdropper is indistinguishable from noise. Hence, the parties cannot expose an eavesdropper just by checking quantum correlations via the shared entangled states. In this case, the amount of information that can be leaked and learned by an eavesdropper from the execution of the protocol should be bounded. As a first step, formal definitions of security, anonymity and privacy in the context of quantum multiparty summation must be established, similar to the field of Quantum Electronic Voting [2].

Although the conclusions from the present study are specific for the investigated protocol, most quantum multiparty protocols rely on the utilisation of entanglement at some stage. Thus, the methodology used here can inspire the analysis of noise robustness in similar protocols.

A   Proofs of Results of Section 4

Lemma 5. *Under depolarising or dephasing noise, the state of the server party right before the application of the Inverse Quantum Fourier Tranform on a* **DQA** *run for n parties can be written as in 4; where* $a = \prod_{i=0}^{n-1}(1 - p_i)^2$ *for dephasing noise and* $a = \prod_{i=0}^{n-1}[1 - (1 - p_i)^2]$ *for depolarising noise, with* p_i *the noise parameter for party i.*

Proof. From the effect that a one qubit *dephasing channel* has on an n-qubit *GHZ* state

$$\rho \mapsto \left(1 - \frac{p_i}{2}\right)\rho + \frac{p_i}{2}Z_i\rho Z_i$$

$$= \frac{1}{2}\left(|0\rangle^{\otimes n}\langle 0|^{\otimes n} + |1\rangle^{\otimes n}\langle 1|^{\otimes n} + (1 - p_i)\left(|0\rangle^{\otimes n}\langle 1|^{\otimes n} + |1\rangle^{\otimes n}\langle 0|^{\otimes n}\right)\right),$$

it can be shown that the application of a noisy non local CNOT between server party and n parties takes their joint state to

$$\rho_{s,1,\dots,n} \mapsto \frac{1}{2}\left(|0\rangle^{\otimes n+1}\langle 0|^{\otimes n+1} + |1\rangle^{\otimes n+1}\langle 1|^{\otimes n}\right.$$

$$\left. + \prod_{i=0}^{n-1}(1 - p_i)\left(|0\rangle^{\otimes n+1}\langle 1|^{\otimes n+1} + |1\rangle^{\otimes n+1}\langle 0|^{\otimes n+1}\right)\right).$$

Then, the parties input their corresponding data through the $\mathcal{Z}$ rotations of an angle θ_j. After that, a second distributed CNOT under dephasing noise with parameter takes the state of the server to state in 4, with $a = \prod_{i=0}^{n-1}(1 - p_i)^2$

For *depolarising noise*, the same expression holds by replacing $\prod_{i=0}^{n-1}(1 - p_i)$ by $1 - (1 - p_i)^2$. Note that for $n > 2$, applying an n-depolarizing channel gives a different result compared to applying a 1-depolarizing channel on all n qubits.

Let ρ_j be the state of the j-th server qubit. After entanglement generation and the first non local CNOT under depolarizing error with parameter $p_{i,1}$, the parties input their corresponding data by applying local $R_Z(\theta_j)$ gates. A second distributed CNOT under depolarizing noise with parameter $p_{i,2}$ gives

$$\rho_j = (1 - p_{i,1})\left((1 - p_{i,2})\rho'_j \otimes |0\rangle\langle 0|^n + \frac{p_{i,1}}{2}I_{n+1}\right) + \frac{p_{i,2}}{2}I_{n+1}$$

$$= \left((1 - p_{i,1})(1 - p_{i,2})\rho'_j \otimes |0\rangle\langle 0|^n + \frac{p_{i,1}(1 - p_{i,2}) + p_{i,1}}{2}I_{n+1}\right)$$

with $\rho_j' = \frac{1}{2}\left(|0\rangle\langle 0| + |1\rangle\langle 1| + e^{-i\theta_j}|0\rangle\langle 1| + e^{i\theta_j}|1\rangle\langle 0|\right)$. Taking $p_{i,1} = p_{i,2} = p_i$, setting $a = 1 - (1-p)^2$ and tracing out the degrees of freedom of the non server parties, we can write

$$\rho_j'' = \mathbf{Tr}_{1\ldots n}[\rho_j] = \frac{1}{2}\left(|0\rangle\langle 0| + |1\rangle\langle 1| + ae^{-i\theta_j}|0\rangle\langle 1| + ae^{i\theta_j}|1\rangle\langle 0|\right). \qquad (10)$$

Theorem 1. *The m-player **DQA** protocol produces the output probability distribution such that for each potential output x*

$$P(x) = \frac{1}{2^n}\prod_{s=0}^{n-1}[1 + a\cos(\theta_s - 2\pi x/2^{n-s})], \qquad (11)$$

with fidelity parameter $a \in [0,1]$ related to the depolarising or dephasing noise level in the shared GHZ states.

Proof. The state before the final inverse quantum Fourier transform is given by

$$|\Psi\rangle = \bigotimes_{s=0}^{n-1}\frac{1}{\sqrt{2}}(|0\rangle + e^{i\theta_s}|1\rangle) = \bigotimes_{s=0}^{n-1}\frac{1}{\sqrt{2}}\left(\sum_{j_s=0,1}e^{i\theta_s j_s}|j_s\rangle\right), \qquad (12)$$

which simplifies to

$$|\Psi\rangle = \frac{1}{\sqrt{2^n}}\sum_{j=0}^{2^n-1}e^{i\sum_{s=0}^{n-1}\theta_s j_s}|j_{n-1}\ldots j_0\rangle. \qquad (13)$$

By Lemma 1, the presence of dephasing or depolarising noise on the quantum edges mixes the state of the server, such that the density matrix takes the form

$$\rho = \bigotimes_{s=0}^{n-1}\frac{1}{2}(|0\rangle\langle 0| + |1\rangle\langle 1| + ae^{-i\theta_s}|0\rangle\langle 1| + ae^{i\theta_s}|1\rangle\langle 0|) \qquad (14)$$

which can be written as

$$\rho = \frac{1}{2^n}\sum_{j=0}^{2^n-1}\sum_{k=0}^{2^n-1}a^{\sum_s |k_s-j_s|}e^{i\sum_s(\theta_s j_s - \theta_s k_s)}|j_{n-1}\ldots j_0\rangle\langle k_{n-1}\ldots k_0|, \qquad (15)$$

where $a = \tilde{a}^2$, as the distributed CNOT gates are performed twice, before and after the rotations.

The last step of the protocol consists of applying the inverse quantum Fourier transform. It maps the state in Eq. (15) to

$$\mathbf{IQFT}_n\rho\mathbf{QFT}_n$$
$$= \frac{1}{2^{2n}}\sum_{x,y=0}^{2^n-1}\sum_{j,k=0}^{2^n-1}a^{\sum_s |k_s-j_s|}e^{i\sum_s(\theta_s-\frac{\pi}{2^s}x)j_s}e^{-i\sum_s(\theta_s-\frac{\pi}{2^s}y)k_s}|x_{n-1}\ldots x_0\rangle\langle y_{n-1}\ldots y_0|.$$
$$(16)$$

As the probability of measuring a computational basis state corresponds to the corresponding diagonal element of the density matrix, we obtain the probability $P(x)$ by setting $x = y$, completing the proof:

$$P(x) = \frac{1}{2^{2n}} \prod_{s=0}^{n-1} \sum_{j_s,k_s=0,1} a^{|k_s-j_s|} e^{i\left(\theta_s - \frac{2\pi}{2^{n-s}}x\right)j_s} e^{-i\left(\theta_s - \frac{2\pi}{2^{n-s}}x\right)k_s}$$

$$= \frac{1}{2^{2n}} \prod_{s=0}^{n-1} \left(1 + ae^{i\left(\theta_s - \frac{2\pi}{2^{n-s}}x\right)} + ae^{-i\left(\theta_s - \frac{2\pi}{2^{n-s}}x\right)} + 1\right)$$

$$= \frac{1}{2^n} \prod_{s=0}^{n-1} \left[1 + a\cos\left(\theta_s - \frac{2\pi}{2^{n-s}}x\right)\right]. \tag{17}$$

References

1. Aleksandrowicz, G., Alexander, T., Barkoutsos, P., Bello, L., Ben-Haim, Y., Bucher, D., et al.: Qiskit: an open-source framework for quantum computing (Feb 2019). https://doi.org/10.5281/zenodo.2562111
2. Arapinis, M., Lamprou, N., Kashefi, E., Pappa, A.: Definitions and security of quantum electronic voting. ACM Trans. Quantum Comput. **2**(1), 1–33 (2021)
3. Barenco, A., et al.: Elementary gates for quantum computation. Phys. Rev. A **52**, 3457–3467 (1995). https://doi.org/10.1103/PhysRevA.52.3457
4. Beals, R., et al.: Efficient distributed quantum computing. Proc. R. Soc. A: Math. Phys. Eng. Sci. **469**(2153), 20120686 (2013). https://doi.org/10.1098/rspa.2012.0686
5. Caleffi, M., et al.: Distributed quantum computing: a survey (2022)
6. van Dam, J.: Analytical model of satellite based entanglement distribution. Master's thesis, TU Delft (2022)
7. DiAdamo, S., Ghibaudi, M., Cruise, J.: Distributed quantum computing and network control for accelerated VQE. IEEE Trans. Quantum Eng. **2**, 1–21 (2021). https://doi.org/10.1109/TQE.2021.3057908
8. Draper, T.G.: Addition on a quantum computer (2000). https://doi.org/10.48550/ARXIV.QUANT-PH/0008033
9. Eisert, J., Jacobs, K., Papadopoulos, P., Plenio, M.B.: Optimal local implementation of nonlocal quantum gates. Phys. Rev. A **62**, 052317 (2000). https://doi.org/10.1103/PhysRevA.62.052317
10. Fedorov, A.K., Gelfand, M.S.: Towards practical applications in quantum computational biology. Nat. Comput. Sci. **1**(2), 114–119 (2021). https://doi.org/10.1038/s43588-021-00024-z
11. Gyongyosi, L., Imre, S.: Scalable distributed gate-model quantum computers. Sci. Rep. **11**(1) (2021). https://doi.org/10.1038/s41598-020-76728-5
12. Harrow, A.W., Hassidim, A., Lloyd, S.: Quantum algorithm for linear systems of equations. Phys. Rev. Lett. **103**, 150502 (2009). https://doi.org/10.1103/PhysRevLett.103.150502
13. Hermans, S.L.N., Pompili, M., Beukers, H.K.C., Baier, S., Borregaard, J., Hanson, R.: Qubit teleportation between non-neighbouring nodes in a quantum network. Nature **605**(7911), 663–668 (2022). https://doi.org/10.1038/s41586-022-04697-y

14. Khabiboulline, E.T., Sandhu, J.S., Gambetta, M.U., Lukin, M.D., Borregaard, J.: Efficient quantum voting with information-theoretic security (2021). https://doi.org/10.48550/ARXIV.2112.14242

15. Li, Z., et al.: Error per single-qubit gate below 10^{-4} in a superconducting qubit. npj Quantum Inform. (2023). https://doi.org/10.1038/s41534-023-00781-x

16. Neumann, Niels M. P.., van Houte, Roy, Attema, Thomas: Imperfect distributed quantum phase estimation. In: Krzhizhanovskaya, V.V., et al. (eds.) ICCS 2020. LNCS, vol. 12142, pp. 605–615. Springer, Cham (2020). https://doi.org/10.1007/978-3-030-50433-5_46

17. Neumann, N.M.P., Wezeman, R.S.: Distributed quantum machine learning. In: Phillipson, F., Eichler, G., Erfurth, C., Fahrnberger, G. (eds.) Innovations for Community Services, vol. 1585, pp. 281–293. Springer International Publishing, Cham (2022). https://doi.org/10.1007/978-3-031-06668-9_20

18. Otero, A.R., Neumann, N.M.P., van der Schoot, W., Wezeman, R.: Noise robustness of a multiparty quantum summation protocol (2023). https://doi.org/10.48550/arXiv.2311.15314

19. Ruiz-Perez, Lidia, Garcia-Escartin, Juan Carlos: Quantum arithmetic with the quantum Fourier transform. Quantum Inf. Process. 16(6), 1–14 (2017). https://doi.org/10.1007/s11128-017-1603-1

20. Shamir, A.: How to share a secret. Commun. ACM 22(11), 612–613 (1979). https://doi.org/10.1145/359168.359176

21. Shor, P.W.: Polynomial-time algorithms for prime factorization and discrete logarithms on a quantum computer. SIAM J. Comput. 26(5), 1484–1509 (1997). https://doi.org/10.1137/S0097539795293172

22. Wu, Q.L., Namekata, N., Inoue, S.: High-fidelity entanglement swapping at telecommunication wavelengths. J. Phys. B: At. Mol. Opt. Phys. 46(23), 235503 (2013). https://doi.org/10.1088/0953-4075/46/23/235503

23. Xu, J.S., Yung, M.H., Xu, X.Y., Tang, J.S., Li, C.F., Guo, G.C.: Robust bidirectional links for photonic quantum networks. Sci. Adv. 2(1), e1500672 (2016). https://doi.org/10.1126/sciadv.1500672

24. Yimsiriwattana, A., Jr., S.J.L.: Distributed quantum computing: a distributed Shor algorithm. In: Donkor, E., Pirich, A.R., Brandt, H.E. (eds.) Quantum Information and Computation II, vol. 5436, pp. 360 – 372. International Society for Optics and Photonics, SPIE (2004). https://doi.org/10.1117/12.546504

25. Yimsiriwattana, A., Lomonaco, S.J.: Generalized GHZ states and distributed quantum computing (2004). https://doi.org/10.48550/ARXIV.QUANT-PH/0402148

26. Šafránek, D., Ahmadi, M., Fuentes, I.: Quantum parameter estimation with imperfect reference frames. New J. Phys. 17(3), 033012 (2015). https://doi.org/10.1088/1367-2630/17/3/033012

Hybrid Approach to Public-Key Algorithms in the Near-Quantum Era

Adrian Cinal[ID], Gabriel Wechta[ID], and Michał Wroński[✉][ID]

NASK National Research Institute, Kolska street. 12, Warsaw, Poland
{adrian.cinal,gabriel.wechta,michal.wronski}@nask.pl

Abstract. Application of post-quantum algorithms in newly deployed cryptosystems is necessary nowadays. In the NIST Post-Quantum Competition several algorithms that seem to be resistant against attacks mounted using quantum computers have been chosen as finalists. However, it is worth noting that one of finalists—SIKE—was catastrophically broken by a classical attack of Castryck and Decru only a month after qualifying for the final round. This shows that absolute trust cannot yet be placed in the algorithms being standardized. And so a proposition was made to use the novel, post-quantum schemes alongside the well-studied classical ones with parameters chosen appropriately to remain secure against quantum attacks at least temporarily, i.e., until a large enough quantum computer is built.

This paper analyzes which classical public-key algorithms should be used in tandem with the post-quantum instances, and studies how to ensure appropriate levels of both classical and quantum security. Projections about the development of quantum computers are reviewed in the context of selecting the parameters of the classical schemes such as to provide quantum resistance for a specified amount of time.

Keywords: post-quantum algorithms · classical algorithms · quantum computing · security level

1 Introduction

Post-quantum cryptography is a relatively new branch of modern cryptology. In 2017 NIST announced their post-quantum cryptography (NIST PQC) competition to which researchers from all around the world could submit their proposals of public-key schemes: key establishment and signature algorithms.

NIST in [7] defined five security levels for assessing the security of post-quantum cryptosystems. Instead of relying on precise estimates of the number of bits of security, these levels take into account both classical and quantum cryptanalysis. Each level is characterized by a reference primitive (AES or SHA), with *its* security forming the basis for subsequent analyses.

Table 1 presents the NIST security levels, providing their definitions and the estimated resources (quantum and classical gates) required to compromise the

© The Author(s), under exclusive license to Springer Nature Switzerland AG 2024
L. Franco et al. (Eds.): ICCS 2024, LNCS 14837, pp. 375–388, 2024.
https://doi.org/10.1007/978-3-031-63778-0_27

reference primitive at each level. As indicated in [7], plausible values for MAXDEPTH range from 2^{40} logical gates (per year) to 2^{64} logical gates (per decade), up to a maximum of 2^{96} logical gates (per millennium).

Table 1. NIST security levels as defined in [7] with estimated complexity of breaking the scheme's security.

Security Level	Security Definition *Must require computational resources comparable to or greater than*	Quantum Gates (estimated)	Classical Gates (estimated)
1	exhaustive key search on a 128-bit key block cipher (e.g. AES-128)	2^{170}/MAXDEPTH	2^{143}
2	collision search on a 256-bit hash function (e.g. SHA3-256)	-	2^{146}
3	exhaustive key search on a 192-bit key block cipher (e.g.AES-192)	2^{233}/MAXDEPTH	2^{207}
4	collision search on a 384-bit hash function (e.g. SHA3-384)	-	2^{210}
5	exhaustive key search on a 256-bit key block cipher (e.g. AES-256)	2^{298}/MAXDEPTH	2^{272}

Once submitted, each algorithm has been analyzed both by NIST specialists and the cryptographic community. These analyses resulted in many weak algorithms being eliminated at an early stage of the competition. Identification and subsequent withdrawal of broken schemes continued, however, all the way to the end. During the third and fourth rounds, respectively, the Rainbow digital signature algorithm and the SIKE key establishment algorithm were compromised in the classical setting. Rainbow was broken completely on security level 1 and significantly weakened on other security levels, whereas SIKE is now known to be breakable in mere 2 h on a classical CPU even at the highest (fifth) security level.

Since post-quantum algorithms have not yet been analyzed thoroughly enough, it is vital from the point of view of security to combine them with classical schemes. Thus each public-key scheme should be a hybrid consisting of at least two parts: a classical instance (secure in the classical setting) and a post-quantum instance (conjectured secure in both classical and post-quantum settings). Should the post-quantum instance prove to be breakable classically, this approach keeps the overall scheme secure against classical attacks and against quantum attacks for some time also (provided the parameters of the classical instance are chosen adequately)—until a powerful enough quantum computer is built. In the long run, if the post-quantum part of the hybrid scheme stands the test of time, the scheme shall remain secure against quantum adversaries despite the classical part having been long obsolesced.

The necessity of using hybrid public-key algorithms has been postulated years ago, even before NIST PQC competition started, but the spectacular failures of the round-three and round-four candidates (finalists) show that the need to combine (largely experimental) post-quantum cryptosystems with conservative, well-studied classical ones is very much real. Such hybrid solutions exist today and are used, for example, in TLS 1.3 [32], where classical algorithms based on elliptic curves over 256-bit prime fields, X25519 and Secp256r1, are used along-side Kyber768, so far believed to be resistant against quantum attacks. TLS 1.3 uses a simple "concatenation approach," where public keys of the two algorithms are concatenated back to back and transmitted as a single value in order to avoid changing the existing data structure and message fields. Similarly, when deriving the session key, two shared secrets are obtained by the two schemes, classical and post-quantum, and are then concatenated to obtain the master shared secret, from which the session key is derived.

This combination, however, is largely ad hoc. What we endeavour to achieve in this paper is a careful analysis of the classical instances (signature and key establishment) and their respective parameters that would match the most closely the security levels of their post-quantum counterparts and accompany them best. We shall also provide estimates about how long these classical companions shall remain secure, based on the current forecasts about the development of quantum computers.

2   Known Attacks Against Post-quantum Instances

2.1   Attack on Rainbow

Beullens in 2022 presented new key recovery attacks against Rainbow [4], one of the three finalist signature schemes in the NIST PQC competition. Previously, it was believed that breaking Rainbow at its lowest security level would take 2^{128} operations. Beullens' attack, however, utilizes differentials to efficiently recover the secret key, thus surpassing all previously known attacks for every parameter set submitted to NIST. Specifically, with a Rainbow public key for the NIST security level 1 parameters from the second-round submission, Beullens' approach can retrieve the corresponding secret key in an average of 53 h (roughly a weekend) using a standard laptop.

2.2   Attack on SIKE

In June of 2022, SIKE advanced to the fourth round of NIST PQC competition as an alternate candidate algorithm for key establishment. Not a whole month afterwards, the algorithm was totally broken by Castryck and Decru [6] on a classical computer. A limitation to their attack is that the endomorphism ring of the base curve must be known (which, however, is already the case in SIKE). Still, not long after this attack, other attacks were devised, with Maino and Martindale [20] presenting a subexponential attack on SIKE, which does not require the knowledge of the endomorphism ring of the base curve. Robert [26] then introduced an algorithm for breaking SIKE, which is, first, polynomial-time, and, second, works no matter the choice of the starting curve.

The attack of Castryck and Decru takes only a few hours on a regular laptop to break SIKE instances at NIST security level 5. It is worth noting that by the time of the attack isogeny-based cryptography had already been studied for about 10 years and no similar attacks had been found. Also, Kani's theorem, lying at the heart of the attack, had been known for over 20 years and was never regarded to be any serious threat. This goes to show that even schemes built on top of well-studied constructions may surprisingly fail, and so extra precautions must be taken as we are entering the near-quantum era.

2.3   Recent Breakthrough Against LWE

In a recent article [8], Chen proposes an efficient quantum algorithm for solving the *learning with errors* (LWE) problem. Hardness of LWE is the assumption underlying the security of lattice-based schemes such as CRYSTALS-Kyber and CRYSTALS-Dilithium standardized already by NIST. As of this writing, Chen's paper is undergoing peer review and the validity of his claims or their practical impact are yet unclear.

3   Quantum Computing

The purpose of this Section is to provide a brief overview of the principles of quantum computation and, above all, introduce the terminology used throughout the paper.

Exploiting quantum mechanics to obtain computational advantage was first proposed by Feynman in 1981 [14], followed by Deutsch defining a quantum Turing machine in 1985 [10]. Thus, the field of quantum computing was born and with publication in 1994 of Shor's seminal paper [31], it gained unprecedented momentum and attracted the attention of cryptographers, as in [31] Shor showed how to leverage quantum computation to break the mathematical problems underpinning contemporary asymmetric cryptography. Two years later, Grover presented an algorithm for searching an unstructured N-element set in time $O(\sqrt{N})$ [17], thus posing further threat to symmetric cryptography.

Similarly to bits in classical computing, a fundamental unit of information in quantum computing is a *qubit* which we associate with a unit vector in $\mathbb{C}^2$:

$$|q\rangle = \alpha_0|0\rangle + \alpha_1|1\rangle \qquad (1)$$

with $|\alpha_0|^2 + |\alpha_1|^2 = 1$ and $|0\rangle = (1,0)^T$, $|1\rangle = (0,1)^T$ column (standard basis) vectors in $\mathbb{C}^2$. We say that a qubit is in a *superposition* of the states $|0\rangle$ and $|1\rangle$. A qubit can be *measured* thus yielding a classical binary value 0 or 1, where 0 is measured with probability $|\alpha_0|^2$ and 1 is measured with probability $|\alpha_1|^2$ according to *Born's rule* (we refer to the numbers $\alpha_0, \alpha_1 \in \mathbb{C}$ as *amplitudes* of their associated basis states $|0\rangle$ and $|1\rangle$, respectively). After a measurement, a qubit is said to have *collapsed* to a classical state and the superposition once present is now destroyed. Before measurement, however, qubits can be manipulated in such a way that once measured they collapse with overwhelming probability to the desired result of computation. This is the basis of quantum computing.

While there are other realizations of quantum computation (cf., for example, *quantum annealing*) the most prevalent model is that of quantum *gates*. A gate represents a *unitary* (reversible and preserving the unit length of the vector $|q\rangle$) transformation to a qubit or a register of n qubits. Conceptually, we design quantum algorithms in this setting as *circuits* with wires going in and out of gates, whereas a physical realization may be completely decoupled from this image (and quantum "circuits" are more temporal than spacial in practice with gates being applied in place one after another).

We say a set of quantum gates is *universal* if any operation possible on a quantum computer (any unitary), or at least a satisfactory approximation thereof, can be expressed as finite sequence of gates from this set. The most common such universal set currently studied is the Clifford+T set and so it shall also be the focus of this paper.

Note that quantum systems are susceptible to noise which has to be accounted for in the quantum computer by implementing extensive error correction. By far the most costly of the Clifford+T gates to implement in a fault-tolerant manner is the T gate [15,27] corresponding to the following unitary matrix:

$$T = \begin{pmatrix} 0 & 0 \\ 0 & e^{i\pi/4} \end{pmatrix}. \tag{2}$$

For this reason complexity of quantum circuits is often expressed in terms of the number of T gates (or T-*count*) or the number of T gates modulo gates which can be "run" in parallel (T-*depth*) [18]. Another useful metric which we shall refer to in this paper is the number of *logical* qubits[1] needed to run the circuit. This is also sometimes referred to as the *width* of the circuit.

Another commonly found measure of complexity is the number of *Toffoli gates* or the associated *Toffoli-depth* [19,27]. Toffoli gates can be implemented using 7 T gates and a T-depth varying between 4 and 1 [1,28], thus we shall translate the Toffoli-count (Toffoli-depth) estimates found in the literature to the T-counts (T-depths) and use these as a common denominator.

---

[1] As a fault-tolerance and error-correction measure a single logical qubit is typically implemented using a number of physical qubits.

4 Analyses

4.1 Methodology

It must be noted that due to noise in the quantum computations, there are success probabilities typically associated with the attacks presented in the literature. For the purposes of this paper, however, we shall work on the assumption that if the quantum resources available are sufficient to mount an attack with non-negligible probability, even if it requires rerunning the computations a number of times, then the relevant cryptosystem is vulnerable.

We shall restrict our attention to RSA as the sole scheme based on integer factorization intractability,[2] and to elliptic-curve schemes (ECDSA, EdDSA, ECDH) based on the discrete logarithm problem. As for the latter, we shall further focus only on elliptic curves over prime fields $\mathbb{F}_p$. The decisions here are motivated by the relevant NIST publications, which deprecate use of DSA [21] as well as ECDSA based on binary curves [21,22]. Also, we are explicitly interested in signature and key establishment algorithms, for which standard implementations always fall into one of the two categories just delineated. While elliptic curves over extension fields (including binary fields) are used in some settings (e.g., in pairing-based cryptography), we intentionally leave them out and focus on schemes of the most fundamental utility and enjoying most standardization and prevalence. Quantum cryptanalysis of binary curves receives thorough treatment in [2].

Our methodology focuses on analyzing progress in the development of quantum computers and the scale of computations able to be run on them. In particular, it is important to determine when we expect to build a quantum computer on which one could run Shor's algorithm for a given problem (integer factorization, discrete logarithm in a finite field, discrete logarithm on an elliptic curve) with given parameters. A crucial factor for this approach is the choice of the time frame and deciding for how long the information (protected by the hybrid scheme) should remain secure (authentic, secret).

Three time periods have to be taken into consideration:

- *implementation time*—time required for the scheme to be globally deployed,
- *usage time*—time when the scheme is actively used,
- *expiration time*—period in which the scheme is being phased out (deprecated), but the information protected by it should still remain secure.

For example, suppose we want to protect a piece of information (encrypted using a key derived from a hybrid key establishment protocol) for 5 years. If we design the system now, implement it by 2025, and intend to use it until 2040, we need to ensure the security of the information until the end of the year 2045

---

[2] The Paillier cryptosystem, found in multi-party computation, uses "de facto" RSA keys so interested parties may use our analyses to evaluate viability of Paillier's encryption for their purposes. Caution is advised here, however, as this scheme is outside the scope of our work.

(the last time any plaintext is encrypted using the hybrid scheme may be in late 2040).

Assuming a declassification period of 25 years[3] and intended usage time until year 2040, we arrive at the conclusion that hybrid schemes should be resistant against quantum attacks until 2065 (on all NIST security levels). As the analyses presented in Sect. 5.1 show, this can be achieved with practical values for security parameters.

Remark 1. When analyzing the complexity of the attack algorithms (Shor's [31], Grover's [17]), it is customary to consider quantum resources usage, not merely time, as is the case for classical algorithms. This is due to the nature of the current state of quantum computing: since the problem of engineering large-scale multi-qubit systems has not yet been solved, these finer points about the algorithms/circuits give better insight into exactly how feasible the attacks are.

4.2 Classical Schemes Based on Factoring

Integers of the form $N = pq$, for p, q—different primes of similar length in bits, are commonly used in the RSA cryptosystem, and thus called *RSA integers* going forward. The security of RSA is based on the assumption that factoring such integers is computationally infeasible.[4] Shor's algorithm [31] efficiently, albeit quantumly, factors RSA integers, thereby solving the underlying computational problem of the RSA cryptosystem. The best known algorithm for factoring RSA integers on classical computers is the General Number Field Sieve [5] which heuristically runs in subexponential time, while the time (or more practically - circuit depth) of Shor's algorithm is polynomial in the size of the input. This is achieved by a reduction to finding the order of an element in $\mathbb{Z}_N$ [31].

Since the introduction of Shor's algorithm, there have been numerous attempts to optimize it in terms of both the number of required logical qubits and the number of quantum gates.

In each run, Shor's factoring algorithm requires $2n$ group operations for an n bit integer. Ekerå and Håstad [13] have shown that, by replacing order finding with short discrete logarithm computation, the number of group operations can be reduced to $\frac{3}{2}n$ without making any trade-offs.

Gidney et al. in [16] have presented a quantum algorithm for factoring RSA integers which, by introducing a number of optimization techniques, has significantly reduced quantum resource costs when compared to the original Shor's algorithm and follow-up works. Additionally, they provide a detailed analysis of the quantum resource requirements of the algorithm in terms of logical and

---

[3] See, e.g., U.S. Executive Order 13526.

[4] Technically, it is based on solving for roots modulo N, but this distinction is not relevant here.

physical qubits as well as Toffoli and T gates. Gidney et al. report over 100x improvement over other top works, which use the same basic cost model as they do. Reported results are presented in Tables 2 and 3.[5]

In 2023, Regev presented an algorithm with lattice reduction post-processing that lowered the number of gates from $\tilde{O}(n^2)$ (original Shor's algorithm) to $\tilde{O}(n^{\frac{3}{2}})$ [25] at the expense of increasing the number of logical qubits from $O(n)$ (optimized Shor's algorithm) to $O(n^{\frac{3}{2}})$.[6] Soon after that, Ragavan and Vaikuntanathan [24] showed how to lower the number of necessary qubits to only $\tilde{O}(n)$, while keeping the circuit size (depth and total number of gates) $\tilde{O}(n^{\frac{3}{2}})$.

Table 2. Expected costs of factoring n-bit RSA integers according to [16].

	Factoring n-bit RSA integer		
	$n = 3072$	$n = 7680$	$n = 15360$
Logical Qubits	9287	23238	46507
T-Count	$1.25 \cdot 2^{27}$	$1.36 \cdot 2^{31}$	$1.47 \cdot 2^{34}$
Circuit Depth	$1.12 \cdot 2^{32}$	$1.76 \cdot 2^{34}$	$1.76 \cdot 2^{36}$

Table 3. Asymptotic costs of factoring n-bit RSA integers according to [16].

	Factoring n-bit RSA integer
Logical Qubits	$3n + 0.002n \lg n$
T-Count	$0.3n + 0.0005n^3 \lg n$
Circuit Depth	$500n^2 + n^2 \lg n$

4.3 Elliptic Curve Cryptography

Alongside the factoring algorithm, Shor also presented an efficient quantum algorithm for solving the discrete logarithm problem in a multiplicative group of a prime field $\mathbb{F}_p$ [31]. This was later made appropriate to the setting of elliptic curves by Proos and Zalka in [23]. It is this latter setting which is relevant to contemporary cryptography. Adapting Shor's algorithm to elliptic curves (or any abelian group) is straightforward provided the group operation can be implemented efficiently. As pointed out in [18], it is the reversible implementation of

---

[5] The metrics outlined in [16] differ from conventional standards, particularly in their nomenclature. To establish a unified basis for comparing various algorithms in both RSA and EC cryptography, we adopt a consistent set of metric names. For a detailed explanation of these metrics, we invite readers to refer to Appendix A in [16].

[6] Although, as we explain in Subsect. 5.1, from a practical point of view, it remains unclear whether such trade-offs lead to faster realization of the attack in practice.

the group operation which contributes the most to the overall cost (in terms of resources) of the quantum circuit.

As per the scope defined in Sect. 4.1, works cited here focus only on elliptic curves over prime fields [11,12,18,23], while neglecting binary fields. Given that curves over binary fields have been deprecated by NIST [21], they shall not receive treatment in this paper either. Interested readers may look to [2] to learn more. Thus we shall henceforth be considering an elliptic curve E over a field $\mathbb{F}_p$ with p prime and n denoting the bit-length of p. Also, without loss of generality, we may assume that E is a Weierstrass curve despite Montgomery and Edwards curves being commonly used. That follows from the fact that there exist birational equivalence relations between (twisted) Edwards curves and Montgomery curves, with every instance of the latter being equivalent to some Weierstrass curve further still [3,9]. We can thus restrict our attention to Weierstrass curves wherever this level of detail is necessary.

Roetteler et al. in [27] have estimated that Shor's algorithm for breaking ECDLP on E would require:

$$(448 \lg(n) + 4090)n^3. \tag{3}$$

Toffoli gates (recall that each Toffoli gate corresponds to 7 T gates). Häner et al. [18] improve on the results of Roetteler et al. by reducing the number of logical qubits and Toffoli gates and providing an asymptotic estimate of the number of T gates:

$$436n^3 + o(n^3). \tag{4}$$

Häner et al. also present various trade-offs possible when implementing Shor's algorithm for ECDLP, optimizing, e.g., for circuit depth (see Table 4) or its T-depth.

Table 4. Expected costs of solving ECDLP according to [18].

	Solving DLP on an n-bit elliptic curve		
	$n = 256$	$n = 384$	$n = 512$
Logical Qubits (optimized for width)	2124	3151	4258
T-Count (optimized for width)	$1.72 \cdot 2^{32}$	$1.51 \cdot 2^{34}$	$1.82 \cdot 2^{35}$
Circuit Depth (optimized for width)	$1.89 \cdot 2^{32}$	$1.77 \cdot 2^{34}$	$1.09 \cdot 2^{36}$
Logical Qubits (optimized for T-count)	2619	3901	5273
T-Count (optimized for T-count)	$1.08 \cdot 2^{31}$	$1.74 \cdot 2^{32}$	$1.00 \cdot 2^{34}$
Circuit Depth (optimized for T-count)	$1.85 \cdot 2^{31}$	$1.31 \cdot 2^{33}$	$1.54 \cdot 2^{34}$
Logical Qubits (optimized for depth)	2871	4278	5789
T-Count (optimized for depth)	$1.34 \cdot 2^{32}$	$1.13 \cdot 2^{34}$	$1.43 \cdot 2^{35}$
Circuit Depth (optimized for depth)	$1.40 \cdot 2^{27}$	$1.48 \cdot 2^{28}$	$1.27 \cdot 2^{29}$

5 Forecasting Evolution of Quantum Computers: When Practical Attacks Will Be Possible

Quantum computing technology, possesses a limited historical track record, and predictions concerning its future development largely rely on quantum experts' educated guesses, occasionally supported by more substantiated arguments. In literature, the performance of quantum computers is frequently monitored via the following quantities:

- average two-qubit-gate error rate,
- number of physical qubits in a system,
- number of logical qubits in a system.

Table 5. Asymptotic costs of solving ECDLP according to [18].

	Solving DLP on an n-bit elliptic curve
Logical Qubits (optimized for width)	$8n + 10.2 \cdot \lfloor \lg n \rfloor - 1$
T-Count (optimized for width)	$436n^3 - 1.05 \cdot 2^{26}$
T-Depth (optimized for width)	$120n^3 - 1.67 \cdot 2^{22}$
Logical Qubits (optimized for T-count)	$10n + 7.4 \cdot \lfloor \lg n \rfloor + 1.3$
T-Count (optimized for T-count)	$1115n^3 / \lg n - 1.08 \cdot 2^{24}$
T-Depth (optimized for T-count)	$389n^3 / \lg n - 1.70 \cdot 2^{22}$
Logical Qubits (optimized for depth)	$11n + 3.9 \cdot \lfloor \lg n \rfloor + 16.5$
T-Count (optimized for depth)	–
T-Depth (optimized for depth)	$285n^2 - 1.54 \cdot 2^{17}$

Remark 2. In our current understanding of quantum computing, the primary bottleneck for executing any of the aforementioned attacks is likely to be the maximal circuit depth. Unfortunately, no methodology has been proposed to forecast the evolution of quantum computers that adequately takes into account the circuit depth. Moreover, researchers working on quantum computers rarely share information concerning this topic. Therefore, we base our analyses on a more freely accessible metric.

5.1 Forecasting Based on a Statistical Model

To the best of our knowledge, the most comprehensive assessment of future quantum computing progress based on statistical modeling was presented in a 2020 article by Sevilla and Riedel [29]. They gathered all available information (scientific articles and enterprises' marketing alike) on quantum computers from 2003 to 2020 [30]. It is vital to note that such data is subject to significant noise and bias primarily because the decision to report findings or failures lies with the researchers.

In light of this, Sevilla et al. devised the *generalized logical qubits* (GLQ) index, which estimates the number of logical qubits that will be available after accounting for the error-correction overheads [29].

The GLQ is expressed as follows:

$$N_{GLQ} = N_{PQ} \left[4 \cdot \frac{\log\left(\sqrt{10}\frac{e_P}{e_L}\right)}{\log\left(\frac{e_{th}}{e_P}\right)} + 1 \right]^{-2} \tag{5}$$

where N_{PQ} represents the number of physical qubits, e_P is the two-qubit gate error rate, $e_{th} = 10^{-2}$ denotes the approximate threshold error under which fault-tolerance becomes viable for the surface code, and $e_L = 10^{-18}$ represents the acceptable logical error rate. To put it simply, this formula utilizes the number of physical qubits and the two-qubit gate error rate to estimate the number of logical qubits available after factoring in the error-correction overhead. In essence, N_{GLQ} models the development of FTQC over time and should represent the quantum computer's "real-world" capability to perform computations. The authors of [29] claim that formula 5 was coined to fit data well for devices that have achieved two-qubit gate error rates below the fault-tolerance threshold mentioned above. They propose a *multivariate log-linear* model that takes a date as input and outputs a distribution for the combination of metrics that quantum computers around that date are likely to represent. The model assumes a linear relation between time and the logarithms of N_{PQ} and e_P, thus giving an exponential relation between time and N_{GLQ}.

Our Estimation. We extended the dataset used in [29] by 25% through the addition of 16 of the most recently published pieces of information regarding quantum processors. By doing so we allowed predictions to represent real-world data more accurately. It is noteworthy that many of the state-of-the-art models currently incorporate technologies other than superconductors, therefore we trained the model on all types of physical realizations. For the bootstrapping process, we used the top 15% of samples with the highest N_{GLQ}. Despite introducing certain refinements to the parameter assumptions of the original model, the resultant curve still exhibits an overestimation of N_{GLQ} for all data points, see Fig. 1. Prediction of N_{GLQ} for the coming years is presented in Table 6.

Table 6. Expected N_{GLQ} for the coming years.

Year	2035	2040	2045	2050	2055	2060	2065	2070
Predicted N_{GLQ}	2	6	24	96	376	1450	5495	20542

6 Recommendations and Closing Remarks

According to the estimations presented in Tables 4 and 5, as well as the results summarized in Table 6, we claim that many classical algorithms currently in use

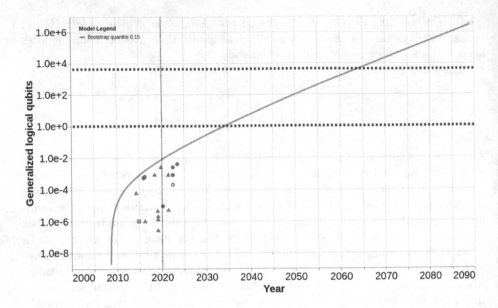

Fig. 1. N_{GLQ} prediction based on a statistical model. Note that the y-axis uses a logarithmic scale. Horizontal dotted lines are, from the bottom, $y = 1$ (one generalized logical qubit) and $y = 4258$ ("breaking" a 512-bit elliptic curve). Different markers denote different physical qubit realizations; for details, see [29].

may continue to be used as the well-studied, conservative fallbacks alongside (relatively experimental) post-quantum instances. Such hybrids are then expected to withstand quantum attacks until the year 2065, even if their post-quantum components have been broken classically by then. The following RSA parameters are recommended:

- at NIST Security Level 1: the RSA modulus N being a 3072-bit integer,
- at NIST Security Level 3: the RSA modulus N being a 7680-bit integer,
- at NIST Security Level 5: the RSA modulus N being a 15360-bit integer.

As far as elliptic curve-based cryptosystems are concerned, forecasts in Fig. 1 show that 512-bit curves should be used to ensure security until year 2065. Smaller elliptic curve groups (e.g., 256-bit) are predicted to withstand quantum attacks until the year 2060. (Note that RSA is more secure against quantum adversaries than elliptic curve-based schemes at a similar classical security level, as already pointed out in [18].) After year 2065 it is nearly impossible to estimate the speed of evolution of quantum computers and so no further predictions are given.

Note that from a number of possible methodologies for studying the pairing of post-quantum schemes with classical ones, we have chosen the one which gives the most practical results (cf. Sect. 4.1), in the sense that it is not so pessimistic as to require exceedingly large group orders (elliptic curve or RSA). We believe, however, that this approach (albeit seemingly best-effort) is the most pertinent to the problem at hand since, as pointed out already, hybridization and parallel use of classical and post-quantum cryptography is to be thought of as a temporary measure for the transition period into the full-blown quantum era. Still, due to much uncertainty in the projections concerning the development of large-scale quantum computers, following the trends regularly and adapting accordingly is imperative. Caution is advised.

References

1. Amy, M., Maslov, D., Mosca, M., Roetteler, M.: A meet-in-the-middle algorithm for fast synthesis of depth-optimal quantum circuits. IEEE Trans. Comput. Aided Des. Integr. Circuits Syst. **32**(6), 818–830 (2013)
2. Banegas, G., Bernstein, D.J., van Hoof, I., Lange, T.: Concrete quantum cryptanalysis of binary elliptic curves. IACR Trans. Cryptogr. Hardw. Embed. Syst. **2021**(1), 451–472 (2020). https://doi.org/10.46586/tches.v2021.i1.451-472. https://tches.iacr.org/index.php/TCHES/article/view/8741
3. Bernstein, D.J., Birkner, P., Joye, M., Lange, T., Peters, C.: Twisted Edwards curves. Cryptology ePrint Archive, Paper 2008/013 (2008). https://eprint.iacr.org/2008/013
4. Beullens, W.: Breaking Rainbow takes a weekend on a laptop. Cryptology ePrint Archive, Paper 2022/214 (2022). https://eprint.iacr.org/2022/214
5. Boudot, F., Gaudry, P., Guillevic, A., Heninger, N., Thomé, E., Zimmermann, P.: The state of the art in integer factoring and breaking public-key cryptography. IEEE Secur. Priv. **20**(2), 80–86 (2022). https://doi.org/10.1109/MSEC.2022.3141918
6. Castryck, W., Decru, T.: An efficient key recovery attack on SIDH. In: Hazay, C., Stam, M. (eds.) EUROCRYPT 2023. LNCS, vol. 14008, pp. 423–447. Springer, Cham (2023). https://doi.org/10.1007/978-3-031-30589-4_15
7. Chen, L., Moody, D., Liu, Y.: NIST post-quantum cryptography standardization. Transition **800**(131A), 164 (2017)
8. Chen, Y.: Quantum algorithms for lattice problems. Cryptology ePrint Archive, Paper 2024/555 (2024). https://eprint.iacr.org/2024/555
9. Costello, C., Smith, B.: Montgomery curves and their arithmetic: the case of large characteristic fields. Cryptology ePrint Archive, Paper 2017/212 (2017). https://eprint.iacr.org/2017/212
10. Deutsch, D.: Quantum theory, the Church-Turing principle and the universal quantum computer. Proc. Roy. Soc. London A **400**, 97–117 (1985)
11. Ekerå, M.: Quantum algorithms for computing general discrete logarithms and orders with tradeoffs. J. Math. Cryptol. **15**(1), 359–407 (2021). https://doi.org/10.1515/jmc-2020-0006
12. Ekerå, M.: Revisiting Shor's quantum algorithm for computing general discrete logarithms (2023)
13. Ekerå, M., Håstad, J.: Quantum algorithms for computing short discrete logarithms and factoring RSA integers. In: Lange, T., Takagi, T. (eds.) PQCrypto 2017. LNCS, vol. 10346, pp. 347–363. Springer, Cham (2017). https://doi.org/10.1007/978-3-319-59879-6_20
14. Feynman, R.P.: Simulating physics with computers. Int. J. Theor. Phys. **21**(6), 467–488 (1982)

15. Fowler, A.G., Mariantoni, M., Martinis, J.M., Cleland, A.N.: Surface codes: towards practical large-scale quantum computation. Phys. Rev. A **86**(3) (2012). https://doi.org/10.1103/physreva.86.032324

16. Gidney, C., Ekerå, M.: How to factor 2048 bit RSA integers in 8 hours using 20 million noisy qubits. Quantum **5**, 433 (2021). https://doi.org/10.22331/q-2021-04-15-433

17. Grover, L.K.: A fast quantum mechanical algorithm for database search (1996)

18. Häner, T., Jaques, S., Naehrig, M., Roetteler, M., Soeken, M.: Improved quantum circuits for elliptic curve discrete logarithms. In: Ding, J., Tillich, J.-P. (eds.) PQCrypto 2020. LNCS, vol. 12100, pp. 425–444. Springer, Cham (2020). https://doi.org/10.1007/978-3-030-44223-1_23

19. Häner, T., Roetteler, M., Svore, K.M.: Factoring using 2n+2 qubits with Toffoli based modular multiplication (2017)

20. Maino, L., Martindale, C.: An attack on sidh with arbitrary starting curve. Cryptology ePrint Archive, Paper 2022/1026 (2022). https://eprint.iacr.org/2022/1026

21. National Institute of Standards and Technology: Digital signature standard (DSS) (2023). https://csrc.nist.gov/pubs/fips/186-5/final

22. National Institute of Standards and Technology: Recommendations for discrete logarithm-based cryptography: elliptic curve domain parameters (2023). https://csrc.nist.gov/pubs/sp/800/186/final

23. Proos, J., Zalka, C.: Shor's discrete logarithm quantum algorithm for elliptic curves (2004)

24. Ragavan, S., Vaikuntanathan, V.: Optimizing space in Regev's factoring algorithm. Cryptology ePrint Archive, Paper 2023/1501 (2023). https://eprint.iacr.org/2023/1501

25. Regev, O.: An efficient quantum factoring algorithm (2023). https://doi.org/10.48550/ARXIV.2308.06572. https://arxiv.org/abs/2308.06572

26. Robert, D.: Breaking SIDH in polynomial time. In: Hazay, C., Stam, M. (eds.) EUROCRYPT 2023. LNCS, vol. 14008, pp. 472–503. Springer, Cham (2023). https://doi.org/10.1007/978-3-031-30589-4_17

27. Roetteler, M., Naehrig, M., Svore, K.M., Lauter, K.: Quantum resource estimates for computing elliptic curve discrete logarithms. In: Takagi, T., Peyrin, T. (eds.) ASIACRYPT 2017. LNCS, vol. 10625, pp. 241–270. Springer, Cham (2017). https://doi.org/10.1007/978-3-319-70697-9_9

28. Selinger, P.: Quantum circuits of t-depth one. Phys. Rev. A **87**(4), 042302 (2013)

29. Sevilla, J., Riedel, C.J.: Forecasting timelines of quantum computing (2020). https://doi.org/10.48550/ARXIV.2009.05045. https://arxiv.org/abs/2009.05045

30. Sevilla, J., Riedel, C.J.: Quantum computing progress - data (2020), 2023. https://docs.google.com/spreadsheets/d/1pwb4gf0FxlxgfVhtXTaqEGS9b7FwsstsJ0v7Zb1naQ0/edit#gid=0

31. Shor, P.: Algorithms for quantum computation: discrete logarithms and factoring. In: Proceedings 35th Annual Symposium on Foundations of Computer Science, pp. 124–134 (1994). https://doi.org/10.1109/SFCS.1994.365700

32. Stebila, D., Fluhrer, S., Gueron, S.: Hybrid key exchange in TLS 1.3. Technical report, Internet Engineering Task Force (2023). https://datatracker.ietf.org/doc/draft-ietf-tls-hybrid-design/09/

Unsafe Mechanisms of Bluetooth, E_0 Stream Cipher Cryptanalysis with Quantum Annealing

Mateusz Leśniak[1][(⊠)], Elżbieta Burek[2], and Michał Wroński[1]

[1] NASK National Research Institute, Kolska Str. 12, Warsaw, Poland
{mateusz.lesniak,michal.wronski}@nask.pl
[2] Military University of Technology, Kaliskiego Str. 2, Warsaw, Poland
elzbieta.burek@wat.edu.pl

Abstract. Due to Shor's and Grover's algorithms, quantum computing has become one of the fastest-evolving areas in computational science. Nowadays, one of the most interesting branches of quantum computing is quantum annealing. This paper presents the efficient method of transforming stream cipher E0 to the QUBO problem and then retrieving the Encryption Key of this cipher using quantum annealing. Comparably to other asymmetric and symmetric cryptographic algorithms, the presented transformation is efficient, requiring only 2,728 (2,751) logical variables for attack with 128 (129) consecutive keystream bits. According to our knowledge, it is the most efficient algorithm transformation with a 128-bit key. Moreover, we show that using current quantum annealers, one can embed the attack for E0 for 58 consecutive bits of keystream, from 128 (129), which are necessary for the attack's first stage (second stage). Therefore, it is likely that it will be possible to embed E0 on available quantum annealers in the next few years.

Keywords: Stream cipher · Bluetooth · E_0 cipher · Cryptanalysis · Quantum annealing

1 Introduction

Quantum computing is an approach that evolves fast nowadays. The (r)evolution in this aspect touches both theoretical and practical aspects, such as building bigger and bigger quantum computers. For cryptological society, several aspects are the most important nowadays. One is: How large are cryptographic problems we can solve using available quantum computers nowadays? This question is also important as quantum annealers are also special cases of quantum computers.

The application of quantum annealing to cryptanalysis of stream ciphers was first presented in [13], in the context of Grain 128 and Grain 128a ciphers. The efficient transformation of algebraic attacks on these two ciphers to the QUBO problem required 5,751 and 6,761 logical variables, respectively. It means that the algebraic attack using quantum annealing is a serious threat because there is a large probability that in several years, it will be possible to build a dedicated quantum annealer on which it will be able to run such algebraic attacks on some stream ciphers.

This paper presents another step in applying quantum annealing to the cryptanalysis of stream ciphers.

The main differences between this paper and [13] are:

1. E0 stream cipher is practically used, for example, in Bluetooth communication protocol, while it is hard to find any practical applications of the Grain cipher.
2. In some cases, which will be described in detail later, it is possible to obtain the whole Encryption key of the E0 cipher running transformation to the QUBO problem twice. Because the same Encryption key is used for 23,5 h, having only 129 consecutive bits of stream, one can retrieve almost the whole 1-day communication using E0 cipher with 50% probability.
3. According to our research and simulations, the creation of a dedicated quantum annealer on which it would be possible to run an algebraic attack on a full E0 cipher should be much easier and more probable than in the case of Grain ciphers family because our attack might be run on the dedicated quantum annealer with 2,751 qubits and 18,156 couplers, for both obtained QUBO problems. On the other hand, the dedicated quantum annealer for Grain 128 (Grain 128a) requires 5,751 (6,761) qubits and 77,496 (94,865) couplers.
4. The QUBO problem for an algebraic attack on the E0 cipher with a 128-bit keystream is too large to embed even in the latest Zephyr architecture. However, smaller problems were generated for the unchanged E0 algorithm but for a shorter keystream. The biggest problem embedded in the Zephyr architecture was generated for a 58-bit keystream.

2   Bluetooth Encryption Overview

Communication over short distances, provided by Bluetooth, protects confidentiality. The security mechanism has evolved over the versions of Bluetooth standards. The following paper focuses on the Legacy mechanism. As presented in [3], Legacy encryption is performed if at least one device does not support Secure Connections and its features. Legacy encryption is performed with the E0 stream cipher, derived from the Massey-Rueppel algorithm.

Before encryption starts, the connection between devices must be established. Each device should carry an initialization phase in order to generate and exchange link keys. The initialization phase consists of five steps:

- generation of an initialization key;
- generation of link key;
- link key exchange;
- authentication;
- generation of Encryption key in each device.

The initialization key K_{init} depends on the identity of the devices. The key is generated with E_{22} algorithm, combining Bluetooth Device Address, PIN code, length of the PIN (in octets), and a random number. It is important to use this key only during initialization.

Link key K_{link} generation and exchange is performed as one procedure. The link key is generated as a combination key, dependent on two devices. Each device has its own random number; this number is used to generate half of the combination key with E_{21} algorithm. A random number is xor-ed with K_{init} and exchanged with another device. After receiving the number, the second half of the combination key is generated. Combination key K_{AB} is created as xor of both halves. If it consists of all zeroes, the key should be discarded. In another case, the new link key $K_{link} = K_{AB}$ is accepted.

Once the key is established, authentication shall be performed. Authentication uses E_1 and additionally outputs an Authenticated Ciphering Offset (ACO).

After this step, encryption key K_{enc} may be generated. The encryption key is the most interesting in terms of attacks presented in this paper. This key is derived by algorithm E_3 as follow:

$$K_{enc} = E_3(K_{link}, EN_{RAND}, COF), \qquad (1)$$

where:

- K_{link} is a 128-bit current link key, common for both devices.
- EN_{RAND} is 128-bit publicly known random number;
- COF is 96-bit Ciphering Offset number. The number is determined in two ways. The offset is derived from the Bluetooth Device Address if the current link key is temporary. Otherwise, an authentic ciphering offset is used.

Algorithm E_3 is called each time encryption is activated. Moreover, K_{enc} shall be periodically refreshed. When E_0 cipher is used, refresh shall be done at least once every 2^{28} ticks of the Bluetooth clock (about 23.5 h).

2.1 Encryption Procedure Concept

The encryption systems consist of three parts: session key generator, keystream generator, and part responsible for encryption and decryption. The first part takes specified inputs, mixes them, and shifts the output stream into the second part. The second part takes the output from the previous part as a seed and generates a keystream. The cipher should be re-initialized for each new packet with a maximum size of 2,745 bits. With each re-initialization, a new session key is generated.

Each device shall have the maximal allowed key length (in octets) $1 \leq L_{MAX} \leq 16$. Before encryption starts, the length of key L should be negotiated. Then, the encryption key is modified as follows:

$$K'_{enc} = g_2^L(x)(K_{enc}(x) \bmod g_1^L(x)), \qquad (2)$$

where $g_1^L(x)$, $g_2^L(x)$ are polynomials defined for specified length L and $\deg(g_1^L(x)) = 8L$, $\deg(g_2^L(x)) \leq 128 - 8L$.

Inputs for the session key generator are as follows:

– 8L-bit encryption key: K'_{enc};
– 48-bit Central's Bluetooth Device Address: ADR;
– 26-bit Central real-time clock: CL.

Such inputs provide that at least one bit is changed between two packets. Thus, each packet is encrypted with the new keystream.

2.2 Stream Generation Algorithm E_0 Description

Algorithm E_0 is illustrated in Fig. 1. Cipher has three major building blocks: linear feedback shift registers, summation combiner logic, and blend register. The generator consists of four different LFSRs with five taps, specified by the following polynomials $f_i(t)$:

– LFSR$_1$: $f_1(t) = t^{25} + t^{20} + t^{12} + t^8 + 1$;
– LFSR$_2$: $f_2(t) = t^{31} + t^{24} + t^{16} + t^{12} + 1$;
– LFSR$_3$: $f_3(t) = t^{33} + t^{28} + t^{24} + t^4 + 1$;
– LFSR$_4$: $f_4(t) = t^{39} + t^{36} + t^{28} + t^4 + 1$.

One bit from specified position of each LFSR, $24, 24, 32, 32$, labeled as x_1, x_2, x_3, x_4 respectively, is connected with second part of generator. Summation Combiner Logic is built with classical xor gate, which produces output bits and function F_1, producing value $s_{t+1} = 2\,s_{t+1}^1 + s_{t+1}^0$, where $s_{t+1}^0, s_{t+1}^1 \in \mathbb{Z}_2$, for last part of generator. Each output bit is generated with the following equation:

$$z_t = x_1 \oplus x_2 \oplus x_3 \oplus x_4 \oplus c_t^0. \tag{3}$$

F_1 function is defined as follows:

$$F_1 : s_{t+1} = \left\lfloor \frac{\sum_{i=1}^4 x_i + c_t}{2} \right\rfloor. \tag{4}$$

Last part of cipher is blend register holding two 2-bit values: $c_t = 2c_t^1 + c_t^0$ and $c_{t-1} = 2c_{t-1}^1 + c_{t-1}^0$, where $c_t^0, c_t^1, c_{t-1}^0, c_{t-1}^1 \in \mathbb{Z}_2$. The next value stored in the blend register is computed with a function F_2 using two different linear bijections, presented with equation (5).

$$\begin{aligned} T_1 &: (x_1, x_2) \rightarrow (x_1, x_0), \\ T_2 &: (x_1, x_2) \rightarrow (x_0, x_1 \oplus x_0). \end{aligned} \tag{5}$$

F_2 function is defined as follows:

$$F_2 : c_{t+1} = s_{t+1} \oplus T_1[c_t] \oplus T_2[c_{t-1}]. \tag{6}$$

Each value stored in registers can be threatened as a vector, then equation (6) can be presented as:

$$c_{t+1}^1 = s_{t+1}^1 \oplus c_t^1 \oplus c_{t-1}^0,$$
$$c_{t+1}^0 = s_{t+1}^0 \oplus c_t^0 \oplus c_{t-1}^1 \oplus c_{t-1}^0. \tag{7}$$

Some simplifications can be applied in (4). This equation can be presented in an alternative form:

$$2\,s_{t+1}^1 + s_{t+1}^0 = \left\lfloor \frac{\sum_{i=1}^4 x_i + 2c_t^1 + c_t^0}{2} \right\rfloor.$$

Moreover, floor can be replaced with additional binary variable $\beta \in \mathbb{Z}_2$ in following way:

$$2\,s_{t+1}^1 + s_{t+1}^0 + \beta \cdot \frac{1}{2} = \frac{\sum_{i=1}^4 x_i + 2c_t^1 + c_t^0}{2}.$$

Finally, equation (4) can be presented as follow:

$$4\,s_{t+1}^1 + 2\,s_{t+1}^0 + \beta = \sum_{i=1}^4 x_i + 2c_t^1 + c_t^0. \tag{8}$$

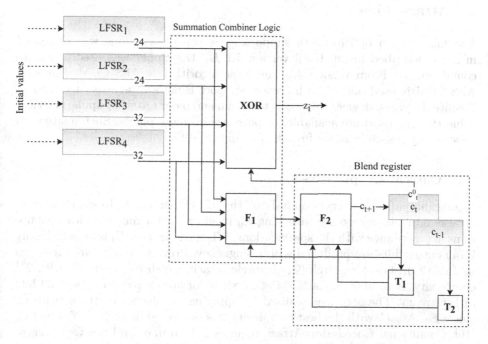

Fig. 1. Bluetooth encryption E_0 scheme. Based on [3].

Before the keystream is generated, initialization must be performed. Inputs are arranged as presented in Table 1. In following table, X_i denoted i-th bit of binary sequence X. As X[i] i-th octet of binary sequence is denoted.

Table 1. Arranging the inputs for session key generator. Based on [3].

	Input						
$LFSR_1$	ADR[2]	CL[1]	$K'_{enc}[12]$	$K'_{enc}[8]$	$K'_{enc}[4]$	$K'_{enc}[0]$	CL_{24}
$LFSR_2$	ADR[3]	ADR[0]	$K'_{enc}[13]$	$K'_{enc}[9]$	$K'_{enc}[5]$	$K'_{enc}[1]$	$CL_3CL_2CL_1CL_0001$
$LFSR_3$	ADR[4]	CL[2]	$K'_{enc}[14]$	$K'_{enc}[10]$	$K'_{enc}[6]$	$K'_{enc}[2]$	CL_{25}
$LFSR_4$	ADR[5]	ADR[1]	$K'_{enc}[15]$	$K'_{enc}[11]$	$K'_{enc}[7]$	$K'_{enc}[3]$	$CL_7CL_6CL_5CL_4111$

Initialization lasts 240 steps. Each input, presented in Table 1, is shifted into the register starting from the rightmost (last column) bit. The state is updated without feedback as long as the first input bit does not reach the rightmost position of the specified LFRS. When the first bit reaches the rightmost position of the last LFRS, blend registers are reset $c_t = c_{t-1} = 0$. From this point, the output sequence is generated. The remaining input bits are shifted; when the last bit is shifted in, then 0 is put as input. After generating 240 output bits, the last 128 bits are loaded without updating the blend register to feedback registers.

3   Attack Idea

The first version of Bluetooth supporting Secure Connections was presented in 2013, described in [2]. Until version 4.1 E_0, the cipher was used to ensure confidentiality. From version 4.1, the basic algorithm is AES-CCM. However, AES-CCM is used only if both devices support Bluetooth at least 4.1 version. Despite the years, devices using Bluetooth 4.0 and older are still popular. Devices using this standard are available for purchase in many places. Such a situation raises many possibilities for breach of confidentiality.

3.1   Current Cryptanalysis

Significant papers on cryptanalysis of the E_0 cipher exist. Research divides attacks into two categories: attacking E_0 itself and attacking E_0 with assumptions in accordance with the specifications. When single level E_0 is attacked, only 2,745 output bits are produced with a single key. Previous results are presented in Table 2. In latest research [9], algebraic attack, which can be made with 2^{84} complexity, is presented. La Scala et al. show an attack requiring only 60 bits of keystream. The attack is realized by applying an algebraic attack twice on single E_0. Attack with the best complexity was presented in 2005 by Yi Lu et al. [10]. Conditional Correlation Attack requires $2^{23.8}$ frames and recovers encryption key with 2^{38} complexity. Despite the large number of packets needed, the attack is still practical, as mentioned in Sect. 2 encryption key is refreshed after 2^{28} clocks.

Table 2. Previous attacks on Bluetooth encryption system.

	Time	Required data	Two level E_0	Additional requirements
Fluhrer, Lucks, 2000, [6]	2^{84}	140 bits	✓	–
	2^{77}	2^{30} bits	✓	
	2^{73}	2^{43} bits	✓	
Golić, Bagini, Morgari, 2002, [7]	2^{70}	45 packets	✓	Pre-computation sorting out a database of 2^{80} 103-bit words
Krause, 2002, [8]	$2^{76.8}$	128 bits	–	–
	$2^{112.95}$	128 bits	✓	–
Courtois, 2003, [5]	2^{49}	$2^{23.4}$ bits	–	2^{28} pre-computation
Armknecht, Krause, 2003, [1]	$2^{67.58}$	$2^{23.07}$ bits	–	–
Lu, Meier, Vaudenay, 2005, [10]	2^{38}	first 24 bits from $2^{23.8}$ packets	✓	–
Shaked, Wool, 2006, [12]	2^{87}	128 bits	✓	–
La Scala, Polese, Tiwari, Visconti, 2022, [9]	2^{84}	60 bits	✓	–

3.2 Proposed Attack Scheme

In this section, the proposed attack is presented. The main idea is to use quantum annealing to solve a system of equations describing the cipher like in [4]. It is important to note that quantum annealing can be used to solve optimization problems in specified form. In this paper, Quadratic Unconstrained Binary Optimization, formulated as Equation (9), is used,

$$\min_{x\in\{0,1\}^N} x^T Q x, \tag{9}$$

where x is binary variable vector, Q is upper-triangular $N \times N$ matrix containing real values. To obtain a cipher-based QUBO problem, algebraic representation of cipher with a system of equations (10) must be converted

$$\begin{cases} f_0(x_0,\ldots,x_{n-1}) \equiv 0 \ (\text{mod } 2), \\ f_1(x_0,\ldots,x_{n-1}) \equiv 0 \ (\text{mod } 2), \\ \vdots \\ f_{m-1}(x_0,\ldots,x_{n-1}) \equiv 0 \ (\text{mod } 2). \end{cases} \tag{10}$$

The above system of equations can be transformed in the following way:

1. Transformation to equations f_i' with binary variables and integer coefficients:

$$f_i' \equiv 0 \ (\text{mod } 2) \rightarrow f_i - 2k_i = 0,$$

where k_i is integer, $k \leq \left\lfloor \frac{f_i^{max}}{2} \right\rfloor$ and f_i^{max} is the maximal value of the polynomial f_i. It is important to note that seven equations are binary in (11), and only the last equation has binary variables and integer values. The above remark simplifies transformation to QUBO problem for E_0 cipher. Only seven remaining equations must be converted.

2. Linearization $f_i' \rightarrow f_{lin}'$, with additional penalty Pen calculation. E_0 design induces no nonlinear equations, so in the case of this paper, $f_i' = f_{lin}'$ and $Pen = 0$.

3. Variables k_i replacement with binary variables $x_0, x_1, \ldots, x_{bl(k_i^{max})-1}$, where $bl(x)$ denotes bit-length of x and

$$k_i = \sum_{j=0}^{bl(k_i^{max})-2} 2^j x_j + (k_i^{max} - 2^{bl(k_i^{max})-1} + 1) \cdot x_{bl(k_i^{max})-1}.$$

4. Determination of polynomial $F_{Pen}' = \sum_{i=0}^{m-1}(f_{lin_i}')^2 + c \cdot Pen$, where c is a constant to prevent achieving the minimum energy for incorrect solutions.

5. Final form of the polynomial $F_{Pen} = F_{Pen}' - C$ calculation, where C is a constant term in F_{Pen}'. This constant corresponds to the minimum energy of the function.

The complexity of quantum annealing mainly depends on a number of variables used. The fewer variables used, the smaller the complexity and the easier to embed the problem.

Bluetooth encryption system design, presented in Sect. 2, requires the use of quantum annealing twice. Figure 2 presents the scheme of attack juxtaposed with the cipher design. The first stage of attack involves initial state recovery. Based on the keystream, a system of equations is generated. To generate equations, 129 bits of output are needed. For each bit, eight equations, presented as (11), are required.

$$\begin{cases} f_0 : z_t = l_{t+1} \oplus m_{t+7} \oplus n_{t+1} \oplus o_{t+7} \oplus c_t^0, \\ f_1 : l_{t+25} = l_t \oplus l_{t+5} \oplus l_{t+13} \oplus l_{t+17}, \\ f_2 : m_{t+31} = m_t \oplus m_{t+7} \oplus m_{t+15} \oplus m_{t+19}, \\ f_3 : n_{t+33} = n_t \oplus n_{t+5} \oplus n_{t+9} \oplus n_{t+29}, \\ f_4 : o_{t+39} = o_t \oplus o_{t+3} \oplus o_{t+11} \oplus o_{t+35}, \\ f_5 : c_{t+1}^1 = s_{t+1}^1 \oplus c_t^1 \oplus c_{t-1}^0, \\ f_6 : c_{t+1}^0 = s_{t+1}^0 \oplus c_t^0 \oplus c_{t-1}^1 \oplus c_{t-1}^0, \\ f_7 : 4 s_{t+1}^1 + 2 s_{t+1}^0 + \beta = l_{t+1} + m_{t+7} + n_{t+1} + o_{t+7} + 2c_t^1 + c_t^0. \end{cases} \quad (11)$$

The entire system of equations can be simplified, similarly to in [13]. The first simplification concerns the equations associated with LFSRs. Last keystream bit is associated with $l_{129}, m_{135}, n_{129}, o_{135}$. This remark allows limiting the number of equations for each keystream bit. Equations for remaining bits are not included in the system. In the final system of equations appears:

- 129 keystream-based equations f_0;
- 105 LFSR-based equations f_1 and f_2;
- 97 LFSR-based equations f_3 and f_4;
- 128 equations f_5, f_6 and f_7.

The second stage is divided into two parts. In the first part, the internal state must be recovered. Based on the initial state from the first stage, a system of equations is made. Successive bits are treated as a 128-bit output according to the order shown in the [3]. With such output, a system of equations, likewise the first stage, is made. Due to the smaller number of keystream bits, the whole system of equations looks slightly different than in the first stage. In the final system of equations appears:

- 128 keystream-based equations f_0;
- 104 LFSR-based equations f_1 and f_2;
- 96 LFSR-based equations f_3 and f_4;
- 127 equations f_5, f_6 and f_7.

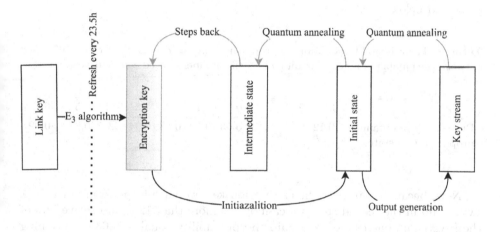

Fig. 2. Attack scenario of the proposed attack.

3.3 Probability of Recovering Used Encryption Key

However, recovering the Encryption key is possible using only a limited number of consecutive bits; even obtaining the ground state of the solved QUBO problem does not necessarily give the proper internal state and, therefore, the proper Encryption key. In this subsection, we focus on estimating how large is the probability of obtaining the proper encryption key using the attack scenario presented in the previous section.

To be able to make any estimations, at the beginning, we have to make the following assumptions:

1. The first assumption is that analyzed stream cipher behaves as the random generator.
2. Even if the analyzed stream cipher behaves as a random generator, it is possible that the same keystream may be obtained for two or more different internal states. It means that there may be two or more proper ground states. It is worth noting that only one, the correct internal state, allows us to obtain the correct Encryption key. However, to make any estimations, it has to be assumed that the probability of obtaining every ground state is the same. In practical implementations, it is not obligatory because the shape of the energy landscape may favor some ground states against other ground states.

So, using the assumptions above, we may estimate the probability of obtaining the correct Encryption key.

Here, we use estimations from [13], where the probability of obtaining a proper internal state was estimated according to our assumptions. The most important thing here is the difference between the size of the internal state n and the number of known keystream bits k. More details may be found in [13]. The table showing the probability of obtaining a proper internal state is presented below (Table 3).

Table 3. The probability of obtaining proper internal state according to the difference between the number of inner state bits n and the number of used consecutive keystream bits k.

$n - k$	3	2	1	0	−1	−2	−3	−4	−5	−6
Probability of obtaining proper ground state	0.12	0.25	0.43	0.63	0.79	0.88	0.94	0.97	0.98	>0.99

Now, because to recover the encryption key, we have to perform our attack twice, and in the second step, we cannot use more than 128 consecutive bits of the keystream, one cannot over gain the probability equal to 0.63 of obtaining the proper encryption key, but in this case in the first step the number of used consecutive bits of keystream should be equal at least to 136. In such a case, the attack complexity increases because, as defined in the first step, the QUBO problem is bigger than the QUBO problem in the second step.

Even though cryptanalysis often assumes that one should focus on attack parameters that give the success probability equal to at least 0.5. In such a case, the situation here is simple: in the first step, one has to use 129 consecutive keystream bits, while in the second step, one has to use 128 consecutive keystream bits. In such a case, the probability of obtaining the proper Encryption key equals 0, 50, which we wanted to obtain.

4   Results

Based on the proposed attack described above and the equations system (11), appropriate optimization problems were created in the form of QUBO for 129 and 128-bit keystreams. The obtained problems consist of 2,751 and 2,728 binary variables for 129 and 128 keystream bits, respectively.

4.1   Graph of the Obtained Optimization Problem

The Q matrix (see Equation (9)) of the obtained optimization problem, for a 129-bit keystream, has a degree of 2,751 and consists of 20,778 non-zero elements, whose values range from -16 to 17. For a 128-bit keystream, the Q matrix of the obtained QUBO problem has a degree of 2,728 and consists of 20,598 non-zero elements, the values of which are also in the range of -16 to 17. For both obtained problems, the number of non-zero elements is only 0.55% of all elements of matrix Q, which indicates that these matrices are sparse. The structure of the Q matrix of the obtained QUBO problem for the 129-bit keystream is presented in Fig. 3, where the axes denote the indexes of the binary variables occurring in the objective function and the black point indicates the non-zero coefficient of a given quadratic monomial.

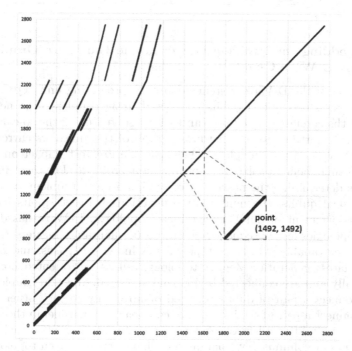

Fig. 3. The structure of the Q matrix of the obtained QUBO problem for the 129-bit keystream.

Since the QUBO problem is defined using an objective function with binary variables, it can be represented as a graph called the problem graph. A vertex represents each binary variable, and each quadratic monomial is represented by an edge between the given vertices.

For the obtained QUBO problems, the number of vertices in the problem graph is equal to the number of binary variables in an optimization problem, while the number of edges is 18,156 for a 129-bit keystream and 17,998 for a 128-bit keystream. For both QUBO problems, the maximum vertex degree is 40, and the remaining vertex degrees and the number of vertices of a given degree are shown in Table 4.

Table 4. Vertex degrees of the resulting graphs of QUBO problems.

Vertex degree	5	6	8	11	12	13	16	17	20	21	22	24
Number of vertices for 128-bit keystream	252	1,323	128	1	135	128	32	1	1	28	173	2
Number of vertices for 129-bit keystream	254	1,335	129	1	136	129	32	1	1	28	174	2

Vertex degree	26	27	28	29	30	32	33	34	36	38	40	
Number of vertices for 128-bit keystream	12	32	36	124	10	19	12	7	142	73	57	
Number of vertices for 129-bit keystream	12	32	36	125	10	19	12	7	144	74	58	

4.2 Embedding the Problem Graph in the Hardware Graph of the D-Wave Quantum Annealer

The structure of the D-Wave system's quantum processing unit (QPU) can be represented as a network of qubits connected using couplers. The network of qubits and the connecting couplers can be represented using the so-called hardware graph. Unfortunately, the hardware graph of the annealers currently provided by D-Wave is not complete, which has the greatest impact on the possibility of using them to solve any optimization problem. The most important parameters determining the possibility of solving a given problem are the number of physical qubits, the number of connections between qubits (couplers), and the arrangement of qubit connections (topology). D-Wave has developed different topologies for subsequent generations of quantum annealers. Currently, annealers are available with three topologies: Chimera, Pegasus, and Zephyr.

In the latest generation Zephyr topology, qubits are oriented horizontally and vertically and are connected by internal, external, and odd couplers. Inner couplers connect a pair of qubits oriented orthogonally to each other, external couplers connect a pair of qubits that are collinear, i.e., parallel in the same row or column, and odd couplers connect a pair of qubits parallel to each other in adjacent rows or columns. Two parameters characterize a given topology:

- nominal length of qubit, meaning the number of orthogonal qubits to which a given qubit is connected using internal couplers,

– degree of qubit, meaning the number of different qubits to which a given qubit is connected using all types of couplers.

The Zephyr topology consists of 71,736 couplers and 7,440 qubits, with a nominal length of 16 and a degree of 20.

To solve an optimization problem using a D-Wave computer, the problem must be embedded in the QPU, i.e., the problem graph must be mapped in the hardware graph, which is called minor-embedding. During this process, each vertex from the problem graph is mapped to a physical qubit of the hardware graph and each edge to a coupler. As mentioned earlier, the hardware graphs of current topologies are not complete; therefore, to map all the connections of a given vertex, it may be necessary to map it not to one but to several physical qubits connected by couplers, creating a chain. All qubits of a given chain must return the same value in solution. Since the embedding problem is an NP-hard problem, its implementation in the currently available D-Wave system is performed using a heuristic tool, which results in the fact that for the same problem, different embedding results can be obtained using a different number of resources. An embedding can be characterized by the number of physical qubits used, the number of all couplers used, and the number and length of chains created.

Since the QUBO problem for an algebraic attack on the E0 cipher with a 128-bit keystream is too large to embed even in the latest Zephyr architecture, smaller problems were generated for the unchanged E0 algorithm but for a shorter keystream. The biggest problem embedded in the Zephyr architecture was generated for a 58-bit keystream. The graph of the embedded problem consists of 1,118 vertices and 6,938 edges. The maximum vertex degree is 38. This problem was mapped to a hardware graph consisting of 6,367 physical qubits and 12,187 couplers and required the creation of 811 chains ranging from 2 to 44 qubits in length. Table 5 shows the embedded problem's parameters and its embedding parameters.

Table 5. Parameters of the problem graph and hardware graph for 58-bit keystream.

Vertex degree	5	6	8	11	12	13	16	17	20	21	22	24	26	27
Number of vertices	112	483	58	1	65	58	40	1	1	31	108	2	12	29

Vertex degree	28	29	30	32	33	36	38							
Number of vertices	19	54	5	15	12	9	3							

Length of chain	2	3	4	5	6	7	8	9	10	11	12	13	14	15
Number of chains	293	117	38	32	37	29	23	23	21	14	19	23	15	19

Length of chain	16	17	18	19	20	21	22	23	24	25	26	27	28	29
Number of chains	13	7	15	12	7	2	5	1	6	6	3	1	3	2

Length of chain	30	31	32	33	34	36	37	38	39	40	41	42	43	44
Number of chains	1	2	3	1	2	3	1	1	2	1	3	2	1	2

4.3   Dedicated Architecture of Quantum Annealer

It is known that D-Wave develops commercial quantum computers that solve optimization problems presented in the BQM (Binary Quadratic Model), CQM (Constrained Quadratic Model), or DQM (Discrete Quadratic Model) models. To enable solving as many different problems as possible, the developed quantum annealer topologies must maintain a kind of compromise between the number of physical qubits, the number of couplers, and their arrangement in the QPU unit, which shows the continuous development of the developed architectures.

In [13] the idea of constructing a quantum annealer only for algebraic attacks on a specific cipher was presented. In such a dedicated annealer, the emphasis would be on mapping the problem graph to the hardware graph as closely as possible, with the ideal situation being that the hardware graph would be the problem graph.

As previously presented, in the case of an algebraic attack on the E0 cipher, we need to solve two QUBO problems, one for a 128-bit keystream and the other for a 129-bit keystream. The QUBO problem for a 129-bit keystream is generated based on 916 equations of the form of Equation (11), of which 908 equations form an equations system based on which the QUBO problem for a 128-bit keystream is generated. Since the objective function of the QUBO problem is derived from the sum of squares, the QUBO problem for a 128-bit keystream is included in the QUBO problem for a 129-bit keystream. It follows that adding further equations generates new quadratic monomials without changing the existing ones, which means that new edges appear in the problem graph without changing the existing ones, but at most increasing the degree of the vertices. This means that the problem graph for a 128-bit keystream is a sub-graph of the problem graph for a 129-bit keystream. Therefore, it is enough to construct one dedicated annealer for the larger problem, to solve both.

A dedicated annealer, one-to-one mapping the problem graph to a hardware graph, would require 2,751 qubits with a maximum degree of 40 and 18,156 couplers. This value of the degree of qubit is higher than the value of the connectivity parameter in current quantum annealers. However, assuming a qubit degree of 20 (the same as in the latest D-Wave annealer topology) and any length and orientation of the qubits, for an algebraic attack on the E0 cipher, the hardware graph would have to consist of 3,542 qubits, 18,947 couplers, and 733 chains, including 675 chains length 2 and 58 chains length 3. It is worth emphasizing here that the amounts of required resources are smaller than the resources currently offered by D-Wave annealers.

5   Conclusion

In this paper was presented the algebraic attack on E0 cipher using quantum annealing. In compare to other attacks on cryptograpihic algorithms using quantum annealing, presented attack requires the smallest number of logical variables from all ciphers with 128-bits of keylength.

It is worth to note, that in some cases, presented attack scenario allows one to retrieve encryption key used for 23,5 h at most, therefore it is possible to decipher the (almost) whole daily communication performing algebraic attack using quantum annealing twice. The first attack uses 129 consecutive keystream bits - this attack allows to obtain 128 bits of inner state, which is also treated as the keystream used in the second attack. If in both attacks one obtains proper ground state as the solution, in such a case there is 50% probability that obtained encryption key will be proper.

However the complexity of algebraic attacks using quantum annealing still is not well known, assuming attack complexity as $O(e^{\sqrt{N}})$ [11], where N is the number of logical variables, the complexity of presented attack for obtaining encryption key is much below the brute force attack complexity and may be estimated as $2^{76.52}$.

References

1. Armknecht, F., Krause, M.: Algebraic attacks on combiners with memory. In: Boneh, D. (ed.) CRYPTO 2003. LNCS, vol. 2729, pp. 162–175. Springer, Heidelberg (2003). https://doi.org/10.1007/978-3-540-45146-4_10
2. Bluetooth Special Interest Group: Bluetooth Core Specification (2013). rev. 4.1
3. Bluetooth Special Interest Group: Bluetooth Core Specification (2021). rev. 5.3
4. Burek, E., Wroński, M., Mańk, K., Misztal, M.: Algebraic attacks on block ciphers using quantum annealing. IEEE Trans. Emerg. Top. Comput. **10**(2), 678–689 (2022)
5. Courtois, N.T.: Fast algebraic attacks on stream ciphers with linear feedback. In: Boneh, D. (ed.) CRYPTO 2003. LNCS, vol. 2729, pp. 176–194. Springer, Heidelberg (2003). https://doi.org/10.1007/978-3-540-45146-4_11
6. Fluhrer, S.R., Lucks, S.: Analysis of the e0 encryption system. In: ACM Symposium on Applied Computing (2001). https://api.semanticscholar.org/CorpusID:2130499
7. Golić, J.D., Bagini, V., Morgari, G.: Linear cryptanalysis of Bluetooth stream cipher. In: Knudsen, L.R. (ed.) EUROCRYPT 2002. LNCS, vol. 2332, pp. 238–255. Springer, Heidelberg (2002). https://doi.org/10.1007/3-540-46035-7_16
8. Krause, M.: BDD-based cryptanalysis of keystream generators. In: Knudsen, L.R. (ed.) EUROCRYPT 2002. LNCS, vol. 2332, pp. 222–237. Springer, Heidelberg (2002). https://doi.org/10.1007/3-540-46035-7_15
9. La Scala, R., Polese, S., Tiwari, S.K., Visconti, A.: An algebraic attack to the Bluetooth stream cipher E0. Finite Fields Appl. **84**, 102102 (2022). https://doi.org/10.1016/j.ffa.2022.102102, https://www.sciencedirect.com/science/article/pii/S1071579722001113
10. Lu, Y., Meier, W., Vaudenay, S.: The conditional correlation attack: a practical attack on Bluetooth encryption. In: Shoup, V. (ed.) CRYPTO 2005. LNCS, vol. 3621, pp. 97–117. Springer, Heidelberg (2005). https://doi.org/10.1007/11535218_7
11. Mukherjee, S., Chakrabarti, B.K.: Multivariable optimization: quantum annealing and computation. Eur. Phy. J. Spec. Top. **224**(1), 17–24 (2015)

12. Shaked, Y., Wool, A.: Cryptanalysis of the Bluetooth E0 cipher using OBDD's. Cryptology ePrint Archive, Paper 2006/072 (2006). https://eprint.iacr.org/2006/072
13. Wroński, M., Burek, E., Leśniak, M.: (in)security of stream ciphers against quantum annealing attacks on the example of the grain 128 and grain 128a ciphers. Cryptology ePrint Archive, Paper 2023/1502 (2023). https://eprint.iacr.org/2023/1502

Towards an In-Depth Detection of Malware Using Multi-QCNN

Tony Quertier and Grégoire Barrué[(✉)]

Orange Innovation, Rennes, France
`tony.quertier@orange.com`, `gregoire.barrue@orange.com`

Abstract. Malware detection is an important topic of current cyberse-curity, and Machine Learning appears to be one of the main considered solutions even if certain problems to generalize to new malware remain. In the aim of exploring the potential of quantum machine learning on this domain using only a few qubits, we implement a new preprocessing of our dataset using Grayscale method, and we couple it with a model composed of five quantum convolutional networks and a scoring function. We get an increase of around 20% of our results, both on the accuracy of the test and its F1-score.

Keywords: Quantum Machine Learning · Malware detection · Quantum convolutional networks

Introduction

Malicious software detection has become an important topic in business, as well as an important area of research due to the ever-increasing number of success-ful attacks using malware. Cybersecurity researchers are recently shifting their attention to Machine Learning (ML) methods to improve malicious files detec-tion [1,2], and they have been incredibly creative in data preprocessing. In fact, this part is essential now that the learning algorithms are already extremely powerful.

In [3], Anderson et al. trained a feature-based malware detection model using a non-optimized LightGBM algorithm, in [4] Nataraj et al. use k-nearest neigh-bors algorithm on image-based malware. Image-based malware detection is a challenge in both classical and quantum computing, but for different reasons. In classical machine learning, one limitation to obtaining results as good as with standard static features is the number of training data. Because malware images are much more complex and less representative than standard images, convolu-tional networks need a lot of images to extract information. However, computing resources are not a problem for two-channel images of size 64×64.

In recent years, research into QML has developed very significantly, whether in the area of learning theory [5], generalisation capabilities [6,7], or more prac-tical use-cases as earth observation study by ESA[1]. In particular, quantum con-volutional neural networks (QCNNs) have proved to be interesting, because they

---

[1] https://eo4society.esa.int/projects/qc4eo-study/.

© The Author(s), under exclusive license to Springer Nature Switzerland AG 2024
L. Franco et al. (Eds.): ICCS 2024, LNCS 14837, pp. 405–412, 2024.
https://doi.org/10.1007/978-3-031-63778-0_29

perform well, and for example do not exhibit Barren Plateaus [8]. They were first introduced in [9], and used for instance in [10] for classical data classification. Their full understanding is still missing, but some works, as [11] start to deeply analyze these models, and we think that it could be a suitable solution for image classification, as is its classical equivalent.

The challenge also comes from the limited number of qubits. On the features, we have restricted the number of qubit to less than 8 [12], but for image-based detection this number is not enough. We still want to give an approach using relatively few qubits, to keep coherence with the current capacities of the quantum computers. It is quite difficult to extract relevant information from a malware image of size 64×64 with just a few qubits, so we turn ourselves to quantum convolutional neural networks, and we split each image of binary file into sub-images, corresponding to specific sections of the binary file.

In this paper, we present an algorithm consisting of five QCNNs trained on the images corresponding to the different sections of a binary file and a scoring function to extract as much information as possible from it.

1   Dataset and Preprocessing

1.1   Data

We rely on two different datasets, Bodmas and PEMachineLearning. More details on these two datasets are available in the paper [13]. We split the dataset into three sub-datasets: one to train the QCNNs on the binary file's sections, one to train the final scoring function and one to test our architecture. Each sub-dataset is composed of 20k benign and 20k malware and the train/validation ratio is 70/30 for the first two.

The PE format is a file format used by Windows operating systems. A PE file is separated into two parts: the header and the sections. The header describes the file and its contents such as the date the file was created, information about loading the file into memory and the number of sections. Each section is described by a specific header containing its name, size and location in virtual memory. Sections generally contain the executable code (.text) and the variables used with their default values (.data, or in read only .rdata). The relocation section (.reloc) contains relocation information and the resource section (.rsrc) contains resources like icons, menus, and other elements. The resource section is often used by malware to evade detection. For example, scripts can be used to inject payloads directly into this section. When the binary file is executed, the embedded payload is extracted and executed. In this paper we focus in the five sections presented above, because they are present in the majority of the binary files, while others sections could be found in less files and thus give a smaller dataset for the training of a QCNN.

1.2   Splitting the Image into Sub-images

When preprocessing our data, we are using the Grayscale method, that transforms the malware into an image [4], but in a more subtle way that is better adapted to our problem. We first explore the PE file using the LIEF library to identify the start and end of each section. Then we transform the content of each section into an image of size 8×8. To begin with, we have identified 5 sections that we consider to be relevant and we focus on these. We train a QCNN on each section thanks to our first sub-dataset. Then we use these trained QCNNs to assign to each file a score per section when the sections are present. If they are not, we give these sections a score of -1. Initially we had chosen a neutral score of 0.5, but this introduced a bias because this is equivalent to considering that the absence of a section is of no importance, which is not true (Figs. 1 and 2).

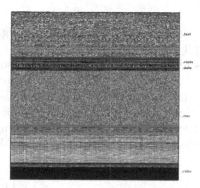

Fig. 1. Sections in a malware

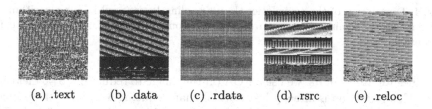

(a) .text   (b) .data   (c) .rdata   (d) .rsrc   (e) .reloc

Fig. 2. Images of different binary's sections

2 Framework

2.1 Description of the QCNN and Its Training on a Section Image

Here we present the architecture of our algorithm. To train a QCNN for each chosen section, we generate images for each section extracted from the file, then

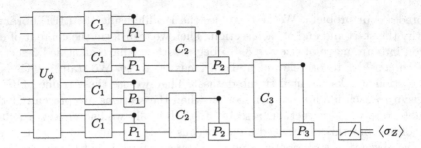

Fig. 3. The architecture of our QCNN, where convolutional and pooling layers alternate up to the measurement by an observable.

we train the QCNNs on the different sections by setting the number of qubits to 8 and therefore using 3 layers, as shown in Fig. 3.

The QCNN consists of a layered architecture as shown in [10]. A layer is composed of a convolutional layer (C_i in Fig. 3), entangling qubits with parameterized gates, and a pooling layer (P_i in Fig. 3), which also entangles qubits but then reduces the number of qubits by tracing out half of them. Figure 4 details the architecture we used for our convolutional and pooling layers. At the end of the quantum circuit, we measure the last qubit in the Z-basis.

(a) Pooling layer

(b) Convolutional layer

Fig. 4. Description of the convolutional and pooling layers used in our algorithm. At the end of the pooling layer we trace out the control qubit in order to reduce the dimension.

Since we use 8 qubits, we use a PCA on the image to reduce $8 \times 8=64$ features to 8. This reduction is already less critical than reducing a $64 \times 64=4096$ image to 8 features. For the encoding of data, we map all input data $x \in \mathbb{R}^n$ in $[0, \frac{\pi}{2}]^n$, and then we apply the encoding map

$$U_\phi : x \mapsto |\phi(x)\rangle = \bigotimes_{i=1}^{n} (\cos(x_i)|0\rangle + \sin(x_i)|1\rangle). \qquad (1)$$

We build five QCNNs with the same architecture, which are trained on images corresponding to a specific section. Scores are available in the Sect. 3. Each QCNN is trained over five epochs. For the optimization method, we use Simultaneous Perturbation for Stochastic Backpropagation (SPSB) [14].

As well as working with a small number of qubits for each QCNN, an interesting advantage is in the interpretation of the results. In addition to the detection score, we are able to see, thanks to the outputs of each QCNN, which section of the binary file is suspicious, and if necessary investigate its content.

2.2 Training a Scoring Function

Once the parameters of each QCNN have been set using the first sub-dataset, we need to find a customized scoring function to treat the scores given by these QCNNs to the files of the second sub-dataset. One could think about using a majority vote, but this would include a bias that suggests that every section is equally important. We use the second sub-dataset, and for each binary file, the file is decomposed into images corresponding to the sections present in the binary. The images are then passed through the QCNNs associated with the sections, and each QCNN returns a score. As a reminder, if the section is missing, the score is set to −1 for that section (Fig. 5).

As an output of this step, we therefore have a vector composed of the five scores, corresponding to the probability that the file is a malware regarding each section, and we want the algorithm to decide if the file is malicious or not as a final output. We test various functions and train some algorithms to try and find the most optimal function for calculating this score. For the final version of the algorithm, we use the third sub-dataset to test all the steps of our model. We keep a XGBoost model as scoring function, which gave us the best results. The various experiments and results can be found in Sect. 3.

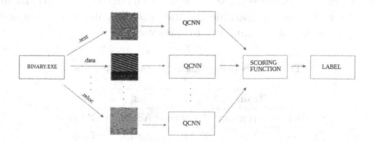

Fig. 5. Architecture of our algorithm. Once everything is trained, we get a model which takes the binary as input and gives its predicted label as output.

3   Experiments and Results

In this section we present the results of our algorithm, which contains three steps. First, we train a QCNN model for each five section of the files. When a section is not present in the file, there is no image corresponding to this section, so the different QCNNs are not trained on exactly the same sizes of dataset, but this guarantees the unbiased property of the training.

Table 1 gathers the training and testing accuracy for each section. We also compute the F1-score, which gives an additional proof of the performances of our results. The results by section are not very good, and actually they are comparable to the model where we train the QCNN on the images of the entire files (see Table 3b). The learning also seems to be better on some sections, which can tell us about their importance for the classification task.

Table 1. QCNN training results for each section

	Train	Test	
Section	Accuracy	Accuracy	F1-score
text	0.67	0.69	0.68
data	0.56	0.55	0.7
rdata	0.66	0.66	0.78
rsrc	0.65	0.63	0.72
reloc	0.69	0.7	0.67

Once the QCNNs are trained, we use them with the second part of the dataset in order to assign to each PE file a vector containing the scores associated to each section. Then we have to train the scoring function on this dataset. We try several scoring functions, which are presented in Table 2. The XGBoost and the Random Forests (RF) models have the best performances, the RF model being better on the train but comparable on the test. We also try a majority vote, which as expected is not relevant in our use-case.

Table 2. Scoring function training results

	Train		Test	
Section	Accuracy	F1-score	Accuracy	F1-score
XGBoost	0.89	0.90	0.84	0.85
LGBM	0.84	0.86	0.82	0.85
RF	0.99	0.99	0.85	0.86
Maj. Vote			0.4	0.27

Finally, once both the QCNNs and the scoring function are trained, we use the third subdataset as a test sample, to evaluate the performances of the whole model. The two compared models in Table 3a, show that the XGBoost model seems to perform slightly better than the RF model.

For comparison, we trained a QCNN on the second sub-dataset and then tested it on the third sub-dataset with the same parameters. Table 3b, shows that training a model with a PCA on the entire image of file is not a suitable solution.

Table 3. Final tests to compare multi-QCNN and single QCNN.

	Accuracy	F1-score
RF	0.82	0.80
XGBoost	0.83	0.83

(a) Testing multi-QCNN with RF and XGBoost on the third dataset.

	Accuracy	F1-score
Train	0.68	0.70
Test	0.60	0.66

(b) Results of QCNN on a complete binary image.

4 Conclusion and Future Work

The power of our proposed algorithm is that we can explore many different directions to enhance its performances. First we can increase the number of sections taken into account, in order to understand their impact on the final score. This could be done very efficiently as we do not have to re-train the QCNNs on the already studied sections, but just train some new QCNNs on the additional sections and take them into account in the input of the scoring function. We can also investigate what scoring function would be the most relevant for our problem. For example, the use of random forests allows to identify which sections are the most likely to help the classification task. Besides, once this information gathered, we could implement a weighted average scoring function to give a percentage of maliciousness of the files, hence giving a more nuanced response to the problem. Finally, we could improve the QCNNs themselves, in particular those trained on the more important sections.

In conclusion, we proposed in this work a multi-QCNN model in order to classify malware and benign files. We identified the different sections of these files, and transformed each section into an image thanks to Grayscale method. Then, we selected five specific sections, and trained a QCNN on each section. We restricted the size of the QCNNs to 8 qubits, to remain consistent with our approach aiming to think how to make better use of the information contained in the data rather than increase the capacities of our models. Once the training done, we gathered the five QCNNs outputs into vectors, that we studied using a scoring function in order to identify the type of each file. Our results show the efficiency of this model compared to a QCNN classifying images of entire files, namely it drastically increased the accuracy and F1 score for our detection task. Besides, the structure of this model makes it easy to modify, adapt, and allows to get more information about the data, increasing our understanding about malware PE files.

References

1. Raff, E., Nicholas, C.: A Survey of Machine Learning Methods and Challenges for Windows Malware Classification (2020)
2. Ucci, D., Aniello, L., Baldoni, R.: Survey of machine learning techniques for malware analysis. Comput. Secur. **81**, 123–147 (2019)

3. Anderson, H., Roth, P.: EMBER: An open dataset for training static PE malware machine learning models (2018)
4. Nataraj, L., Karthikeyan, S., Jacob, G., Manjunath, B.S.: Malware images: visualization and automatic classification. In: ACM International Conference Proceeding Series (2011)
5. Cerezo, M., Verdon, G., Huang, H.-Y., Cincio, L., Coles, P.J.: Challenges and opportunities in quantum machine learning. Nat. Comput. Sci. **2**(9), 567–576 (2022)
6. Abbas, A., Sutter, D., Zoufal, C., Lucchi, A., Figalli, A., Woerner, S.: The power of quantum neural networks. Nat. Comput. Sci. **1**(6), 403–409 (2021)
7. Larocca, M., Ju, N., García-Martín, D., Coles, P.J., Cerezo, M.: Theory of over-parametrization in quantum neural networks (2021)
8. Pesah, A., Cerezo, M., Wang, S., Volkoff, T., Sornborger, A.T., Coles, P.J.: Absence of barren plateaus in quantum convolutional neural networks. Phys. Rev. X **11**(4), 041011 (2021)
9. Cong, I., Choi, S., Lukin, M.D.: Quantum convolutional neural networks. Nat. Phys. **15**(12), 1273–1278 (2019)
10. Hur, T., Kim, L., Park, D.K.: Quantum convolutional neural network for classical data classification. Quantum Mach. Intell. **4**(1), 3 (2022)
11. Umeano, C., Paine, A.E., Elfving, V.E., Kyriienko, O.: What can we learn from quantum convolutional neural networks? (2023)
12. Barrué, G., Quertier, T.: Quantum machine learning for malware classification (2023). arXiv preprint arXiv:2305.09674
13. Marais, B., Quertier, T., Morucci, S.: Ai-based malware and ransomware detection models. In: Conference on Artificial Intelligence for Defense (2022)
14. Hoffmann, T., Brown, D.: Gradient estimation with constant scaling for hybrid quantum machine learning (2022)

Correction to: A Numerical Feed-Forward Scheme for the Augmented Kalman Filter

Fabio Marcuzzi

Correction to:
Chapter 10 in: L. Franco et al. (Eds.): *Computational Science – ICCS 2024*, **LNCS 14837,**
https://doi.org/10.1007/978-3-031-63778-0_10

In the originally published version of chapter 10, the reference 4 had been rendered incorrectly. This has been corrected.

The updated version of this chapter can be found at
https://doi.org/10.1007/978-3-031-63778-0_10

Author Index

L. Franco et al. (Eds.): ICCS 2024, LNCS 14837, pp. 413–414, 2024.
https://doi.org/10.1007/978-3-031-63778-0

Printed in the United States
by Baker & Taylor Publisher Services